W9-AUP-588

Water Quality

Guidelines, Standards and Health: Assessment of risk and risk management for water-related infectious disease

Water Quality

Guidelines, Standards and Health: Assessment of risk and risk management for water-related infectious disease

Edited by

Lorna Fewtrell
Centre for Research into Environment and Health, Aberystwyth, Wales

and

Jamie Bartram
World Health Organization, Geneva, Switzerland

Published on behalf of

World Health Organization

SMITTSKYDDSINSTITUTET
Swedish Institute for Infectious Disease Control

Published on behalf of the World Health Organization by
IWA Publishing, Alliance House, 12 Caxton Street, London SW1H 0QS, UK
Telephone: +44 (0) 20 7654 5500; Fax: +44 (0) 20 7654 5555; Email: publications@iwap.co.uk
www.iwapublishing.com

First published 2001

Printed by TJ International (Ltd), Padstow, Cornwall, UK

British Library Cataloguing-in-Publication Data
A CIP catalogue record for this book is available from the British Library

Library of Congress Cataloging-in-Publication Data
A catalog record for this book is available from the Library of Congress

ISBN 1 900222 28 0 (IWA Publishing)
ISBN 92 4 154533 X (World Health Organization)

Contents

	Foreword	*vii*
	Acknowledgements	*ix*
	List of Contributors	*x*
	Disclaimer	*xiv*
1	Harmonised assessment of risk and risk management for water-related infectious disease: an overview *Jamie Bartram, Lorna Fewtrell and Thor-Axel Stenström*	1
2	Guidelines: the current position *Arie Havelaar, Ursula J. Blumenthal, Martin Strauss, David Kay and Jamie Bartram*	17
3	The Global Burden of Disease study and applications in water, sanitation and hygiene *Annette Prüss and Arie Havelaar*	43
4	Endemic and epidemic infectious intestinal disease and its relationship to drinking water *Pierre Payment and Paul R. Hunter*	61
5	Excreta-related infections and the role of sanitation in the control of transmission *Richard Carr (with contributions from Martin Strauss)*	89
6	Disease surveillance and waterborne outbreaks *Yvonne Andersson and Patrick Bohan*	115
7	Epidemiology: a tool for the assessment of risk *Ursula J. Blumenthal, Jay M. Fleisher, Steve A. Esrey and Anne Peasey*	135
8	Risk assessment *Chuck Haas and Joseph N.S. Eisenberg*	161

9	Quality audit and the assessment of waterborne risk *Sally Macgill, Lorna Fewtrell, James Chudley and David Kay*	185
10	Acceptable risk *Paul R. Hunter and Lorna Fewtrell*	207
11	A public health perspective for establishing water-related guidelines and standards *Joseph N.S. Eisenberg, Jamie Bartram and Paul R. Hunter*	229
12	Management strategies *Dan Deere, Melita Stevens, Annette Davison, Greg Helm and Al Dufour*	257
13	Indicators of microbial water quality *Nicholas J. Ashbolt, Willie O.K. Grabow and Mario Snozzi*	289
14	Risk communication *Sue Lang, Lorna Fewtrell and Jamie Bartram*	317
15	Economic evaluation and priority-setting in water and sanitation interventions *Guy Hutton*	333
16	Implementation of guidelines: some practical aspects *Marcos von Sperling and Badri Fattal*	361
17	Regulation of microbiological quality in the water cycle *Guy Howard, Jamie Bartram, Stephen Schaub, Dan Deere and Mike Waite*	377
18	Framework for guidelines development in practice *David Kay, Dan Deere, Marcos von Sperling and Martin Strauss*	395
	Index	*413*

Foreword

The quality of water, whether it is used for drinking, irrigation or recreational purposes, is significant for health in both developing and developed countries worldwide. Water quality can have a major impact on health, both through outbreaks of waterborne disease and by contributing to the background rates of disease. Accordingly, countries develop water quality standards to protect public health. Recognising this, the World Health Organization (WHO) has developed a series of normative "guidelines" that present an authoritative assessment of the health risks associated with exposure to health hazards through water and of the effectiveness of approaches to their control. The three principal guidelines are intended to assist countries in establishing effective national or regional strategies and standards and are:

- *Guidelines for drinking-water quality.*[1]
- *Guidelines for the safe use of wastewater and excreta in agriculture and aquaculture.*[2]
- *Guidelines for safe recreational water environments.*[3]

These guidelines are updated as scientific and managerial developments occur, to ensure that they continue to be based on the best available evidence.

The assessment and management of the health risks associated with exposure to microbial hazards through water present special challenges, for example:

— not all of the microbial hazards (pathogens) are recognised and many cannot be readily enumerated or studied;
— adverse health effects may arise after a single exposure, yet water quality varies widely and rapidly;
— management actions are rarely of consistent effectiveness, and their outcome may be difficult to predict; and
— when water is unsafe, conventional testing indicates this only after exposure has occurred, i.e. too late to contribute to disease prevention.

---

[1] *Guidelines for drinking-water quality*, 2nd ed. (Addendum, in press). *Volume 1: recommendations*, 1993 (Addendum, 1998); *Volume 2: health criteria and other supporting information*, 1996 (Addendum, 1998); *Volume 3: surveillance and control of community supplies*, 1997. Geneva, World Health Organization.

[2] Mara D, Cairncross S. *Guidelines for the safe use of wastewater and excreta in agriculture and aquaculture.* Geneva, World Health Organization, 1989 (update in preparation).

[3] *Guidelines for safe recreational water environments.* Geneva, World Health Organization, in preparation.

To date, the various WHO guidelines relating to water have been developed in isolation from one another. Their primary water quality concern is for health hazards derived from excreta. Addressing their specific areas of concern together will tend to support better health protection and highlight the value of interventions directed at sources of pollution, which may otherwise be undervalued.

The potential to increase consistency in approaches to assessment and management of water-related microbial hazards was discussed by an international group of experts between 1999 and 2001. The group included professionals in the fields of drinking-water, irrigation, wastewater use and recreational water with expertise in public health, epidemiology, risk assessment/management, economics, communication, and the development of standards and regulations. These discussions led to the development of a harmonised framework, which was intended to inform the process of development of guidelines and standards. Subsequently, a series of reviews was progressively developed and refined, which addressed the principal issues of concern linking water and health to the establishment and implementation of effective, affordable and efficient guidelines and standards. This book is based on these reviews, together with the discussions of the harmonised framework and the issues surrounding it.

In its simplest form, the framework consists of an iterative cycle, comprising: an assessment of risk; health targets linked to the wider public health context; and risk management, with these components being informed by aspects of environmental exposure and tolerable ("acceptable") risk. A key component of the harmonised framework is the use of an inclusive range of tools for the assessment of risk, including epidemiology and information collected during the investigation of outbreaks of waterborne disease, as well as the formal risk assessment process (Chapters 6–8). Simultaneously, WHO is developing detailed guidelines on the characterization of hazards associated with exposure to both food and water, which will further aid the process of harmonisation. Another important development is the move towards integrated risk management strategies (Chapter 12). Information needs to be made available to managers in a timely manner, so that they can take appropriate action to prevent exposure to microbial hazards. Present approaches to end-product quality testing for microbial indicators are inadequate for this. Improved management of water safety therefore requires development, validation and use of more process-oriented indicators and testing methods (Chapter 13). This issue is being examined by WHO in collaboration with the Organization for Economic Co-operation and Development (OECD).

It is hoped that this book will be useful to all those concerned with issues relating to microbial water quality and health, including environmental and public health scientists, water scientists, policy-makers and those responsible for developing standards and regulations.

Acknowledgements

The World Health Organization wishes to express its appreciation to all those whose efforts made the production of this book possible. An international group of experts from a diverse range of backgrounds provided the material for the book and also submitted the material to a process of mutual review and endorsement. The contributors are listed on the following pages.

Thanks are also due to James Chudley, for assistance in the preparation of the illustrations, to Andy Fewtrell for technical advice, and to the Centre for Research into Environment and Health for its continued support.

Special thanks are due to the Karolinska Institute, the Ministry of Health of Sweden, the city of Stockholm, the Swedish Institute for Infectious Disease Control and the United States Environmental Protection Agency, which provided financial support for the meeting and this book.

List of Contributors

Yvonne Andersson
Swedish Institute for Infectious Disease Control, Stockholm, Sweden.
Email: yvonne.andersson@smi.ki.se

Nicholas J. Ashbolt
University of New South Wales, Sydney, New South Wales, Australia.
Email: n.ashbolt@unsw.edu.au

Jamie Bartram
World Health Organization, Geneva, Switzerland.
Email: bartramj@who.int

Ursula J. Blumenthal
London School of Hygiene and Tropical Medicine, London, England.
Email: ursula.blumenthal@lshtm.ac.uk

Patrick Bohan
Centers for Disease Control and Prevention, Atlanta, GA, USA.
Email: pfb3@cdc.gov

Richard Carr
World Health Organization, Geneva, Switzerland.
Email: carrr@who.int

James Chudley
School of the Environment, University of Leeds, Leeds, England.
Email: jameschudley@hotmail.com

Annette Davison
Australian Water Technologies, Sydney, New South Wales, Australia.[1]
Email: adavison@dlwc.nsw.gov.au

---

[1] Current address: Department of Land and Water Conservation, Sydney, New South Wales, Australia.

Dan Deere
Sydney Catchment Authority, Sydney, New South Wales, Australia.
Email: daniel.deere@sca.nsw.gov.au

Al Dufour
United States Environmental Protection Agency, Cincinnati, OH, USA.
Email: dufour.alfred@epa.gov

Joseph N.S. Eisenberg
School of Public Health, University of California, Berkeley, CA, USA.
Email: eisenber@socrates.berkeley.edu

Steve A. Esrey
United Nations Children's Fund, New York, NY, USA.
Email: sesrey@unicef.org

Badri Fattal
Hebrew University of Jerusalem, Jerusalem, Israel.
Email: badri@shum.cc.huji.ac.il

Lorna Fewtrell
Centre for Research into Environment and Health, Aberystwyth, Wales.
Email: lorna@creh.demon.co.uk

Jay M. Fleisher
Eastern Virginia Medical School, Norfolk, VA, USA.
Email: fleishe1@ix.netcom.com

Willie O.K. Grabow
University of Pretoria, Pretoria, South Africa.
Email: wgrabow@icon.co.za

Chuck Haas
Drexel University, Philadelphia, PA, USA.
Email: haascn@post.drexel.edu

Arie Havelaar
National Institute of Public Health and the Environment, Bilthoven, Netherlands.
Email: arie.havelaar@rivm.nl

Greg Helm
Sydney Water, Sydney, New South Wales, Australia.
Email: greg.helm@sydneywater.com.au

Guy Howard
Water, Engineering and Development Centre, Loughborough University, Loughborough, England.
Email: a.g.howard@lboro.ac.uk

Paul R. Hunter
Public Health Laboratory, Chester, England (current address Medical School, University of East Anglia, Norwich, England) paul.hunter@uea.ac.uk

Guy Hutton
Swiss Tropical Institute, Basel, Switzerland.
Email: guy.hutton@unibas.ch

David Kay
Centre for Research into Environment and Health, Aberystwyth, Wales.
Email: dave@crehkay.demon.co.uk

Sue Lang
South-East Water, Moorabbin, Victoria, Australia.
Email: sue.lang@sewl.com.au

Sally Macgill
School of the Environment, University of Leeds, Leeds, England.
Email: s.m.macgill@leeds.ac.uk

Pierre Payment
Armand-Frappier Institute, National Institute of Scientific Research, University of Quebec, Montreal, Quebec, Canada.
Email: pierre.payment@inrs-iaf.uquebec.ca

Anne Peasey
London School of Hygiene and Tropical Medicine, London, England.[2]
Email: apeasey@quetzal.innsz.mx

---

[2] Current address: National Institute of Medical Sciences and Nutrition, Mexico City, Mexico.

Annette Prüss
World Health Organization, Geneva, Switzerland.
Email: pruessa@who.int

Stephen Schaub
United States Environmental Protection Agency, Washington, DC, USA.
Email: schaub.stephen@epamail.epa.gov

Mario Snozzi
Swiss Federal Institute for Environmental Science and Technology, Dübendorf, Switzerland.
Email: snozzi@eawag.ch

Thor-Axel Stenström
Swedish Institute for Infectious Disease Control, Stockholm, Sweden.
Email: thor-axel.stenstrom@smi.ki.se

Melita Stevens
Melbourne Water Corporation, Melbourne, Victoria, Australia.
Email: melita.stevens@melbournewater.com.au

Martin Strauss
Swiss Federal Institute for Environmental Science and Technology, Dübendorf, Switzerland.
Email: strauss@eawag.ch

Marcos von Sperling
Federal University of Minas Gerais, Belo Horizonte, Brazil.
Email: marcos@desa.ufmg.br

Mike Waite
Drinking-Water Inspectorate, London, England.
Email: mike_waite@detr.gov.uk

Disclaimer

The opinions expressed in this publication are those of the authors and do not necessarily reflect the views or policies of the World Health Organization, United States Environmental Protection Agency or the Swedish Institute for Infectious Disease Control. In addition, the mention of specific manufacturer's products does not imply that they are endorsed or recommended in preference to others of a similar nature that are not mentioned. Errors and omissions excepted, the names of proprietary products are distinguished by initial capital letters.

1

Harmonised assessment of risk and risk management for water-related infectious disease: an overview

Jamie Bartram, Lorna Fewtrell and Thor-Axel Stenström

This chapter examines the need for a harmonised framework for the development of guidelines and standards in terms of water-related microbiological hazards. It outlines the proposed framework and details the recommendations derived from an expert meeting held to examine these issues. In its simplest form the framework consists of an iterative cycle, comprising an assessment of public health, an assessment of risk, health targets and risk management, with these components being informed by aspects of environmental exposure and acceptable risk.

 Water Quality: Guidelines, Standards and Health. Edited by Lorna Fewtrell and Jamie Bartram. Published by IWA Publishing, London, UK. ISBN: 1 900222 28 0

1.1 INTRODUCTION

In both developing and developed countries worldwide principal starting points for the setting of water quality standards, including microbiological standards, are World Health Organization Guidelines (Box 1.1).

These guidelines are, in large part, health risk assessments and are based upon scientific consensus, best available evidence and broad expert participation. The use of the term 'guidelines' is deliberate since they are not international standards. Rather, the intention is to provide a scientific, rational basis from which national standards are developed. It is specifically recognised that the process of adaptation requires that account be taken of social, economic and environmental factors and that the resulting standards may differ, sometimes appreciably, from the original guidelines. The guidelines advocate that a risk-benefit approach, whether quantitative or qualitative, be taken to the control of public health hazards associated with water.

Box 1.1. World Health Organization guidelines concerned with water quality

Guidelines for Drinking-water Quality
First published in 1984 in three volumes to replace earlier international standards. The guidelines are divided into three volumes:
Volume 1: Recommendations
Volume 2: Health Criteria and other Supporting Information
Volume 3: Surveillance and Control of Community Supplies.
Second editions of the three volumes were released in 1993, 1996 and 1997. Addenda to volumes 1 and 2 covering selected chemicals were released in 1998 and 1999 and a microbiological addendum is expected in 2001.

Guidelines for the Safe use of Wastewater and Excreta in Agriculture and Aquaculture
These were published in 1989 based upon the Engelberg guidelines and associated consultations and consensus. They replaced an earlier technical note (1973).

Guidelines for Safe Recreational Water Environments
These have been prepared progressively from 1994. Volume 1: Coastal and Freshwaters was released as a draft to the public domain for comment in 1998 and Volume 2: Swimming pools, spas and similar recreational water environments was released to the public domain for comment in 2000. Finalisation is envisaged in 2001. Volume 1 of the guidelines *per se* is supported by the text 'Monitoring Bathing Waters'.

In relation to chemical hazards, the guidelines for drinking-water quality (which provide the clearest example) are principally hazard characterisations in the context of the now 'classic' conception of risk assessment and risk management applied to chemical hazards. Delimiting the position of the guidelines to the rational scientific component of standard setting and advocating the role of national authorities in adapting guidelines to specific circumstances has proven a valuable means of supporting countries at all levels of socio-economic development and also a means of providing a common basis among them for activities protective of public health. While the guidelines are not international standards they are frequently referred to in international fora (such as the Codex Alimentarius Commission) as international points of reference for water quality, as well as supporting national standard setting.

In relation to microbiological hazards the sharp distinction between risk assessment and risk management that characterises approaches to chemical hazard is not maintained. This reflects a series of factors, most important among which are:

- The recognition that the hazards of greatest concern are multiple and share a common source - human excreta (and indeed that unrecognised hazards from the same source exist).
- The recognition that important health effects (both acute and delayed) may occur as a result of short-term exposure.
- The approach (derived from traditional 'hygiene' but reflected in modern risk management such as the hazard analysis and critical control point (HACCP) principles used in the food industry) that because the pathogens of concern are widespread and because their occurrence varies widely and rapidly in time and space, the absence of (a) safeguard(s) in itself constitutes a hazard.

As a result, all three of the WHO water quality-related guidelines include requirements for what may loosely be described as 'adequate safeguards' or 'good practice', in addition to stipulating numerical values for water quality measures. Whereas in the case of chemical hazards, the principal outcome is a guideline value expressed as a concentration of the substance of concern (i.e. a direct measurement of the human health hazard), in the case of microbiological hazards, the guideline is expressed in terms of measures not of the hazard itself, but of indicators that would assist in confirming that adequate safeguards were in place and operating within reasonable performance requirements (Table 1.1). Such measures include both analytical measurements and inspection-based procedures.

Table 1.1. Indicators and good practice requirements by guideline area

Guideline area	Indicators	Good practice requirements
Drinking-water quality	Value stipulated for faecal coliforms, with recommendations on turbidity, pH and disinfection (chlorination)	Groundwater source protection Treatment proportional to (surface) water quality Sanitary inspection as part of surveillance and control
Safe use of wastewater and excreta in agriculture and aquaculture	Faecal coliforms (unrestricted irrigation) Intestinal helminth counts (restricted and unrestricted irrigation) Trematode egg counts (aquaculture)	Involvement of adequate treatment chains
Safe recreational water environments	Numerical values for indicators (faecal streptococci/enterococci) related to defined levels of risk	'Annapolis Protocol' proposes a series of interventions

The three guidelines differ appreciably from one another, reflecting the state of scientific advance in the three distinct areas that they cover at the time they were produced (see Chapter 2). As a result, it is unlikely that they provide equivalence in terms of the degree of health protection provided by each.

1.2 THE NEED FOR A HARMONISED FRAMEWORK

In the areas of drinking water and wastewater and excreta reuse substantial new epidemiological evidence has become available since the time of the original development of the corresponding WHO guidelines. In parallel, the science of microbiological risk assessment has advanced and continues to advance rapidly, and substantial developments have occurred in the science and application of integrated water resource management. In the broader sphere of public health:

- There has been increasing acceptance that hazards previously managed in isolation should be understood as aspects of a whole.
- There has been an increasing demand for evidence-based decision making.
- There has been an increasing demand for information to support cost-benefit analysis.

In relation to microbiological aspects of water quality it is clear that the three areas of guidelines discussed here are joined by a common source of the hazard of primary concern – human (and to a lesser extent animal) excreta. They are therefore inseparable from the issue of adequate sanitation to contain, inactivate and control the pathogens derived from such excreta (Chapter 5). Dealing with the three aspects in isolation will tend to discriminate against interventions close to the source of the hazard (which is therefore contrary to the general principle of containing and treating pollution close to source).

Demands for an improved environment and health evidence base have tended to focus on the need to describe the response of communities (and individuals) to specific exposures to pollutants of concern. The evidence base for what is in effect 'population dose–response' is often weak. It is derived, directly or indirectly, from four principal sources of information:

- Epidemiological study of disease occurring under 'normal' situations of exposure. (Such studies may be better or worse controlled; exposure may be reasonably described. The study size is limited principally by financial considerations and the ability to define suitable study groups. Such studies reflect real populations under real conditions of exposure and are therefore of unique value.)
- Study of outbreaks of disease. (Such studies also reflect real populations under real conditions of exposure but the utility of information generated is often constrained by the inability to retrospectively estimate exposure and the physical constraints of the natural event and by necessarily reactive investigation.)
- Human volunteer studies (highly controlled but artificial exposures amongst real human populations).
- Microbiological risk assessment (which provides a framework through which data from multiple sources may be combined and used more effectively than in isolation).

It should be noted that the first two of these provide not only information concerning population dose–response but also information concerning the effectiveness of preventive measures.

When considering only health-related outcomes of environmental interventions, difficult choices have to be made regarding the relative priority that should be given to multiple interventions competing for limited available resourcing (even where the financial resourcing for the intervention is outside the health sector *per se,* as is commonly the case). During the earlier part of the

'Water Decade' (1981–90), for example, it was suggested that an intervention that was acting on a cause of less than 5% of diarrhoeal disease burden should not be justified on health grounds but, rather, interventions acting on greater proportions should be prioritised. The problem is analogous (although not equivalent) to that of 'apportionment' of exposure to chemical hazards through multiple routes. Such simplifications, while illustrative of real concerns, have tended to be superseded by demands for more comprehensive cost-benefit analysis – itself extremely difficult to apply to environmental interventions with health benefits.

Costs of interventions may be high and substantial benefits may accrue not only to health but also to, for example, diverse economic sectors (see Chapter 15). Both health and non-health benefits may be delayed. Care is therefore required in promoting one area of intervention (or indeed one specific intervention) on the basis of health gain and there is an increasingly recognised need for representatives of the health sector to engage more effectively as participants in intersectoral planning and decision-making.

The limited inter-guideline consistency, new advances, and the need to take a more holistic approach to risk management logically lead to the need for a harmonised approach to the development of guidelines for water-related exposures to microbiological hazards.

This issue was tackled by a group of experts at a meeting in Stockholm held in September 1999. The output from the meeting was the proposal of a harmonised framework to inform guideline development and revision, along with a series of recommendations for the adoption of the framework. The remainder of this chapter describes the framework and the principal reasons underpinning its elements. It also outlines the important issues that are covered in greater detail in other chapters of this book.

1.3 THE OVERALL FRAMEWORK

Experts at the meeting in Stockholm agreed that future guidelines should integrate assessment of risk, risk management options and exposure control elements within a single framework with embedded quality targets. The normative part of the end product of the guidelines would therefore constitute the requirement to define, adopt and implement a strategy and measures to adequately protect human health appropriate to specific conditions. While this would require the embedding of water quality targets (in turn justified on the basis of targets for health protection) and also the development of measures and limit values for measures of water quality, the experts recommended strongly that such measures and values were a part of, and supportive, to the requirement to define and exercise good management. The harmonised framework put a mechanism in place to achieve this goal, which would be applicable within and between the three areas of present concern (drinking water, wastewater and recreational water). It also allows the guidelines to be considered within the overall context of public health policy and transmission of disease through other routes.

In its simplest form the framework can be conceptualised as shown in Figure 1.1. It is essentially an iterative process linking assessment of risk with risk management via the definition of health targets and the assessment of health outcomes. While health targets and outcomes are inevitably local or national in character, the former can be informed by 'acceptable risk' which provides a means to support the development of internationally-relevant guidelines which can, in turn, be adapted to specific national and local conditions.

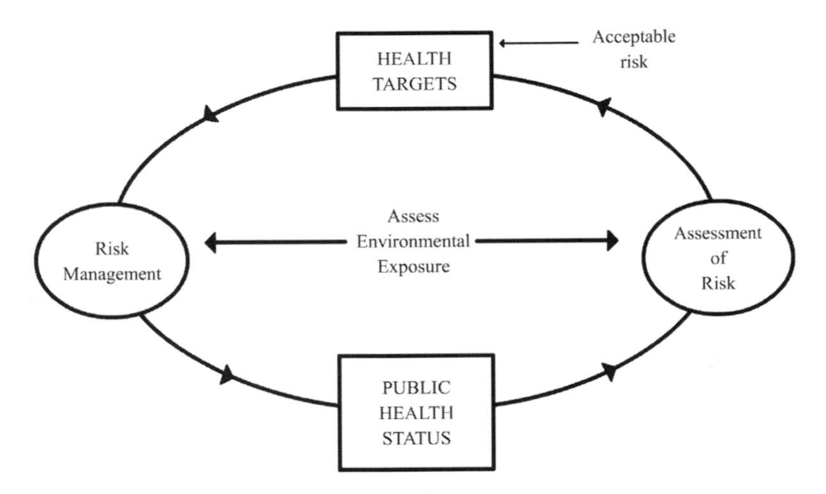

Figure 1.1. A simplified framework.

1.3.1 Assessment of risk in the overall framework

In this framework, the assessment of risk is not a goal in its own right but rather a basis for decision-making and in the first iteration of the process it is the starting point. For the purposes of WHO guidelines the exclusive emphasis is upon health and, as such, the assessment is an assessment of health risk. In applying the guidelines to specific circumstances one may wish to take into account other non-health factors and in practice these may have a considerable impact upon both costs and benefits.

For the purposes of microbiological hazards, the health risk is the risk of disease, which in turn translates into the risk of infection. The group recommended that the guidelines utilise a best estimate of risk and not overlay conservative or safety factors as a means to accommodate uncertainty. This was recommended in order to better inform decision-making and especially the prioritisation of interventions and cost-benefit analysis. It was recognised that this would in turn lead to an iterative process within the guidelines themselves and progressive adjustment to take account of new information. Assuming equivalence between risk of infection and risk of disease may appear to be a measure of conservatism. It is also, however, a means to specifically reflect the health concerns of more sensitive members of the normal population, such as children who in the absence of previous exposure have not developed immunity. As such it is similar to the approach taken towards chemical hazards in the 'guidelines for drinking-water quality'.

Given the diverse range of possible infections which may be water-related, the range in severity of immediate health outcome and also the existence of, sometimes important, delayed effects associated with some of the infections concerned, a common exchange unit (such as Disability Adjusted Life Years (DALYs)) was considered essential to account for acute, delayed and chronic effects (including both morbidity and mortality) in order to maximise relevance to policy making and decision-taking.

The guidelines should operate from the assumptions that pathogens do occur in the environment (unless there is specific reason to exclude a particular pathogen, such as its absolute absence from the area under consideration) and that there is a susceptible population. These assumptions are strongly supported by the evidence outlined in Chapters 3–6, and by the continued occurrence of water borne disease outbreaks in countries, at all levels of socio-economic development, worldwide.

Full use should be made of the vast array of information sources, studies and tools to inform the assessment. Where available and appropriate, information sources should include outbreak investigation (Chapter 6), epidemiological studies (Chapter 7) and microbiological risk assessment (Chapter 8) as well as studies on behaviour of microbes in the environment (and their inactivation, removal and addition/multiplication through resource and source management and in water abstraction and use). Some of these sources provide information on exposure-response, some on the effectiveness of interventions and some on both. Bringing together information on these two aspects of health protection was considered important.

Explicit attention should be paid to the quality of studies and of data and information from them (Chapter 9). In general, publication in the internationally accessible peer-reviewed literature serves as an initial screen for quality but is not a guarantee of it. Coherence among multiple studies (including differences with rational explanation) should be seen as an important element in determining the quality of evidence. Ideally a simple ranking scheme should be developed to assist in assessing the quality of available evidence in terms of its suitability for demonstrating cause-effect and (separately) for supporting quantitative study (including guidelines derivation).

Considerable discussion at the meeting of experts related to the importance of short-term deviations in quality to health, to the extent that overall health risk may be dominated not by the 'typical' or 'average' water quality but water quality in short periods of sub-optimal performance (even where these may in fact comply with conventional 'standards'). The overall agreement was that specific measures were required to enable identification and management response to such events and also that such events should be properly accounted for in estimating human health risk.

1.4 THE ELEMENTS OF THE FRAMEWORK

This section describes the individual elements of the framework in more detail. Figure 1.2 shows an expanded version of the framework shown in Figure 1.1.

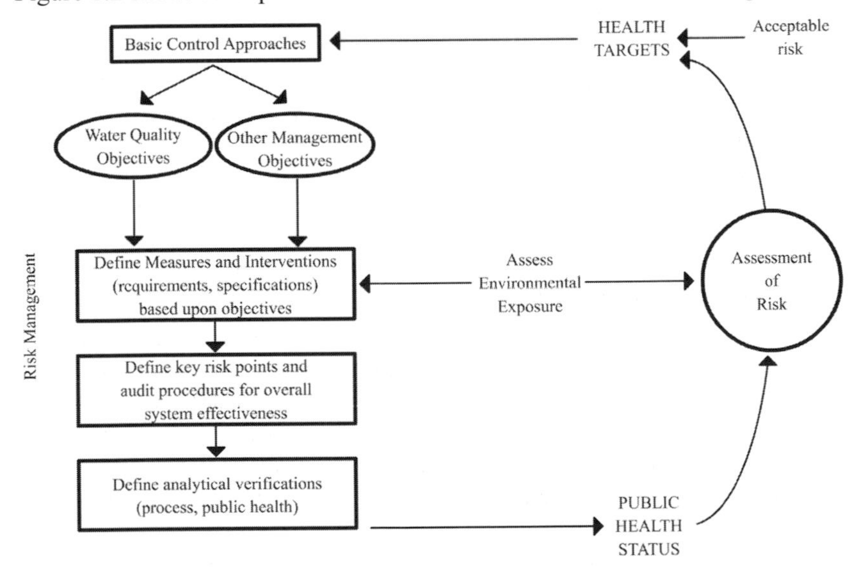

Figure 1.2. Expanded framework.

1.4.1 Environmental exposure assessment

Environmental exposure assessment is an important input to both the assessment of risk and to risk management. Exposure assessment is a formal component of the risk assessment process (Chapter 8).

Exposure assessment is a required input for microbiological risk assessment. As noted earlier, the expert group that met in Stockholm agreed that the harmonised process should be based upon the assumption that pathogens occur in the environment. However, representative quantified assumptions will have to be made in the development of guidelines and these may then be one of the fields for adaptation in passing from guidelines to national and/or local standards. In such a process of adaptation, both pathogen occurrence *per se* and, indirectly, weighting factors applied to pathogens of greater concern should be taken into account. Paradoxically this might imply the need for greater stringency in protective measures and safeguards in less developed countries where capacities to apply such measures are least.

An important role for environmental exposure assessment is in prioritisation among potential interventions in the context of overall environmental exposure to pathogenic micro-organisms. Thus, for example, if most exposure to a given pathogen occurs from non-water related sources and, say, only 5% of the burden of disease is associated with (for example) drinking water, then it may reasonably be argued that greater public health benefit is likely to be achieved by intervening in the other routes of exposure. Such simple analysis in practice is conditioned by factors such as the availability of interventions in the various exposure routes and their cost. Furthermore, prioritisation of this type is normally applied to at the local and national levels and is not applicable within the context of global guidelines, where representative assumptions must be made that may then be amended by local and national authorities to take account of specific conditions.

1.4.2 Acceptable risk and health targets

In its Guidelines for Drinking-water Quality (1993), WHO suggests that:

> The judgement of safety – or what is an acceptable level of risk in particular circumstances – is a matter in which society as a whole has a role to play. The final judgement as to whether the benefit resulting from the adoption of any of the Guideline Values … justifies the cost is for each country to decide.

While the general public may prefer the idea of 'zero risk', in a world of limited resources and competing demands some idea of tolerable risk is vital in order that health targets are sensible and achievable and that measures to pursue them are cost-effective.

There is increasing recognition, especially among the policy-making and scientific communities, of the concept of 'acceptable risk'. The term 'tolerable risk' is preferred by some workers to recognise that the risk is not truly acceptable but may be tolerated, either absolutely, or in deference to greater or more highly perceived priorities.

Different agencies have begun to explore what might constitute a tolerable disease burden. WHO, for example, calculates its guideline values for genotoxic carcinogens (for which there is no threshold concentration below which there is zero risk) as equivalent to the upper bound estimate of the one in 100,000 lifetime excess risk (of cancer). For other toxic chemicals, where a threshold does exist, guideline values are set in relation to this. The present state of knowledge suggests that infection and disease can be initiated by a single micro-organism and can therefore show non-threshold properties. The consequence, given that sterility is not a feasible goal, is the need to recognise the issue of 'tolerable' risk (see Chapter 10). The United States surface water treatment rule is concerned with minimising health risks from pathogenic micro-organisms occurring in surface waters and originally established a goal that fewer than one person in 10,000 per year would become infected from exposure to the protozoan *Giardia* in drinking water (and this was assumed to be protective against other diseases at the time).

All present descriptions of tolerable disease burden in relation to water are expressed in terms of specific health outcomes (such as cancer, diarrhoeal disease, etc.). The expert group in Stockholm was concerned that such approaches would prove problematic in relating some common water-related diseases to one another, whether because of their diverse acute effects (cholera, dysentery, typhoid, infectious hepatitis, intestinal worms) or because of their varied severity weightings (mild self-limiting diarrhoea through to significant case mortality rates) or because of delayed effects (such as the association of Guillain-Barré syndrome with campylobacteriosis). The group therefore recommended that a reference level of acceptable risk be adopted which should be expressed in DALYs with an appropriate accompanying explanation to assist non-expert readers in interpreting its significance.

Unnecessarily strict guidelines and standards may militate against beneficial uses of water and therefore prevent society from enjoying their benefits. Recreational water use leads to significant benefits to the individual and to society as a whole (rest, recreation, hygiene) and guidelines and standards should be established that are protective of public health without unnecessarily hampering the enjoyment of these benefits. The use of wastewater in irrigation can similarly contribute to food security, the closing of nutrient cycles in agriculture and improved conservation and protection of aquatic ecosystems. Such benefits should be considered alongside the requirements for the protection of human health.

Wealthy and poor countries are united by increasing prevalence of sensitive sub-populations, particularly those that are immunocompromised, in addition to the young, elderly and pregnant. The issue of immunocompromised populations has been especially highlighted because of HIV/AIDS but in some (especially more industrially developed) regions other causes (notably therapy) may also be significant. Questions remain regarding water quality requirements to protect specific sensitive sub-populations and the Stockholm group therefore recommended that guidelines normally be set so as to offer protection throughout a lifetime, acknowledging the different sensitivities and susceptibilities within that timeframe (i.e. to include the young, elderly and pregnant). For more specific sub-groups, the prevalence of which may vary widely between countries and whose water quality requirements may not be achievable through available measures, additional guidance should be included where adequate evidence allows this.

Health targets are to be based upon the outcome of the assessment of risk and on information concerning levels of acceptable risk. Although health targets have not, as yet, been used in WHO water-related guidelines they have been used very successfully in other areas. Table 1.2 outlines some of their benefits.

Table 1.2. Benefits deriving from the use of health targets

Target development stage	Benefit
Formulation	Gives insight into the health of the population
	Reveals gaps in knowledge
	Gives insight into consequences of alternative strategies
	Supports the priority-setting process
	Increases the transparency of health policy
	Ensures consistency among several health programmes
	Stimulates debate
Implementation	Inspires and motivates partners to take action
	Improves commitment
	Fosters accountability
	Guides the allocation of resources
Monitoring and evaluation	Supplies concrete milestones for evaluation and adjustments
	Provides opportunities to test feasibility of the targets
	Provides opportunities to take actions to correct deviations
	Exposes data needs and discrepancies

WHO guidelines should be relevant to the widely varying socio-cultural, economic and environmental conditions that prevail in different countries and regions. Use of a reference level would facilitate the adaptation of guidelines to enable account to be taken of such conditions. In consequence, it was felt to be important that guidelines make explicit reference to and provide guidance on issues associated with the adaptation of guidelines to standards.

1.4.3 Risk management

Consideration of the risk management process leads to the expanded version of the framework as shown in Figure 1.2. Based on the defined health targets acceptable risk water quality targets are defined. Ideally, such health targets will employ a selected index pathogen (see Chapter 13) that combines both control challenges and health significance in terms of health hazard and, ideally, the availability of other relevant data. In practice, more than one pathogen will normally be required in order properly to reflect diverse challenges to the safeguards available. While water quality targets may be expressed in terms of exposure to specific pathogens, care is required in relating this to overall population exposure, which may be concentrated into small periods of time. Further care is required to account properly for potentially 'catastrophic' events (leading to large-scale outbreaks of disease) rather than only for background rates of disease during normal cycles of performance and efficiency. Both relate to the recognised phenomenon of short periods of very decreased efficiency in many processes and provide a logical justification for the long established 'multiple barrier principle' in water safety. It is important to note that the inclusion of water quality targets expressed in terms of human exposure to pathogens does not imply that those pathogens should be directly measured, nor even that the capacity for such measurement should be within the analytical capacity of normal ('routine') monitoring laboratories, nor that measuring their reduction to below the water quality target necessarily implies safety. This is because the reference pathogens act as surrogates for other pathogens in determining safe practices but may not necessarily occur in the environment when other pathogens of concern occur.

Information concerning the efficiency of processes combined with data on the occurrence of pathogens in source waters and water quality targets enables definition of operating conditions that would reasonably be expected to achieve those targets. In this, information on process efficiency and pathogen occurrence should take account of steady-state performance and performance during maintenance and periods of unusual load. While the indicator systems required to verify adequate performance may require the use of 'conventional' laboratory-based analytical measures, it was seen that overall a greater relative emphasis would be given to periodic inspection/auditing and to simple measurements that could be rapidly and frequently made and directly inform management. Greater emphasis on measures to confirm that processes are operating as expected is required to protect public health and this will create challenges for the form of present approaches to monitoring.

Within each set of guidelines, water quality objectives and their associated management controls will need to respond not only to 'steady-state' conditions but also the possibility of short-term events (such as variation in environmental water quality, system challenges and process problems) in order to minimise the likelihood of outbreaks of disease.

The overall package of appropriate measures will vary between countries and localities. In order that guidelines be relevant and supportive, the experts recommended that representative scenarios including description of assumptions, management options, critical control points and indicator systems for verification be included (see Chapter 12). It was envisaged that these would be supported by general guidance regarding the identification of priorities and regarding progressive implementation that would be of special, but not unique, relevance to less industrially developed countries, thereby helping to ensure that best use is made of limited resources.

The expert group suggested that the management strategy adopted within the risk management process, whilst being adapted to the specific needs of the respective guidelines, should be based on the extensive and accumulating experience with Hazard Analysis and Critical Control Points (HACCP). An examination of various management tools, including details of HACCP, is made in Chapter 12.

1.4.4 Implementation

A range of tools and approaches may be deployed in seeking implementation. These may include incentives, legal enforcement, education (both professional and public) and so on (see Chapters 14–17). They may be linked to wider level management (e.g. integrated basin or coastal zone management) or may fall largely outside traditional water sector management (certification of materials, chemicals, operators, consumer protection, and so on). While general comment on the available measures and experience with their effective application is important, detailed guidance on such aspects (which vary widely with social, political, economic and cultural factors) is not universally applicable and should not therefore constitute a part of the guidelines.

The issue of progressive implementation is however a prime concern for the guidelines and is of universal relevance. WHO guidelines should provide explicit guidance on step-wise implementation. Advice, in the form of a procedure, on gradation and likely speed of achievement will reduce false expectations and should increase incentives for compliance. The need for stepwise implementation based upon public health priority is especially great in developing countries, a point which is well illustrated in Chapter 16.

1.4.5 Public health status

There has been an increasing trend to reappraise the 'linear' presentation of risk assessment and associated risk management into a more circular format, recognising both the need to respond to advances and general developments and to explicitly address the incremental nature of most environment and health decision making and the need to identify and to respond to both successes and failures through specific feedback. Such a circular process better accommodates the need to identify opportunities for public participation.

The final stage before re-entering the process is, therefore, logically to examine the public health outcome (see Chapter 11). Are the measures being put into place having the desired effects in the required time frame? The first iteration or iterations may lead to water quality objectives and management objectives being met without the desired public health outcome, or contrariwise that a greater response is achieved than expected. Equally, it may be found that the 'management and implementation' side of the circle requires further attention in order that the measures applied lead to the desired management changes. Without explicitly addressing these aspects it is impossible to see if the processes put into place are effective. Failure to achieve stated health targets in early stages should not be seen as a weakness of the approach but as part of the process, enabling best use to be made of resources, and also a source of experience and information with which to inform future stages.

Approaches to reliably estimating the disease burden (Chapter 3) are under development and, if reliable and adequately sensitive, will be important at this stage as they will allow changes to be monitored. Measurement of public health outcomes will vary between countries and it is recognised that present approaches and capacities for both surveillance and for outbreak detection and investigation are typically inadequate for this purpose.

1.5 FURTHER DEVELOPMENT

The proposed harmonised framework has not yet been subjected to the acid test of implementation. Groups of experts, however, have tested the process in a desk exercise examining hypothetical studies from each of the guideline areas. These are detailed in Chapter 18.

It is likely that there will be extensive data requirements to support the application of guidelines of all types at country level. While some of this information will be presented in the guidelines *per se*, WHO could also be instrumental in collating, synthesising and making more readily available such information and this was considered a priority by experts at the Stockholm meeting.

Outcomes, especially health-related outcomes, deriving from the implementation of the guidelines within the three areas of concern are, and will continue to be, very important in disease reduction in terms of global burden of disease in both developing and developed countries. However, until recently, there has been a trend in some quarters to believe that drinking water in more industrially developed countries was the cause of little disease and that infectious disease in particular was of largely historical interest. The experience with a single recently recognised pathogen significantly associated with water borne disease (i.e. *Cryptosporidium*) has shattered that optimistic assessment and focused interest on this area of universal concern (Chapter 6).

Experts noted that the experience of bringing together individuals from three sub-sectors (drinking water, recreational water and wastewater reuse) and from different disciplinary areas (risk assessment, epidemiology, engineering, regulatory affairs and economics) has highlighted the need for care in the use of terms that may be used with subtle or grossly different meanings, and recommended that all guidelines be accompanied with a simple glossary of terms to minimise misunderstanding.

2

Guidelines: the current position

Arie Havelaar, Ursula J. Blumenthal, Martin Strauss, David Kay and Jamie Bartram

The setting of guidelines is a key normative function of the World Health Organization. This chapter examines the development of the current water-related WHO guidelines. Within the area of water, microbiology and guideline setting there are three distinct but related areas, namely:

- drinking water;
- wastewater reuse; and
- recreational water.

The following explores the background to the current guidelines, highlighting the different pathways to their formation.

2.1 INTRODUCTION

The aim of the water-related WHO guidelines is the protection of public health. They are intended to be used as the basis for the development of national standards and as such the values recommended are not mandatory limits, but are designed to be used in the development of risk management strategies which may include national or regional standards in the context of local or national environmental, social, economic and cultural conditions. The main reason for not promoting the adoption of international standards is the advantage provided by the use of a risk-benefit approach to the establishment of specific national standards or regulations. This approach is thought to promote the adoption of standards that can be readily implemented and enforced and should ensure the use of available financial, technical and institutional resources for maximum public benefit.

2.2 GUIDELINES FOR DRINKING-WATER QUALITY

The WHO Guidelines for Drinking-Water Quality (GDWQ) have a long history and were among the first environmental health documents published by the Organization. The first WHO publication dealing specifically with drinking-water quality was published in 1958 as International Standards for Drinking-Water. It was subsequently revised in 1963 and 1971 under the same title. To encourage countries of advanced economic and technological capabilities in Europe to attain higher standards, and to address hazards related to industrial development and intensive agriculture, the European Standards for Drinking-Water Quality were published in 1961 and revised in 1970. In the mid-1980s the first edition of the WHO guidelines for Drinking-Water Quality was published in three volumes:

- Volume 1: Recommendations
- Volume 2: Health criteria and other supporting information
- Volume 3: Surveillance and control of community water supplies.

The second editions of the three volumes were published in 1993, 1996 and 1997 respectively. In 1995, a co-ordinating committee decided that the GDWQ would be subject to rolling revision, and three working groups were established to address microbiological aspects, chemical aspects and aspects of protection and control of drinking water quality.

As with all the water-related guidelines the primary aim of the GDWQ is the protection of human health, and to serve as a basis for development of national water quality standards. The guideline values recommended for individual constituents are not mandatory limits but if they are properly implemented in light of local circumstances will ensure the safety of drinking water supplies through the elimination, or reduction to a minimum concentration, of constituents of water that are known to be hazardous to health.

The GDWQ cover chemical and physical aspects of water quality as well as the microbiological aspects which are the focus of this publication. Within the GDWQ it is emphasised that the control of microbiological contamination is of paramount importance and must never be compromised. Likewise, it is stated that disinfection should not be compromised in attempting to control chemical by-products.

Chemical, physical and radiological contaminants are extensively covered by critical review and summary risk assessment documents published by international bodies such as the International Programme on Chemical Safety (IPCS), the International Agency for Research on Cancer (IARC), Joint FAO/WHO Meetings on Pesticide Residues (JMPR) and Joint FAO/WHO Expert Consultation on Food Additives (JECFA). These documents are mainly based on animal studies. For most chemicals, the risk assessment results in the derivation of a threshold dose below which no adverse effects are assumed to occur. This value is the basis for a Tolerable Daily Intake (TDI), which can be converted into a guideline value for a maximum allowable concentration in drinking water using a series of assumptions and uncertainty factors. For genotoxic carcinogens a threshold value is not assumed to exist, and the guideline value is based on extrapolation of the animal dose–response data to the low dose region typically occurring through drinking water exposure. Concentrations associated with an excess lifetime cancer risk of 10^{-5} are presented as guideline values. For both types of chemical substances, with and without threshold values, the guidelines take the form of end-product standards, which can be evaluated by chemical analysis of the finished water or the water at the point of consumption. However, guideline values are not set at concentrations lower than the detection limits achievable under routine laboratory operating conditions and are recommended only when control techniques are available to remove or reduce the concentration of the contaminant to the desired level.

Microbiological risks are treated very differently. In Volume 2, reviews are available of the characteristics of many different pathogenic micro-organisms, and an Addendum covering new information on a number of important pathogens is in preparation (Table 2.1).

Table 2.1. Pathogens reviewed in GDWQ (Volume 2, 1996 and Addendum, in preparation)

Bacteria	Viruses	Protozoa and Helminths
Salmonella	Picornaviruses (inc. Hep A)	*Giardia*
Yersinia	Adenoviruses	*Cryptosporidium*
Campylobacter	Parvoviruses	*Entamoeba histolytica*
Vibrio cholera	Small round structured viruses	*Balantidium coli*
Shigella	Hepatitis E virus	*Naegleria* + *Acanthamoeba*
Legionella	Papovaviruses	*Dracunculus medinensis*
Aeromonas		*Schistosoma*
Ps. Aeruginosa		*Cyclospora cayatenensis*
Mycobacterium		
Cyanobacterial toxins		

However, the information on pathogens is barely used in the derivation of guidelines for the production of safe drinking water. Instead, the guidelines are based on tried and tested principles of prevention of faecal pollution and good engineering practice. This approach results in end product standards for faecal indicator organisms and operational guidelines for source water protection and adequate treatment. These aspects are complementary but only loosely connected.

2.2.1 Faecal indicator organisms

The rationale for using faecal indicator organisms as the basis for microbiological criteria is stated as follows:

It is difficult with the epidemiological knowledge currently available to assess the risk to health presented by any particular level of pathogens in water, since this risk will depend equally on the infectivity and invasiveness of the pathogen and on the innate and acquired immunity of the individuals consuming the water. It is only prudent to assume, therefore, that no water in which pathogenic micro-organisms can be detected can be regarded as safe, however low the concentration. Furthermore, only certain waterborne pathogens can be detected reliably and easily in water, and some cannot be detected at all. (WHO 1996 p. 93)

Escherichia coli and to a lesser extent thermotolerant coliform bacteria are considered to best fulfil the criteria to be satisfied by an ideal indicator. These are:

- universally present in large numbers in the faeces of humans and warm-blooded animals;
- readily detected by simple methods;
- do not grow in natural waters; and
- persistence in water and removal by water treatment similar to waterborne pathogens.

It is recommended that when resources are scarce it is more important to examine drinking-water frequently by means of a simple test than less often by several tests or a more complicated one. Hence, the recommendations are mainly based on the level of *Escherichia coli* (or thermotolerant coliform organisms). Basically, the criterion is that *E. coli* must not be detectable in any 100 millilitre (ml) sample. For treated water entering, or in, the distribution system the same recommendation is also given for total coliform bacteria, with a provision for up to 5% positive samples within the distribution system. The rationale for this additional criterion is the greater sensitivity of total coliforms for detecting irregularities (not necessarily faecal contamination) in treatment and distribution. The concept of indicators is covered in detail in Chapter 13.

In many developing countries, high quality water meeting the *E. coli* criterion is not readily available, and uncritical enforcement of the guideline may lead to condemnation of water sources that may be more appropriate or more accessible than other sources, and may even force people to obtain their water from more polluted sources. Under conditions of widespread faecal contamination, national surveillance agencies are recommended to set intermediate goals that will eventually lead to the provision of high quality water to all, but will not lead to improper condemnation of relatively acceptable supplies (this is expanded upon in Volume 3 of the GDWQ).

2.2.2 Operational guidelines

The GDWQ do not specify quantitative criteria for virus concentrations in drinking water. Estimates of health risks linked to the consumption of contaminated drinking-water are not considered sufficiently developed to do so, and the difficulties and expense related to monitoring viruses in drinking water preclude their practical application. Similar considerations preclude the setting of guideline values for pathogenic protozoa, helminths and free-living (parasitic) organisms. Instead, the importance of appropriate source water protection and treatment related to the source water quality are emphasised. Recommended treatment schemes include disinfection only for protected deep wells and protected, impounded upland waters. For unprotected wells and impounded water or upland rivers, additional filtration is recommended and more extensive storage and treatment schemes are recommended for unprotected watersheds. Different treatment processes are described in Volume 2 (WHO 1996) in some detail. Performance objectives for typical treatment chains are also outlined, including, for example, the recommendation that turbidity should not exceed 1 Nephelometric Turbidity Unit (NTU) under average loading conditions, and 5 NTU under maximum loading.

The experience gained in surveillance and improvement of small-community supplies through a series of WHO-supported and other demonstration projects is reflected in Volume 3 (WHO 1997). This gives detailed guidance on all aspects of planning and executing surveillance programmes, emphasising the importance of sanitary inspection as an adjunct to water quality analysis. There is also guidance on technical interventions to improve water quality by source protection, by affordable treatment and disinfection and by household water treatment and storage.

2.3 SAFE USE OF WASTEWATER AND EXCRETA IN AGRICULTURE AND AQUACULTURE

All around the world, people both in rural and urban areas have been using human excreta for centuries to fertilise fields and fishponds and to maintain the soil organic fraction. Use of faecal sludge in both agriculture and aquaculture continues to be common in China and south-east Asia as well as in various African countries. In the majority of cases, the faecal sludge collected from septic tanks and unsewered family and public toilets is applied untreated or only partially treated through storage.

Where water-borne excreta disposal (sewerage) was put in place, the use of the wastewater in agriculture became rapidly established, particularly in arid and seasonally arid zones. Wastewater is used as a source of irrigation water as well as a source of plant nutrients, allowing farmers to reduce or even eliminate the purchase of chemical fertiliser. Recent wastewater use practices range from the piped distribution of secondary treated wastewater (i.e. mechanical and biological treatment) to peri-urban citrus fruit farms (e.g. the city of Tunis) to farmers illegally accessing and breaking up buried trunk sewers from which raw wastewater is diverted to vegetable fields (e.g. the city of Lima). Agricultural reuse of wastewater is practised throughout South America and in Mexico and is also widespread in Northern Africa, Southern Europe, Western Asia, on the Arabian Peninsular, in South Asia and in the US. Vegetable, fodder and non-food crops as well as green belt areas and golf courses are being irrigated. In a few countries (such as the US and Saudi Arabia) wastewater is subjected to advanced treatment (secondary treatment, filtration and disinfection) prior to use.

The use of human wastes contributes significantly to food production and income generation, notably so in the fast-growing urban fringes of developing countries. Yet, where the waste is used untreated or health protection measures other than treatment are not in place, such practice contributes to the 'recycling' of excreted pathogens among the urban/peri-urban populace. Farmers and their families making use of untreated faecal sludge or wastewater, as well as consumers, are exposed to high risks of disease transmission.

2.3.1 History of wastewater reuse guideline development

The wastewater reuse guidelines enacted in California in 1918 may have been the first ones of their kind. They were modified and expanded and now stipulate a total coliform (TC) quality standard of 2.2/100 ml (seven-day median) for wastewater used to irrigate vegetable crops eaten uncooked (State of California 1978). This essentially means that faecal contamination should be absent and there should be no potential risk of infection present (although low coliform levels do not necessarily equate to low pathogen levels). The level of 2.2 TC/100ml is virtually the same as the standard expected for drinking water quality and was based on a 'zero risk' concept. The standard set for the irrigation of pastures grazed by milking animals and of landscape areas with limited public access is also quite restrictive, and amounts to 23 total coliforms/100 ml. Such levels were thought to be required to guarantee that residual irrigation water attached to vegetables at the time of harvest would not exceed drinking water quality limits. However, vegetables bought on open markets that are grown with rainwater or freshwater (which is often overtly or covertly polluted with raw or partially treated wastewater) may exhibit faecal indicator counts much higher than this. The Californian standards were influential in the formulation of national reuse guidelines by the US Environmental Protection Agency (US EPA/USID 1992), which are designed to guide individual US states in the formulation of their own reuse regulations. They also influence countries which export wastewater-irrigated produce to the US, as the exporting country is under some pressure to meet the water quality standards of the US.

The formulation of the 'California' standards was strongly influenced by the wastewater treatment technologies in use in industrialised countries at the time. This comprised secondary treatment (activated sludge or trickling filter plants) for the removal of organic contaminants, followed by chlorination for removal of bacteria. Such technology can result in very low coliform levels, especially if heavy chlorination is used, allowing the standard to be achievable. Coliforms, as indicators of faecal bacteria, were the only microbiological criterion used (Hespanhol 1990).

California-type standards were adopted in a number of countries including developing countries, as this constituted the only guidance available at the time. However, the very strict coliform levels were not achievable in developing countries due to the lack of economic resources and skills to implement and operate the rather sophisticated treatment technology in use, or thought to be available, at the time. Hence, standards in these countries existed on paper only and were not enforced. Although the standards set by the State of California had limited applicability on a worldwide scale, they were probably instrumental in enhancing the acceptance of wastewater reuse among planners, engineers, health authorities, and the public in industrialised countries.

WHO published wastewater reuse guidelines for the first time in 1973 (WHO 1973). The group drafting the guidelines felt that to apply drinking water-type standards (2.2 coliforms/100 ml) for wastewater reuse was unrealistic and lacked an epidemiological basis. Moreover, recognition was given to the fact that few rivers worldwide used for irrigation carry water approaching such quality. The group was further convinced that few, if any, developing countries could meet such standards for reused wastewater. As a result of these deliberations, a guideline value of 100 coliforms/100 ml for unrestricted irrigation was set. The guidelines also made recommendations on treatment, suggesting secondary treatment (such as activated sludge, trickling filtration or waste stabilisation ponds (WSP)) followed by chlorination or filtration and chlorination. However, the implementation of such wastewater treatment technologies (with the exception of WSP) remained unattainable for most developing countries and, in some circumstances, this led to authorities tolerating the indirect reuse of untreated wastewater. Indirect reuse being the abstraction of water for irrigation from a water body containing wastewater (the quality of which may vary markedly as dilution depends on the seasonal flow regime in the receiving water body).

In the past two decades, recycling of urban wastewater for agricultural use has been receiving increasing attention from decision makers, planners and external support agencies, largely as a result of the rapid dwindling of easily accessible freshwater sources (groundwater in particular) and the consequent sharp rise in cost of procuring irrigation water. Reduction in environmental pollution caused by wastewater disposal was seen as a benefit from the recycling of human waste. With this change of paradigm in (urban) water resources management, a renewed need for informed guidance on health protection arose. As a result, WHO, United Nations Development Programme (UNDP), the World Bank, United Nations Environment Programme (UNEP), Food and Agriculture Organisation (FAO), and bilateral support agencies commissioned reviews of credible epidemiological literature related to the health effects of excreta and wastewater use in agriculture and aquaculture. The results are documented in Blum and Feachem (1985) and in Shuval *et al.* (1986). The above stakeholders, with the aid of independent academic institutions and experienced scientists, aimed to develop a rational basis for the formulation of updated health guidelines in wastewater reuse, which would be applicable in many different settings, i.e. in economically less developed as well as in industrialised countries. Reviews of the relationships between health, excreted infections and measures in environmental sanitation (Feachem *et al.* 1983), on survival of excreted pathogens on soils and crops (Strauss 1985) were conducted at the same time.

Earlier regulatory thinking was guided largely by knowledge of pathogen detection and survival in wastewater and on irrigated soils and crops, i.e. by what constitutes the so-called potential risk. In the light of the reviews undertaken, it was concluded that potential risk should not, alone, automatically be interpreted as constituting a serious public health threat. This can be estimated only by determining actual risks, which are a result of a series of complex interactions between different factors (Figure 2.1), and which can be measured using epidemiological studies.

A relative ranking of health risks from the use of untreated excreta and wastewater was determined from the review of epidemiological studies (Shuval *et al.* 1986). Use of untreated or improperly treated waste was judged to lead to:

- a high relative excess frequency of intestinal nematode infection;
- a lower relative excess frequency of bacterial infections; and
- a relatively small excess frequency for viruses.

For viruses, direct (i.e. person-to-person) transmission is the predominant route and immunity is developed at an early age in endemic areas. The excess frequency for trematodes (e.g. *Schistosoma*) and cestodes (e.g. tapeworms) vary from high to nil, according to the particular excreta use practice and local circumstances. A major factor determining the relative ranking is pathogen survival on soil and crops. Figure 2.2 (derived from Feachem *et al.* 1983 and Strauss 1985) shows this for selected excreted pathogens. Pathogen die-off following the spreading of wastewater or faecal sludge on agricultural land acts as an important barrier against further transmission, and results in a diminished risk of infection for both farmers and consumers.

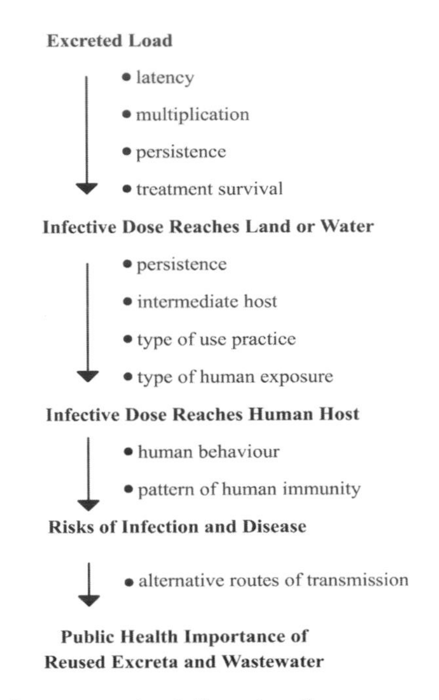

Figure 2.1. Pathogen–host properties influencing the sequence of events between the presence of a pathogen in excreta and measurable human disease attributable to excreta or wastewater reuse (Blum and Feachem 1985; reproduced by permission of the International Reference Centre for Waste Disposal).

Waste stabilisation ponds had, meanwhile, been proven to be a low-cost, sustainable method of wastewater treatment, particularly suited to the socio-economic and climatic conditions prevailing in many developing countries. Well-designed and operated WSP schemes, comprising both facultative ponds (to remove organic contaminants) and maturation ponds (to inactivate pathogenic micro-organisms), can reliably remove helminth eggs and consistently achieve faecal coliform effluent levels of <1000/100 ml. No input of external energy or disinfectants is, therefore, needed. This means that the production of effluent that is likely to satisfy reasonable quality standards has become within the reach of developing countries.

Representatives from UN agencies, including the World Bank, and various research institutions convened in 1985 (IRCWD 1985) and in 1987 to discuss and propose a new paradigm to quantify the health impacts of human waste utilisation. The meetings recommended the formulation of new guidelines for the reuse of human waste. A document was produced, pertaining to both wastewater and excreta use and also addressing the planning aspects of waste utilisation schemes (Mara and Cairncross 1989). The meetings resulted in the formation of a WHO Scientific Group, which was mandated to recommend revised wastewater reuse guidelines. WHO published the current guidelines in 1989 (WHO 1989).

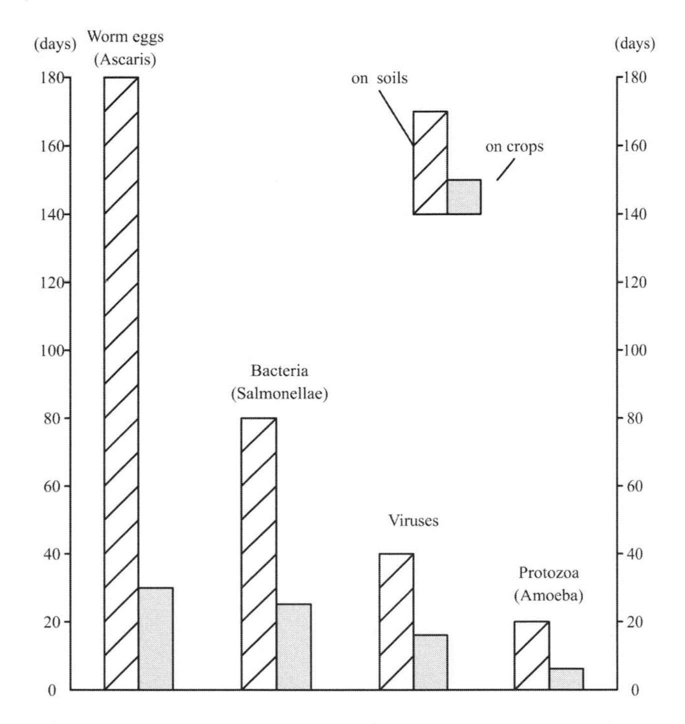

Figure 2.2. Survival of excreted pathogens on soils and crops in a warm climate.

2.3.2 How the current WHO (1989) guidelines were derived

The purpose of the guidelines was to guide design engineers and planners in the choice of waste treatment technologies and waste management options. The guideline levels were derived from the results of the available epidemiological studies of wastewater use, along with a consideration of what was achievable by wastewater treatment processes. A great deal of evidence was available on the risk of exposure to raw wastewater and excreta, and on the risks to farm workers and populations living nearby spray-irrigated areas of use of partially-treated wastewater (Shuval *et al.* 1986). However, there was less evidence of the effect of use of treated wastewater, particularly in relation to consumption of vegetable crops. Where epidemiological evidence was not sufficient to allow the definition of a level (microbiological quality) at which no excess risk of infection would occur, data on pathogen removal by wastewater treatment processes and pathogen die-off in the field, and prevailing guidelines for water quality were taken into account.

The recommended microbiological quality guidelines are shown in Table 2.2.

Table 2.2. Recommended microbiological quality guidelines for wastewater use in agriculture[a] (WHO 1989)

Cat.	Reuse conditions	Exposed group	Intestinal nematodes[b] (/litre*[c])	Faecal coliforms (/100ml**[c])	Wastewater treatment expected to achieve required quality
A	Irrigation of crops likely to be eaten uncooked, sports fields, public parks[d]	Workers, consumers, public	≤1	≤1000	A series of stabilisation ponds designed to achieve the microbiological quality indicated, or equivalent treatment
B	Irrigation of cereal crops, industrial crops, fodder crops, pasture and trees[e]	Workers	≤1	None set	Retention in stabilisation ponds for 8-10 days or equivalent helminth removal
C	Localised irrigation of crops if category B exposure of workers and the public does not occur	None	n/a	n/a	Pre-treatment as required by the irrigation technology, but not less than primary sedimentation

[a] In specific cases, local epidemiological, sociocultural and environmental factors should be taken into account, and the guidelines modified accordingly
[b] *Ascaris* and *Trichuris* species and hookworms
[c] During the irrigation period
[d] A more stringent guideline (≤200 faecal coliforms/100ml) is appropriate for public lawns with which the public may come into direct contact
[e] In the case of fruit trees, irrigation should cease two weeks before the fruit is picked and none should be picked off the ground
* Arithmetic mean
** Geometric mean

An intestinal nematode egg guideline was introduced for both unrestricted (category A) and restricted (category B) irrigation because epidemiological evidence showed a significant excess of intestinal nematode (*Ascaris*, *Trichuris*, hookworm) infections in farm workers and consumers of vegetable crops irrigated with untreated wastewater. A high degree of helminth removal was therefore proposed, especially as there were some data indicating that rates of infection were very low when treatment of wastewater occurred. The level was set at ≤1 egg per litre, equivalent to a removal efficiency of up to 99.9% (3 log removal). This level is achievable by waste stabilisation pond treatment (with a retention time of 8–10 days) or equivalent treatment options. The intestinal nematode egg guideline was also meant to serve as an indicator for other pathogens, such as helminth eggs and protozoan cysts.

A bacterial guideline of ≤1000 faecal coliforms (FC) per 100ml (geometric mean) was recommended for unrestricted irrigation (category A). Epidemiological evidence, particularly from outbreaks, indicated the transmission of bacterial infections such as cholera and typhoid through use of untreated wastewater. It was thought that transmission was less likely to occur through treated wastewater, considering the degree of bacterial removal achievable through treatment and the relatively high infectious dose for some bacterial infections. Data on pathogen removal from well-designed waste stabilisation ponds showed that at an effluent concentration of 1000 FC/100ml (reflecting >99.99% removal) bacterial pathogens were absent and viruses were at very low levels (Bartone *et al.* 1985; Oragui *et al.* 1987; Polpraset *et al.* 1983). Natural die-off of pathogens in the field, amounting to 90–99% reduction over a few days, represented an additional safety factor that was taken into consideration when formulating the guidelines. In addition, the level set was similar to guidelines for irrigation water quality and bathing water quality adopted in industrialised countries. These were 1000 FC/100ml for unrestricted irrigation with surface water promulgated by the US EPA (US EPA 1973) and 2000 FC/100ml for bathing water stipulated by the EU (CEC 1976). No bacterial guideline was recommended for restricted irrigation (category B) as there was no epidemiological evidence for the transmission of bacterial infections to farm workers when wastewater was partially treated.

Health protection measures were also considered. They included:

crop selection
wastewater application measures
human exposure control.

These are management practices, the aim of which is to reduce exposure to infectious agents. The concept was based on the principle of interrupting the flow of pathogens from the wastewater to the exposed worker or consumer, and the measures described act as barriers to pathogen flow whereas the use of treatment achieves removal of the pathogens. In this way, crop restrictions would reduce consumers' exposure to contaminated raw vegetables, wastewater application through drip irrigation would reduce contamination of low-growing crops and farm worker exposure, and wearing protective clothing would reduce the risk for farm workers. Integration of these measures and adoption of a combination of several protection measures was encouraged. A number of possible combinations are shown in the model of choices of health protection measures (Figure 2.3) (Blumenthal *et al.* 1989); for example, partial treatment of wastewater to a level less stringent than that recommended in the guidelines would be adequate if combined with other measures e.g. crop restriction.

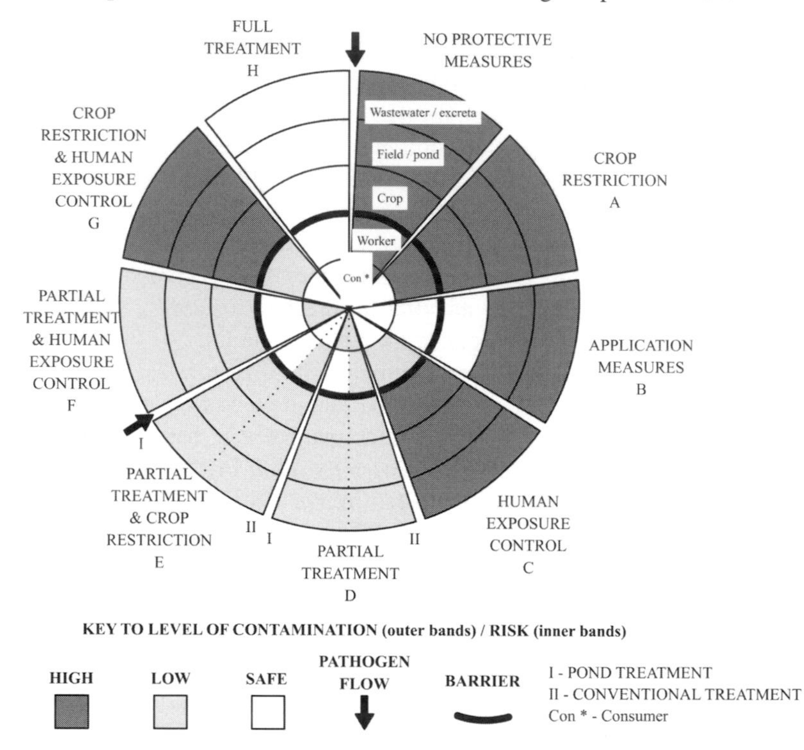

Figure 2.3. Generalised model illustrating the effect of different control measures in reducing health risks from wastewater reuse (adapted from Blumenthal *et al.* 1989; WHO 1989).

Combinations of measures could be selected to suit local circumstances. For example, where there was a market for cereal crops and good institutional capacity but insufficient resources to treat wastewater to category A quality, crop restrictions with partial wastewater treatment could be used. In situations where wastewater treatment could not be provided for a number of years, combinations of management options could be used in the interim (e.g. crop restrictions and human exposure control). The model of combinations of management practices and treatment processes drew on experience of reuse practices in the field (Strauss and Blumenthal 1990).

The main features of the WHO (1989) guidelines for wastewater reuse in agriculture are therefore as follows:

- Wastewater is considered as a resource to be used, but used safely.
- The aim of the guidelines is to protect against excess infection in exposed populations (consumers, farm workers, populations living near irrigated fields).
- Faecal coliforms and intestinal nematode eggs are used as pathogen indicators.
- Measures comprising good reuse management practice are proposed alongside wastewater quality and treatment goals; restrictions on crops to be irrigated with wastewater; selection of irrigation methods providing increased health protection, and observation of good personal hygiene (including the use of protective clothing).
- The feasibility of achieving the guidelines is considered alongside desirable standards of health protection.

Similar principles were applied to the derivation of guidelines for the use of excreta in agriculture and aquaculture (Mara and Cairncross 1989), and to tentative guidelines for the use of wastewater in aquaculture (WHO 1989). The latter are based on, among other things, extensive wastewater-fed aquaculture field studies (Edwards and Pullin 1990).

2.3.3 How WHO (1989) guidelines have been incorporated into standards

In the WHO (1989) guidelines, it was specified that in specific cases of standard setting, 'local epidemiological, socio-cultural and environmental factors should be taken into account and the guidelines modified accordingly'. The microbiological quality guidelines have been used as the basis for standard setting in several countries and regional administrations. In some situations, the microbiological quality guideline levels have been adopted unchanged as standards, e.g. the Balearic Islands and Catalonia in Spain (Bontoux 1998). In other situations the quality guideline levels have been adopted, but within a more cautious approach where management practices and restrictions are closely specified. In France, for example, sanitary recommendations for the use of wastewater for the irrigation of crops and landscapes, drawing on the WHO guidelines, were published in 1991. These recommendations are used to guide wastewater reuse projects. Standards will be formulated and enacted, following evaluation of these projects (Bontoux and Courtois 1998). The French recommendations stipulate additional safety measures besides restricting the use of wastewater according to the quality of the treated effluents (for which WHO microbiological guideline values are used). Special measures include the protection of groundwater and surface waters, distribution networks for treated wastewater, hygiene regulations at treatment and irrigation facilities, and the training of operators and supervisors.

Standard setting in other countries has been influenced by the WHO guidelines, but often with some modification of the microbiological guidelines before adoption as standards. In Mexico, large areas are irrigated with untreated wastewater and crop restrictions are enforced. A standard of ≤5 eggs per litre has been set for restricted irrigation (Norma Oficial Mexicana 1997). The revised standards for unrestricted irrigation are 1000 FC/100ml (monthly mean) and ≤1 helminth ova per litre (similar to WHO). The rationale for this relates to what is practicable through currently available or planned treatment technology, and it was believed that a stricter helminth standard for restricted irrigation would require the use of filters in treatment plants, which would be unaffordable (Peasey et al. 1999). In Tunisia, the WHO guideline for restricted irrigation has been adopted (≤1 helminth ova per litre) but irrigation of vegetables to be eaten raw with reclaimed wastewater is prohibited (Bahri 1998; République Tunisienne 1989). The effluent of secondary treatment plants (supplemented by retention in ponds or reservoirs where necessary) is mainly used to irrigate fruit trees, fodder crops, industrial crops, cereals and golf courses.

2.3.4 Controversy over WHO guidelines on wastewater reuse

Controversy arose over the WHO guidelines on wastewater reuse shortly after their introduction in 1989. The criticism raised was that they were too lenient and would not sufficiently protect health, especially in developed countries. The rationale for the opposing views may well originate from a difference in underlying paradigm. Views critical of the WHO recommendations appear to be based largely on a 'zero-risk' concept (an idea explored in more detail in Chapter 10) which results in guidelines or standards where the objective is to eliminate pathogenic organisms in wastewater. WHO guidelines, however, are based on the objective that there should be no excess infection in the population attributable to wastewater reuse and that risks from reuse in a specific population must be assessed relative to risks of enteric infections from other transmission routes. Achieving wastewater quality close to drinking water standards is economically unsustainable and epidemiologically unjustified in many places.

2.4 SAFE RECREATIONAL WATER ENVIRONMENTS

In 1998, WHO published 'Guidelines for Safe Recreational Water Environments' in Consultation Draft form (Anon 1998). These guidelines deal with many different hazards including drowning, spinal injury, excess ultra-violet (UV) and so on. However, this section will consider the material relating to faecal contamination of coastal and freshwater. The publication followed a series of four expert meetings held between 1989 and 1997. Amongst broader management issues, these meetings considered:

- epidemiological protocol design and data quality
- appropriate data for use in guidelines design
- statistical treatment of data
- alternative guideline systems.

The following outlines the stages in guideline derivation for this aspect.

2.4.1 The process of microbiological guideline design for recreational waters

Ideally, a scientifically supportable guideline value (or numerical standard) is defined to provide a required level of public health protection, measured either in terms of 'acceptable' disease burden and/or some percentage attack rate of illness in the population which, again, is felt to be acceptable.

Derivation of such a numerical standard depends on the existence of:

- A dose–response curve linking some microbial concentration in the recreational waters with the 'outcome' illness, generally gastroenteritis.
- An understanding of the probability that a defined population would be exposed to a given water quality.

2.4.1.1 Epidemiology

Very few microbiological standards currently in force could claim good data on the first of these requirements, let alone the second. For example, current European Union mandatory standards for recreational waters are based on Directive 76/160/EEC (CEC 1976) which does not appear to have a firm epidemiological foundation. Subsequent attempts to revise these, now dated, European standards (Anon 1994a) have met with resistance from the competent authorities in member countries due to the lack of epidemiological evidence to underpin proposed changes (Anon 1994b, 1995a,b,c).

In the US, new standards were derived in 1986 (US EPA 1986), based on the work of Cabelli *et al.* (1982) which resulted in a dose–response relationship linking microbiological water quality and disease outcome (principally gastroenteritis). However, these studies have received a strong methodological critique (Fleisher 1990a,b, 1991; Fleisher *et al.* 1993) which has cast some doubt on the validity of the dose–response relationships reported.

In effect, the problem facing the WHO expert advisers was the plethora of epidemiological investigations in this area which had:

- adopted different protocols
- measured different exposure variables
- employed different sampling protocols for environmental and health data
- applied different case definitions to quantify the outcome variables
- assessed and controlled differently for potential confounding variables.

Thus, precise comparison between studies was difficult. However, a consistent finding of the body of evidence presented by these investigations was that significant illness attack rates were observed in populations exposed to levels of water quality well within existing standard parametric values and that a series of dose–response relationships were evident, suggesting increased illness from increasingly polluted waters.

To clarify the utility of available epidemiological evidence for guideline design, WHO commissioned an internal review of epidemiological investigations in recreational water environments (Prüss 1998). Following an exhaustive literature search and a pre-defined set of criteria, this paper classified some 37 relevant studies and concluded that the most precise dose–response should derive from the studies which had applied a randomised trial design because this approach:

- facilitated acquisition of more precise exposure data, thus reducing misclassification bias; and
- allowed better control of, and data acquisition describing, potential confounding factors.

Published data from studies of this nature were, however, only available (at the time) from government-funded studies in the UK conducted between 1989 and 1993 (Fleisher *et al.* 1996; Kay *et al.* 1994) and a pilot study conducted in the Netherlands by Asperen *et al.* (1997).

2.4.1.2 Water quality data

A key problem in using microbiological data to define standards is the inherent variability of microbiological concentrations in environmental waters. Many workers have reported changes of several orders of magnitude occurring over short time intervals of a few hours (e.g. McDonald and Kay 1981; Wilkinson *et al.* 1995; Wyer *et al.* 1994, 1996). However, analysis of 'compliance' data (and special survey information) from recreational waters suggested that the bacterial concentrations approximated to a $\log_{10}$-normal probability density function (pdf) which could be characterised by its geometric mean value and $\log_{10}$ standard deviation. This was true of UK coastal beaches (Kay *et al.* 1990) and EU-identified bathing waters.

Thus, the bacterial probability density function could be used to calculate the probability of exposure to any given water quality for any specific bathing water. Clearly, this assumes that historical 'compliance' data adequately characterises current water quality to which bathers are exposed.

2.4.1.3 Combining epidemiological and environmental data

The first stage in guideline design can be characterised by disease burden estimation. This requires the combination of the dose–response curve with the probability of exposure to different levels of water quality predicted by the probability density function of bacterial distribution. Figures 2.4–2.6 illustrate this process using UK compliance data and the dose–response curve linking faecal streptococci and gastroenteritis published in Kay *et al.* (1994), assuming a population exposed of 1000 individuals and a resultant disease burden of 71 cases of gastroenteritis.

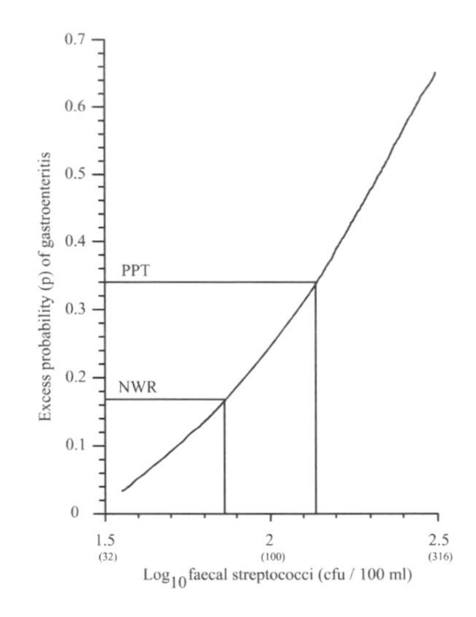

Figure 2.4. Dose–response curve linking faecal streptococci with excess probability of gastroenteritis (reproduced from Kay *et al.* 1999 with permission of John Wiley and Sons Limited). PPT: person to person transmission; NWR: non-water related.

Assuming universally applicable relationships, the policy maker could simply define the 'acceptable' level of illness in the exposed population and use this to derive a feasible region of the probability density function geometric mean and standard deviation values to comply with the accepted disease attack rate.

The approach adopted used the disease burden model outlined in Figure 2.6 and the concept of an 'acceptable' number of gastroenteritis incidents in a 'typical' bather. For example, one case in 20 exposures, one case in 80 exposures and one case in 400 exposures. These were derived from the theoretical proposition that, on average:

- the bather experiencing 20 exposures in a season might experience one case of gastroenteritis
- the family of 4 experiencing 20 bathing events might experience one case of gastroenteritis
- the family of 4 experiencing 20 bathing events per year for 5 years might experience 1 case of gastroenteritis.

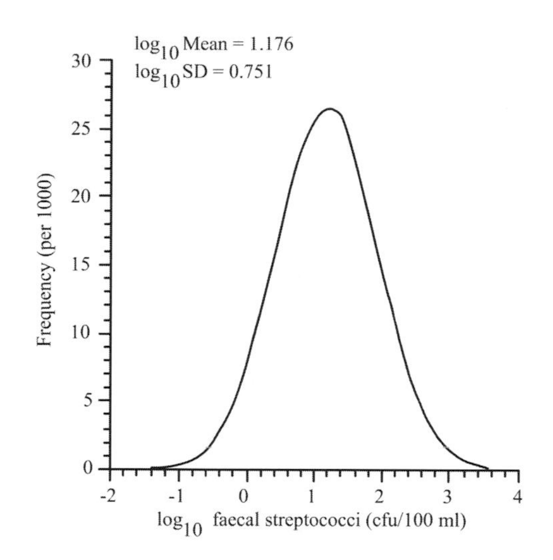

Figure 2.5. Probability density function of faecal streptococci in bathing waters – curve adjusted to have a total area of 1000 (reproduced from Kay *et al.* 1999 with permission of John Wiley and Sons Limited).

Using the average $\log_{10}$ standard deviation for over 500 EU bathing locations, these disease burden levels were used to define the 95 percentile points of the theoretical probability density function that would produce this risk of exposure. These correspond approximately to the 200, 50 and 10 faecal streptococci cfu/100ml levels.

The final guideline is not a 95 percentile but an absolute level of 1,000 faecal streptococci cfu/100ml, which if exceeded should lead to immediate investigation and follow-up action. This level was derived from the 1959 Public Health Laboratory Service (PHLS) investigation of serious illness in the UK, which suggested that paratyphoid might be possible where total coliform concentrations exceeded 10,000 cfu/100ml (PHLS 1959). Converting to faecal streptococci concentrations, this gave an approximate level of 1000 which the WHO committee considered should represent a maximum acceptable concentration because of the risk of serious illness.

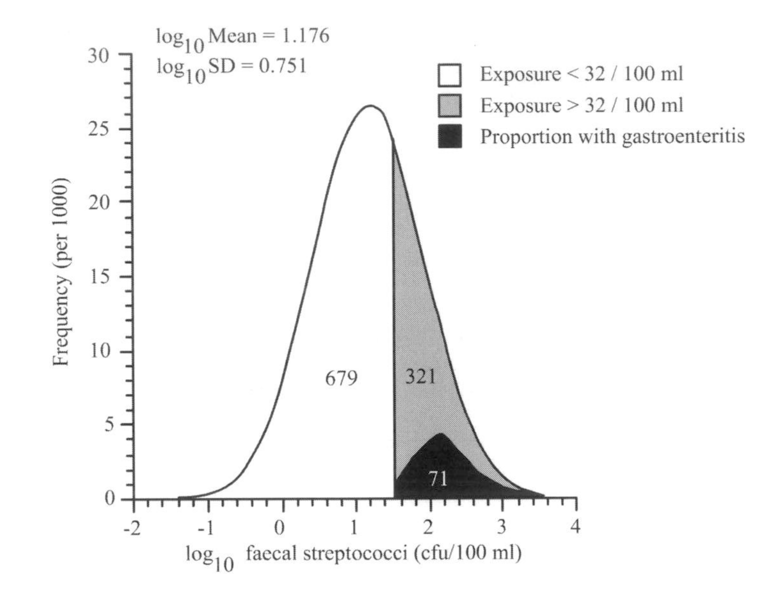

Figure 2.6. Example of an estimated disease burden (reproduced from Kay *et al.* 1999 with permission of John Wiley and Sons Limited).

2.4.1.4 Problems with this approach

The epidemiological database is very narrow and potentially culturally specific. It derives from the UK marine investigations and was chosen because of the greater accuracy in dose–response curve construction produced by randomised studies. However, its application worldwide must be questioned. This highlights the urgent need for further implementations of the randomised trial protocol to the quantification of recreational water dose–response relationships in other water types (e.g. fresh waters), in other regions (e.g. Mediterranean and tropical) and with other risk groups (e.g. canoeists, surfers etc.).

The nature of the randomised trial can mean that the exposed population is restricted. For example, the UK studies used healthy adult volunteers, and children were excluded because they were not considered able to give informed consent. Thus, significant risk groups that the standards seek to protect can be systematically excluded. However, this problem was not encountered in the studies of Asperen *et al.* (1997) in the Netherlands.

If a single number is required to define the guideline, e.g. a geometric mean or 95 percentile, then some assumption must be made concerning the other parameters of the probability density function. In this case a uniform $\log_{10}$ standard deviation was assumed. However, it is known that this parameter changes at compliance points in response to, for example, non-sewage inputs such as rivers and streams. The standard deviation of the probability density function certainly affects the probability of exposure to polluted waters and thus the disease burden.

2.5 IMPLICATIONS FOR INTERNATIONAL GUIDELINES AND NATIONAL REGULATIONS

It can be seen from the outline of the three guideline areas that although there are similarities, they have very different histories and there is little commonality in the way they have been derived. Key to all three areas is the hazard of primary concern, namely human (and animal) excreta. These three areas should not, ideally, be considered in isolation but should be examined together and subject to integrated regulation and management. The harmonised framework should allow further development and future revisions of the guideline areas to be carried out in a consistent way, allowing the consideration of the water environment in general rather than components of it in isolation. It is important to bear in mind that guidelines represent the international evidence base and they require adaptation prior to implementation in order to be appropriate for individual national circumstances.

2.6 REFERENCES

Anon (1994a) Proposal for a Council Directive concerning the Quality of Bathing Water. *Official Journal of the European Communities* No C112, 22 April, pp. 3–10.

Anon (1994b) Select Committee on the European Communities, *Bathing Water,* House of Lords Session 1994–5, 1st Report with evidence 6 December, HMSO, London.

Anon (1995a) Select Committee on the European Communities, *Bathing Water Revisited,* House of Lords Session 1994–5, 7th Report with evidence 21 March, HMSO, London.

Anon (1995b) Parliamentary Debates, House of Lords Official Report, Hansard **564**(90), 18 May, pp. 684–708.

Anon (1995c) Cost of compliance with proposed amendments to Directive 76/160/EEC. Presented to the House of Lords Select Committee Enquiry. Prepared for the DoE by Halcrow plc.

Anon (1998) Guidelines for safe recreational water environments: coastal and freshwaters. Consultation Draft, World Health Organization, Geneva.

Asperen, I.A. van, Medema, G.J., Havelaar, A.H. and Borgdorff, M.W. (1997) *Health effects of fresh water bathing among primary school children*, Report No. 289202017, RIVM, Bilthoven, the Netherlands.

Bahri, A. (1998) Wastewater reclamation and reuse in Tunisia. In *Wastewater Reclamation and Reuse*, (ed. T. Asano), Water Quality Management Library, Vol. 10, pp. 877–916.

Bartone, C.R., Esparza, M.L., Mayo, C., Rojas, O. and Vitko, T. (1985) *Monitoring and maintenance of treated water quality in the San Juan lagoons supporting aquaculture,* Final Report of Phases I and II, UNDP/World Bank/GTZ Integrated Resource Recovery Project GLO/80/004, CEPIS.

Blum, D. and Feachem, R.G. (1985) *Health aspects of nightsoil and sludge use in agriculture and aquaculture – part III: An epidemiological perspective,* IRCWD Report No. 04/85, SANDEC, CH-8600, Dubendorf, Switzerland.

Blumenthal, U.J., Strauss, M., Mara D.D. and Cairncross, S. (1989) Generalised model of the effect of different control measures in reducing health risks from waste reuse. *Water Science and Technology* **21**, 567–577.

Bontoux, J. and Courtois, G. (1998) The French wastewater reuse experience. In *Wastewater Reclamation and Reuse* (ed. T. Asano), Water Quality Management Library Vol. 10, pp 1193–1210.

Bontoux, L. (1998) The regulatory status of wastewater reuse in the European Union. In *Wastewater Reclamation and Reuse* (ed. T. Asano), Water Quality Management Library Vol. 10, pp. 1463–1475.

Cabelli, V.J., Dufour, A.P., McCabe, L.J. and Levin, M.A. (1982) Swimming associated gastroenteritis and water quality. *Am. J. Epi.* **115**, 606–616.

CEC (1976) Council Directive 76/160/EEE concerning the quality of bathing water. Official Journal of the European Communities **L31**, 1–7.

Edwards, P. and Pullin, R.S.V. (eds) (1990) Wastewater-Fed Aquaculture. Proceedings, International Seminar on Wastewater Reclamation and Reuse for Aquaculture, Calcutta, India, 6–9 December 1988.

Feachem, R.G., Bradley, D.J., Garelick, H. and Mara, D.D. (1983) *Sanitation and Disease – Health Aspects of Excreta and Wastewater Management,* John Wiley & Sons, Chichester, UK.

Fleisher, J.M., Jones, F., Kay, D. and Morano, R. (1993) Setting recreational water quality criteria. In *Recreational Water Quality Management: Fresh Water,* vol II, pp. 123–126 (eds D. Kay and R. Hanbury), Ellis Horwood, Chichester.

Fleisher, J.M. (1990a) Conducting recreational water quality surveys. Some problems and suggested remedies. *Marine Pollution Bulletin* **21**(2), 562–567.

Fleisher, J.M. (1990b) The effects of measurement error on previously reported mathematical relationships between indicator organism density and swimming associated illness: a quantitative estimate of the resulting bias. *International Journal of Epidemiology* **19**(4), 1100–1106.

Fleisher, J.M. (1991) A reanalysis of data supporting the US Federal bacteriological water quality criteria governing marine recreational waters. *Journal of the Water Pollution Control Federation* **63**, 259–264.

Fleisher, J.M., Kay, D., Salmon, R.L., Jones, F., Wyer, M.D. and Godfree, A.F. (1996) Marine waters contaminated with domestic sewage, non-enteric illnesses associated with bather exposure in the United Kingdom. *American Journal of Public Health* **86**(9), 1228–1234.

Hespanhol, I. (1990) Guidelines and integrated measures for public health protection in agricultural reuse systems. *J Water SRT – Aqua* **39**(4), 237–249.

IRCWD (1985) Health aspects of wastewater and excreta use in agriculture and aquaculture: the Engelberg report. *IRCWD News* **23**, 11–19.

Kay, D., Fleisher, J.M., Salmon, R., Jones, F., Wyer, M.D., Godfree, A., Zelanauch-Jaquotte, Z. and Shore, R. (1994) Predicting the likelihood of gastroenteritis from sea bathing: results from randomised exposure. *Lancet* **34**, 905–909.

Kay, D., Wyer, M.D., McDonald, A.T. and Woods, N. (1990) The application of water quality standards to United Kingdom bathing waters. *Journal of the Institution of Water and Environmental Management* **4**(5), 436–441.

Kay, D., Wyer, M.D., Crowther, J., O'Neill, J.G., Jackson, G, Fleisher, J.M. and Fewtrell, L. (1999) Changing standards and catchment sources of faecal indicators in nearshore bathing waters. In *Water Quality Processes and Policy* (eds S. Trudgill, D. Walling and B. Webb), John Wiley & Sons, Chichester, UK.

Mara, D.D. and Cairncross, S. (1989) Guidelines for the Safe Use of Wastewater and Excreta in Agriculture and Aquaculture – Measures for Public Health Protection. WHO/UNEP, Geneva.

McDonald, A.T. and Kay, D. (1981) Enteric bacterial concentrations in reservoir feeder streams: baseflow characteristics and response to hydrograph events. *Water Research* **15**, 861–868.

Norma Oficial Mexicana (1997) NOM-001-ECOL-1996 Que establece los limites maximos permisibles de contaminantes en las descargas de aguas residuales en aguas y bienes nacionales. Diario Oficial de la Federation Enero 01 de 1997, pp. 68–85. (Official Mexican quality standards (1997) STANDARD-001-ECOL-1996 for effluent discharge into surface waters and on national property. Official Federal Newspaper; in Spanish).

Oragui, J.I., Curtis, T.P., Silva, S.A. and Mara, D.D. (1987) The removal of excreted bacteria and viruses in deep waste stabilization ponds in Northeast Brazil. *Water Science and Technology* **19**, 569–573.

Peasey, A., Blumenthal, U.J., Mara, D.D. and Ruiz-Palacios, G. (1999) A review of policy and standards for wastewater reuse in agriculture: a Latin American perspective. WELL study Task 68 part II, London School of Hygiene and Tropical Medicine and WEDC, Loughborough University, UK.

PHLS (1959) Sewage contamination of coastal bathing waters in England and Wales: a bacteriological and epidemiological study. *Journal of Hygiene, Cambs.* **57**(4), 435–472.

Polpraset, C., Dissanayake, M.G. and Thanh, N.C. (1983) Bacterial die-off kinetics in waste stabilization ponds. *J. WPCF* **55**(3), 285–296.

Prüss, A. (1998) Review of epidemiological studies on health effects from exposure to recreational water. *International Journal of Epidemiology* **27**, 1–9.

République Tunisienne (1989) Decree No. 89-1047. Journal officiel de la République Tunisienne. (In French.)

Shuval, H.I., Adin, A., Fattal, B., Rawitz, E. and Yekutiel, P. (1986) *Wastewater Irrigation in Developing Countries – Health Effects and Technical Solutions. Integrated Resource Recovery,* UNDP Project Management Report No. 6, World Bank Technical Paper No. 51.

State of California (1978) Wastewater Reclamation Criteria. California Administrative Code Title 22, Division 4, Environmental Health, Dept. of Health Services, Sanitary Engineering Section (also cited in *Wastewater Reclamation and Reuse* (ed. T. Asano), Water Quality Management Library Vol. 10, pp. 1477–1490.

Strauss, M. (1985) *Health Aspects of Nightsoil and Sludge Use in Agriculture and Aquaculture – Pathogen Survival*, IRCWD Report No. 04/85, SANDEC, CH-8600 Duebendorf, Switzerland.

Strauss, M. and Blumenthal, U.J. (1990) *Human Waste Use in Agriculture and Aquaculture – Utilisation Practices and Health Perspectives,* IRCWD Report No. 08/90, SANDEC, CH-8600 Duebendorf, Switzerland.

US EPA (1973) Water quality criteria. Ecological Research Series, EPA R-3-73-033. US Environmental Protection Agency, Washington, DC.

US EPA (1986) *Ambient water quality criteria for bacteria – 1986*. EPA440/5-84-002. Office of Water Regulations and Standards Division, Washington DC.

US EPA/USID (1992) *Manual – Guidelines for Water Reuse*, Report EPA/625/R-92/004, US EPA (Office of Water) and USID, Washington DC.

WHO (1973) *Reuse of Effluents: Methods of Wastewater Treatment and Public Health Safeguards. Report of a WHO Meeting of Experts,* Technical Report Series No. 517, WHO, Geneva.

WHO (1989) *Health Guidelines for the Use of Wastewater in Agriculture and Aquaculture. Report of a WHO Scientific Group,* Technical Report Series No. 778, WHO, Geneva.

WHO (1993) *Guidelines for Drinking Water Quality. Volume 1: Recommendations*. Geneva.

WHO (1996) *Guidelines for Drinking Water Quality. Volume 2: Health criteria and other supporting information.* Geneva.

WHO (1997) *Guidelines for Drinking Water Quality. Volume 3: Surveillance and control of community supplies*. Geneva.

Wilkinson, J., Jenkins, A., Wyer, M.D. and Kay, D. (1995) Modelling faecal coliform dynamics in streams and rivers. *Water Research* **29**(3), 847–855.

Wyer, M.D., Jackson, G.F., Kay, D., Yeo, J. and Dawson, H.M. (1994) An assessment of the impact of inland surface water input to the bacteriological quality of coastal waters. *Journal of the Institution of Water and Environmental Management* **8**, 459–467.

Wyer, M.D., Kay, D., Dawson, H.M., Jackson, G.F., Jones, F., Yeo, J. and Whittle, J. (1996) Delivery of microbial indicator organisms to coastal waters from catchment sources. *Water Science and Technology* **33**(2), 37–50.

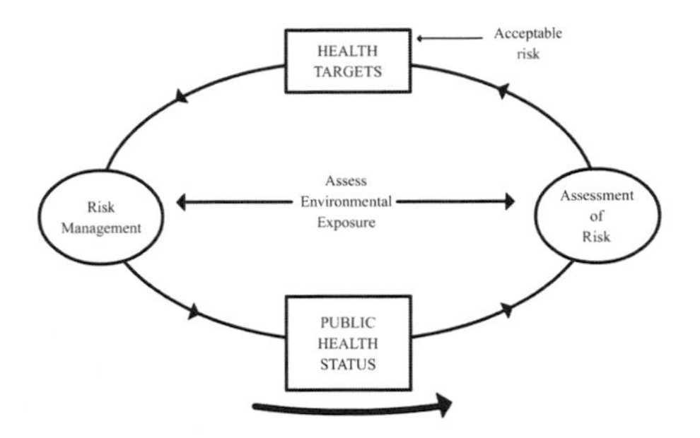

3

The Global Burden of Disease study and applications in water, sanitation and hygiene

Annette Prüss and Arie Havelaar

This chapter introduces the concept of the global burden of disease and its key measure, the Disability-Adjusted Life Year (DALY). It illustrates the use of DALYs both to integrate the effects of a single agent and also to compare the health effects of different agents. It also examines their role in informing the development of guidelines.

3.1 INTRODUCTION

In 1996, in a landmark publication, the Global Burden of Disease and Injury series appeared on our shelves, representing the culmination of over eight years of work (Murray and Lopez 1996 a,b). These volumes outline the Global Burden of Disease (GBD) Study and the associated global health statistics, and represent the first global and internally consistent collection of epidemiology information on disease burden. The volumes describe the burden from 107 diseases and injuries and 10 major risk factors or risk groups for various age groups and geographical regions. It represents a unique achievement describing the world's disease burden status and trends in the health of populations.

The project was undertaken in a number of stages, with the first stage initiated by the World Bank in 1988. The initial aims were to assess the significance to public health of individual diseases (or related clusters of disease) and what was known about the cost and effectiveness of relevant interventions for their control (Jamieson 1996). This first phase led to the introduction of a new common measure for examining diverse disease outcomes, the DALY or Disability Adjusted Life Year. Phase two extended the effort by attempting to provide a comprehensive set of estimates for total disease burden by including disability as well as number of deaths. The publication of the Global Burden of Disease and Injury series represents the third phase of the project. The publication of these volumes was undertaken to inform policy analysis, particularly assessment of priorities in terms of health research and development in developing countries (Jamieson 1996). The initial estimates outlined in the Global Burden of Disease and Injury series are constantly undergoing a process of updating and development (WHO 1999).

3.2 MEASURING POPULATION HEALTH

Of key importance to the GBD study was the introduction of a common unit of currency to allow comparisons to be made between different health outcomes and allowing quantification of non fatal outcomes. This section details the development of the DALY. While not without their problems (Anand and Hanson 1997; Barendregt *et al.* 1996; Williams 1999) DALYs and other summary measures of population health do go at least some way towards providing a level playing field from which comparisons can be made.

For the purpose of integrating the health burden of different health effects of one agent, or comparing the effects of different agents, a common measure is necessary. Traditionally, public health policy has concentrated on mortality, and the severity of disease was expressed in death rates or the number of life years lost due to a certain cause. However, many diseases do not lead to premature mortality, but may be a significant cause of morbidity. Healthy life expectancy is increasingly becoming the focus of public health policy (Van der Maas and Kramers 1997). As outlined in the introduction, Murray (1994) and Murray and Lopez (1996a) have developed the DALY. The DALY is part of a family of population health summary measures. It is based on measuring health gaps, as opposed to measuring health expectancies (Murray and Lopez 1999), and as such it measures the difference between current conditions and a selected target, for example an ideal health state. This integrated measure combines years of life lost by premature mortality (YLL) with years lived with a disability (YLD), standardised by means of severity weights. Thus:

$$DALY = YLL + YLD \quad (3.1)$$

3.2.1 Years of life lost

To estimate YLL on a population basis, the age-specific mortality rates must be combined with the life expectancy of the fatal cases, had they not developed the disease. If mortality affects the population in a random fashion, the life expectancy can be derived from standard life tables. Murray (1996) proposed a table based on the highest observed national life expectancy (for Japanese women), taking into account differences in life expectancy between men and women. The standard life expectancy at birth is 80.0 years for men and 82.5 years for women. For comparison, the life expectancy in the Netherlands in 1994 was 74.6 years for men and 80.3 years for women (Van der Maas and Kramers 1997) while that in Zimbabwe in 1998 was 39 years. If mortality affects a susceptible sub-population, the use of standard life expectancy would lead to a gross overestimation of YLL. In this case, disease-specific information is necessary to estimate the additional loss of life years by the disease under consideration. The total loss of life years is calculated as:

$$YLL = \sum_i e^*(a_i) \sum_j d_{ij} \quad (3.2)$$

where i is an index for different age-classes, d_{ij} is the number of fatal cases per age-class, j is an index for different disease categories and $e^*(a_i')$ is the mean life expectancy in that age class.

3.2.2 Years lived with disability

To estimate YLD on a population basis, the number of cases must be multiplied by the average duration of the disease and a weight factor that reflects the severity of the disease on a scale from 0 (perfect health) to 1 (dead). If necessary, the disease process can be subdivided into several stages according to duration or severity. Thus,

$$YLD = \sum_j N_j \, L_j \, W_j \quad (3.3)$$

where j is an index for different disease categories, N is the number of patients, L is the duration of disease and W is the severity weight.

3.2.3 Measuring disability

Disability needs to be assessed in three different domains: the physical, psychological and social domains. Each of these domains is an aggregate of a number of dimensions, which are usually measured by means of questionnaires. There are three main types of questionnaire for health status measurement: generic, disease-specific and domain-specific (Essink-Bot 1995). Generic instruments cover the three domains of health in a non-disease specific way, assuming that different diseases can be characterised as patterns of physical, psychological and social dysfunction. Several generic instruments have been developed, which differ in the emphasis that each places on each domain. Disease-specific instruments are developed to study changes in health as a consequence of (treatment for) a specific disease. Domain-specific instruments concentrate on the consequences of disease in a specific domain of health or, more specifically, on a specific symptom.

The choice between these three types of instruments depends on the purpose and the perspective of the study. In this case, the objective of the study is to integrate and compare the health effects of very different diseases, which leads naturally to the choice for generic instruments. This choice is further supported by the societal perspective of the study: the objective is to evaluate the impact of disease on a public health level, which leads to the need for non-disease-specific and comprehensive, i.e. generic, measurements.

Information from questionnaires gives a descriptive evaluation of health status, which must be evaluated for further analysis. Different valuation methods are available, such as Standard Gamble (SG), Time Trade Off (TTO), Person Trade Off (PTO) and Visual Analog Scale (VAS) (Brooks 1996; Murray 1996; Torrance 1986). For public health analyses the Person Trade Off and the Time Trade Off methods are the most natural approaches. The Person Trade Off protocol has two variants. In the PTO1-variant, respondents are asked to choose between an intervention that prolongs the life of 1000 individuals in perfect health and an intervention that prolongs the life of N individuals with less than perfect health. In the PTO2-variant, the alternative is to cure N individuals in less than perfect health. The value of N at which the respondent cannot make a choice (the indifference point) is used to calculate the disability weight of the health state under consideration. In the Time Trade Off protocol, respondents are asked to weigh the benefits of an immediate 'cure' against possible later loss of health. Nord (1995) has outlined that the PTO protocol is by its nature most suitable for evaluation of health care programmes from a societal perspective.

Societal perspective also requires that the values used be based on public perception rather than on the opinion of patients or health professionals. However, in the GBD study (Murray and Lopez 1996a,b) and in the VTV study (Van der Maas and Kramers 1997), the panels were composed of medical experts, because they were expected to be best able to compare a large number of diseases in an objective manner.

In the GBD study (Murray and Lopez 1996a,b), a set of 22 indicator conditions was described, representing different grades of disability in the dimensions of physical functioning, neuro-psychological conditions, social functioning, pain and sexual/reproductive functions. In a formal procedure, these indicator conditions were assigned disability weights and classified into seven disability classes. In the next step, several hundred outcomes were evaluated with respect to the distribution of each condition across the seven disability classes. From these data, a composite disability weight for each condition was calculated (Table 3.1).

Table 3.1. Disability classes and indicator diseases (Murray 1996)

Class	Weight	Examples
1	0.00–0.02	Vitiligo on face, low weight
2	0.02–0.12	Watery diarrhoea, sore throat
3	0.12–0.24	Infertility, arthritis, angina
4	0.24–0.36	Amputation, deafness
5	0.36–0.50	Down's syndrome
6	0.50–0.70	Depression, blindness
7	0.70–1.00	Psychosis, dementia, quadriplegia

3.3 MAJOR OUTCOMES OF THE GBD STUDY

In the Global Burden of Disease Study (Murray and Lopez 1996a,b), DALYs have been calculated with age-weighting and a three-percent discount rate. The leading causes of mortality and burden of disease for 1990 are shown in Table 3.2.

Table 3.2 shows the importance of accounting for non-fatal outcomes, as can be seen from the change in ranking position for a number of causes and the appearance of illnesses such as unipolar major depression when disability, not just death, is accounted for. A more recent estimation of mortality and disease burden (WHO 1999) shows a similar pattern but with HIV/AIDs taking up fourth position for both deaths and DALYs and malaria being an important cause in terms of DALYs (Table 3.3).

Table 3.2. Leading causes of death and burden of disease estimates for 1990 (adapted from Murray and Lopez 1996a)

Rank	Cause	% of total	Deaths or DALYs (1000s)
Deaths			
1	Ischaemic heart disease	12.4	6260
2	Cerebrovascular disease	8.7	4381
3	Lower respiratory infections	8.5	4299
4	Diarrhoeal diseases	5.8	2946
5	Perinatal conditions	4.4	2443
6	Chronic obstructive pulmonary disease	3.9	2211
7	Tuberculosis	2.1	1960
8	Measles	2.1	1058
9	Road traffic accidents	1.9	999
10	Cancer of trachea/bronchus/lung	1.9	945
DALYs			
1	Lower respiratory infections	8.2	112,898
2	Diarrhoeal diseases	7.2	99,633
3	Perinatal conditions	6.7	92,313
4	Unipolar major depression	3.7	50,810
5	Ischaemic heart disease	3.4	46,699
6	Cerebrovascular disease	2.8	38,523
7	Tuberculosis	2.8	38,426
8	Measles	2.7	36,520
9	Road traffic accidents	2.5	34,317
10	Congenital abnormalities	2.4	32,921

Table 3.3. Leading causes of death and burden of disease estimates for 1998 (adapted from WHO 1999)

Rank	Cause	% of total	Deaths (1000s)
1	Ischaemic heart disease	13.7	7375
2	Cerebrovascular disease	9.5	5106
3	Lower respiratory infections	6.4	3452
4	HIV/AIDS	4.2	2285
5	Chronic obstructive pulmonary disease	4.2	2249
6	Diarrhoeal diseases	4.1	2219
7	Perinatal conditions	4.0	2155
8	Tuberculosis	2.8	1498
9	Cancer of trachea/bronchus/lung	2.3	1244
10	Road traffic accidents	2.2	1171

Table 3.3 (cont'd)

Rank	Cause	% of total	DALYs (1000s)
1	Lower respiratory infections	6.0	82,344
2	Perinatal conditions	5.8	80,564
3	Diarrhoeal diseases	5.3	73,100
4	HIV/AIDS	5.1	70,930
5	Unipolar major depression	4.2	58,246
6	Ischaemic heart disease	3.8	51,948
7	Cerebrovascular disease	3.0	41,626
8	Malaria	2.8	39,267
9	Road traffic accidents	2.8	38,849
10	Measles	2.2	30,255

Table 3.4 examines the DALY data shown in Table 3.2 by developed versus developing region. As might be expected, there are some notable differences between developed and developing as well as between the overall world picture.

Table 3.4. Causes of DALYs by developed and developing regions, 1990 (adapted from Murray and Lopez 1996a)

Rank	Developed regions		Developing regions	
	Cause	%	Cause	%
1	Ischaemic heart disease	9.9	Lower respiratory infections	9.1
2	Unipolar major depression	6.1	Diarrhoeal disease	8.1
3	Cerebrovascular disease	5.9	Perinatal conditions	7.3
4	Road traffic accidents	4.4	Unipolar major depression	3.4
5	Alcohol use	4.0	Tuberculosis	3.1
6	Osteoarthritis	2.9	Measles	3.0
7	Cancer of trachea/bronchus/lung	2.9	Malaria	2.6
8	Dementia etc.	2.4	Ischaemic heart disease	2.5
9	Self-inflicted injuries	2.3	Congenital abnormalities	2.4
10	Congenital abnormalities	2.2	Cerebrovascular disease	2.4

Disease burden is also being assessed at national and regional levels, and for specific purposes such as analysing the importance of certain diseases or risk factors in population subgroups. The disease burden caused by an environmental problem, and the preventable part of it, are major elements in driving the field of decision-making for priority setting and resource allocation in health and the environment. The global burden of disease attributable to various risk factors is shown in Table 3.5.

Table 3.5. Global burden of disease and injury attributable to selected risk factors, 1990 (adapted from Murray and Lopez 1996a)

Risk factor	Deaths (1000s)	As % total deaths	DALYs (1000s)	As % total DALYs
Malnutrition	5881	11.7	219,575	15.9
Poor water supply, sanitation and personal and domestic hygiene	2668	5.3	93,392	6.8
Unsafe sex	1095	2.2	48,702	3.5
Tobacco	3038	6.0	36,182	2.6
Alcohol	774	1.5	47,687	3.5
Occupation	1129	2.2	37,887	2.7
Hypertension	2918	5.8	19,076	1.4
Physical inactivity	1991	3.9	13,653	1.0
Illicit drugs	100	0.2	8467	0.6
Air pollution	568	1.1	7254	0.5

Quantitative assessment of the burden, together with information on effectiveness and cost-effectiveness of interventions within a social and ethical framework, provide a rational basis for research, implementation and policy development. The attributable burden would usually be based upon the burden that would have been observed if the past exposure of concern had been absent or reduced to a plausible level. The preventable burden would be the burden that could be avoided if current levels of exposure were reduced to a minimum or eliminated.

3.4 GBD ESTIMATE APPLICATIONS

GBD estimates can be used in assessing the performance of a country or region in terms of health-supporting conditions and actions, to map out geographical or population-specific differences, and to monitor trends. GBD information is therefore a tool for identifying overall inequalities in a population. It also allows for comparison between regions or comparison with the developmental status of a region.

GBD information may also be used as a basis for identifying control priorities. Alongside information on effectiveness of interventions and their costs, it helps to prioritise action to prevent or reduce problems associated with a high disease burden. Disease burden measurements become essential when an effort will have a benefit proportional to the size of the problem being addressed. This is the case with political attention, allocation of time in training curricula or, to a certain extent, allocation of resources to research and development. GBD trends permit planning for a shift in priorities rather than reacting to signs of change.

3.5 GBD AND GUIDELINES

Traditionally, guideline values for environmental media (the Drinking-Water Quality Guidelines, WHO 1993, for example) aim to provide the answer to the question:

At which value can we reasonably expect that no or only negligible health impacts will occur in an exposed population?

The question:

How much disease burden will be reduced in a population if the guidelines are implemented?

cannot be answered without additional information. This means that, although the costs of implementation could be estimated, the efficiency of such an intervention in terms of health status improvement of the population concerned remains unresolved.

Ignorance of the effectiveness of an intervention in terms of disease burden can be acceptable provided that the intervention is affordable, and that resource allocation is not in competition with other interventions (or that other aspects such as ethical or considerations are involved). When resource allocation is a problem, informed choices have to be made, at least in the short term. This is not necessarily a problem only in developing countries, but is also a common problem in developed country situations. For example, a significant number of bathing beaches do not meet the bathing water quality requirements. The Recreational Water Quality Guidelines (WHO 1998) therefore propose various levels of recreational water quality, described by the associated burden a population would experience if exposed (see Chapters 2 and 11). The policy maker can, on the basis of such information and population exposure, not only estimate the current burden of disease caused by such an exposure in the population, but also the reduction of the burden if improvement action was taken.

Looking at the normative function from a wider angle, disease burden measurement also provides information on the relative importance of a problem involving an environmental condition. It therefore puts the normative function of a specific type of exposure into a certain perspective concerning the priority that its development deserves.

3.5.1 Use of DALYs in guideline derivation

DALYs can be utilised in a variety of ways. They can be used to integrate the effects of a single agent, compare the health effects of different agents or conditions and to inform the debate on levels of acceptable risk.

3.5.1.1 Disease development – gastrointestinal disease

The first step in disease burden estimation requires an understanding of the natural history of the disease. This is best illustrated by the use of a diagram (see Figure 3.1) and it allows disease development to be broken down into various health outcomes or end points. The host can be in any of a number of possible health states, and the transitions between these states can be described by a set of conditional probabilities, i.e. the chance of moving to a health state, given the present health state.

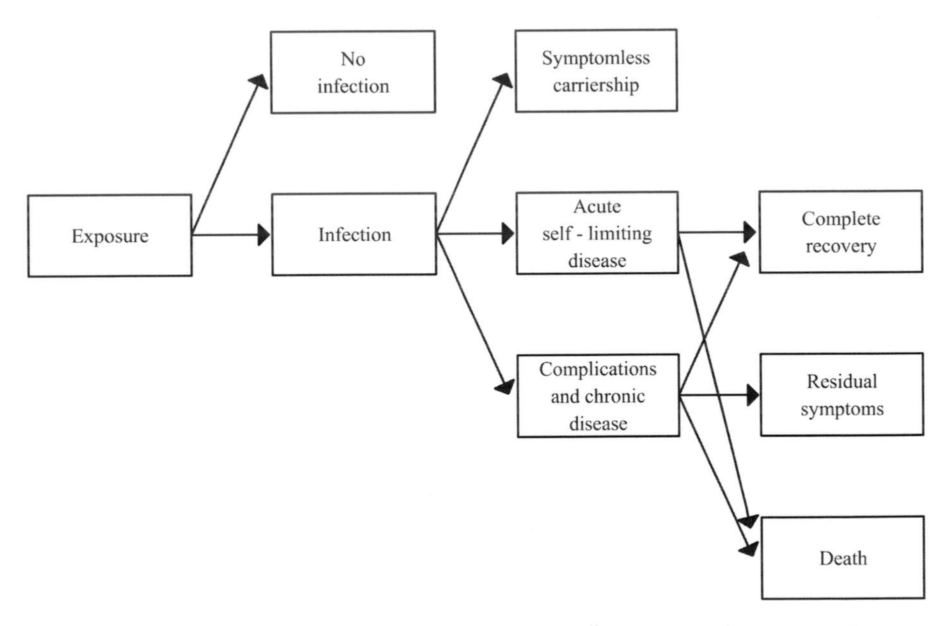

Figure 3.1. Chain model of infectious gastrointestinal disease.

The probability of infection (that is, the ability of the pathogen to establish and multiply within the host) depends on the level of exposure to the organisms in food, water or other environmental factors. Based on data from human feeding studies, statistical dose–response models have been developed to quantify the relationship between the number of ingested organisms and the probability of infection (Havelaar and Teunis 1998; Teunis *et al.* 1996). These models are empirical and do not explicitly identify the factors that may influence the process of infection. Such factors include:

- the physiological status of the pathogen
- the matrix in which it is presented to the host
- the microbial dynamics in the host
- the aspecific host resistance (e.g. gastric acid, enzymes, bile, peristalsis)
- the specific (cellular and humoral immunity) host resistance.

Thus, generalisation of dose–response models is only possible to a limited extent. There are also experimental data on the probability of acute, gastrointestinal disease after infection. In most human feeding studies, clinical symptoms are also described, but the relationship with the ingested dose is less uniform than for infection (Teunis *et al.* 1997). Additional data may be derived from epidemiological studies, such as outbreak investigations or prospective cohort studies.

Usually, gastroenteritis is a self-limiting disease and the host will generally recover within a few days to a few weeks without any residual symptoms (although this may not be true of susceptible individuals, those with weakened immune systems and those in developing countries). In most cases, symptomatic or asymptomatic infection confers immunity that may protect from infection and/or disease upon subsequent exposure. Usually, immunity against enteric pathogens is short-lived and the host will again enter a susceptible state within a period of months to years. In a small fraction of infected persons (with or without acute gastroenteritis), chronic infection or complications may occur. Some pathogens, such as salmonellae, are invasive and may cause bacteraemia and generalised infections. Other pathogens produce toxins that may be transported by the blood to susceptible organs, where severe damage may occur. An example is the haemolytic uremic syndrome, caused by damage to the kidneys from Shiga-like toxins of some *E. coli* strains. Complications may also arise by immune-mediated reactions, where the immune response to the pathogens is then also directed against the host tissues. Reactive arthritis and Guillain-Barré syndrome are well-known examples of such diseases. The complications from enteritis normally require medical care, and frequently result in hospitalisation. There may be a substantial risk of mortality, and not all patients may recover fully, but may suffer from residual symptoms, which may be life-long. Therefore, despite the low probability of complications, the public health burden may be significant.

3.5.1.2 Integrating the health effects of exposure to one agent

This application of the DALY scale, in terms of exposure to a single agent, is illustrated by the example of the health burden of infection with thermophilic *Campylobacter* spp. in the Dutch population. *Campylobacter* infection may lead to a great diversity of symptoms, but most important in terms of incidence and severity are acute gastroenteritis (in the general population and leading to a general practitioner visit), Guillain-Barré syndrome (clinical phase as well as residual symptoms) and reactive arthritis.

Table 3.6. Health burden due to infection with thermophilic *Campylobacter* spp. in the Netherlands, assuming no age-weighting or discounting* (adapted from Havelaar *et al.* 2000a)

Population	Number of cases	Duration (years)	Severity weight	YLD/ YLL
Morbidity				
General population: gastroenteritis	311,000	0.014	0.067	291
General practitioner: gastroenteritis	17,500	0.023	0.393	159
Clinical phase Guillain-Barré	58.3	1	0.281	16
Residual symptoms: Guillain-Barré	57.0	37.1	0.158	334
Reactive arthritis	6570	0.115	0.210	159
Mortality				
Gastroenteritis	31.7	13.2	1.00	419
Guillain-Barré	1.3	18.7	1.00	25
TOTAL				1403

* based on mean values of the estimated annual incidence, the severity weight and the duration

Table 3.6 shows a summary of results, indicating an annual loss of approximately 1400 DALYs per year in the Dutch population of 15 million. The most significant impact on public health is from gastroenteritis-related mortality and the residual symptoms of Guillain-Barré Syndrome, despite the fact that the incidence is low. Acute gastroenteritis (both patients who do and do not visit their GP), is an additional important source of disease burden.

3.5.1.3 Integrating health effects from exposure to different agents

DALYs can also be used to compare the effects of different agents and allow a balancing of risks. Disinfection of drinking water reduces the risk of infectious disease but oxidants such as chlorine and ozone react with water constituents to produce a wide range of disinfection by-products, with toxic and carcinogenic properties. The dilemma of how to balance these positive and negative health effects has long hampered decision making with regard to implementing or modifying drinking-water disinfection processes (Craun *et al.* 1994a,b). The use of DALYs as a tool to quantify all effects in one single metric has simultaneously been suggested in the Netherlands (Havelaar *et al.* 2000b) and by the United States Environmental Protection Agency (US EPA). Havelaar *et al.* (2000b), using a hypothetical case study, examined the reduction of the risk of infection with *Cryptosporidium parvum* following ozonation of drinking water in comparison to the concomitant increase in the risk of renal cell cancer arising from the formation of bromate. It was found that the health benefits of preventing gastroenteritis in the general population and premature death in AIDS patients outweighed health losses by premature death from renal cell cancer by a factor of more than ten.

3.5.1.4 Defining a level of acceptable risk

The approach used above can be extended to derive a target value for acceptable risk from pathogens in water that offers a similar level of protection as current standards for genotoxic carcinogenic compounds.

The definition of acceptable risk used in the guidelines for drinking water quality (WHO 1993) for genotoxic carcinogens is: '*less than one excess cancer case per 10^{-5} consumers after lifetime exposure*'. If a cohort of one million people experienced this risk, there would be ten excess cancer cases in this cohort. Renal cell cancer (RCC) caused as a result of exposure to bromate (as discussed above) will be used as an example. RCC occurs at a median age of 65 years (standardised life expectancy 19 years) and has a case-fatality ratio of 60%. If the relatively minor effects of morbidity are ignored, the health burden of one case of RCC is equal to the number of Life Years Lost, which is $1 \times 60\% \times 19 = 11.4$ years (Havelaar *et al.* 2000b). Averaged over the total life expectancy of this population at birth (80 years), the annual (acceptable) loss of healthy life years is a fraction of $10 \times 11.4/80 \times 10^6 = 1.4 \times 10^{-6}$. Compare this fraction with the annual health burden of *Campylobacter*-associated infections in the Netherlands of almost 1500 DALYs per year per 15 million inhabitants. This is a fraction of 10^{-4}, or more than 70 times higher than deemed acceptable for genotoxic carcinogens. Note that in the Netherlands, as in many other industrialised countries, the acceptable level of risk is set at 10^{-6}, making the current health burden of campylobacteriosis 700 times higher than the equivalent risk limit for carcinogens. It should be noted, however, that cancer types other than RCC may occur at different ages and may have a different prognosis. Therefore, the level of protection of one standard for acceptable risk from all types of cancer does not lead to a common level of protection.

3.6 PROBLEMS IN ASSESSING DISEASE BURDEN IN RELATION TO WATER QUALITY

As for many environmental exposures, the links between disease burden and specific water-related exposures have been difficult to identify. In recent years, with the development of more sophisticated epidemiological methods, increased evidence has been compiled for the health impacts related to water. This is the case for exposure to recreational water (Prüss 1998), and also the impacts of microbiological aspects of drinking water (Payment 1997).

The main difficulties in assessing the water-related disease burden lie in the following points:

- Exposure often occurs at household or small community level and can only be measured with major expenditure and, therefore, cannot be determined on a routine basis. This means that exposures such as drinking-water quality cannot realistically be measured on a large scale, because contamination may vary between adjacent households. Although drinking-water quality is routinely assessed at the point of distribution, it has been shown that the quality at point of consumption may differ significantly.
- The diseases transmitted by water are mostly non-specific, such as the large cluster of 'diarrhoeal diseases'. The problem associated with this relates to the difficulty in attributing a disease to a specific exposure, especially when it is difficult to assess this exposure.
- In settings where the water-related disease burden is greatest (in certain developing country situations and in small community supplies), exposures to disease-causing organisms are frequent and occur often through various 'competing' pathways. These pathways can include exposure to drinking-water, contaminated food, person-to-person contact and lack of hygiene, and it is difficult to determine the relative contributions of these various causes.
- For many water-related exposures, the risks have not been clearly established in terms of exposure–risk relationships. Without an established linkage between exposure and disease outcome, and the difficulties, outlined above, in attributing outcome to specific exposures, disease burden cannot be estimated with any degree of confidence.

Given the importance of the disease burden related to water supply, sanitation and hygiene (two to three million deaths per year), it is imperative that further investigations be made to improve our knowledge about the relative importance of pathways of transmission and the relation between population exposure and disease burden. This information is necessary to construct the picture that will allow efficient and equitable allocation of resources and efforts in order to achieve the greatest improvement of population health status.

3.7 IMPLICATION FOR INTERNATIONAL GUIDELINES AND NATIONAL REGULATIONS

Developments in the global burden of disease and the use of DALYs will play an important role in prioritising risk factors, determining levels of acceptable risk, setting health targets and appraising effectiveness through examining public health outcome. Their use is, therefore, key to the development of future guidelines driven by the harmonised framework. Because international guidelines should be tailored to the public health needs and conditions of individual countries, DALYs are also likely to play an important role in that process of adaptation.

3.8 REFERENCES

Anand, S. and Hanson, K. (1997) Disability-adjusted life years: a critical review. *Journal of Health Economics* **16**, 695–702.

Barendregt, J.J., Banneux, L. and Van de Maas, P.J. (1996) DALYs: the age weights on balance. *Bulletin of the World Health Organization* **74**, 439–443.

Brooks, R. with the EuroQol Group (1996) EuroQol: the current state of play. *Health Policy* **37**, 53–73.

Craun, G.F., Bull, R.J., Clark, R.M., Doull, J., Grabow, W., Marsh, G.M. *et al.* (1994a) Balancing chemical and microbiological risks of drinking water disinfection, part I. Benefits and potential risks. *J. Water SRT-Aqua* **43**, 192–199.

Craun, G.F., Regli, S., Clark, R.M., Bull, R.J., Doull, J., Grabow, W. *et al.* (1994b) Balancing chemical and microbiological risks of drinking water disinfection, part II. Managing the risks. *J. Water SRT-Aqua* **43**, 207–218.

Essink-Bot, M-L. (1995) Health status as a measure of outcome of disease and treatment. Rotterdam: Erasmus Universiteit.

Havelaar, A.H. and Teunis, P.F.M. (1998) Effect modelling in quantitative microbiological risk assessment. In *Proceedings of the 11th Annual Meeting of the Dutch Society for Veterinary Epidemiology* (eds A.M. Henken and E.G. Evers), Bilthoven, the Netherlands.

Havelaar, A.H., De Wit, M.A.S. and Van Koningsveld, R. (2000a) Health burden of infection with thermophilic Campylobacter species in the Netherlands. Report no. 284550004, Bilthoven: National Institute of Public Health and the Environment.

Havelaar, A.H., De Hollander, A.E.M., Teunis, P.F.M., Evers, E.G., Van Kranen, H.J., Versteegh, J.F.M., Van Koten, J.E.M. and Slob, W. (2000b) Balancing the risks and benefits of drinking water disinfection: DALYs on the scale. *Environmental Health Perspectives* **108**, 315–321.

Jamieson, D.T. (1996) Forward to the global burden of disease and injury series. In *The Global Burden of Disease and Injury Series. Volume I. The Global Burden of Disease* (eds C.J.L. Murray and A.D. Lopez), pp. xv–xxiii, Boston: Harvard School of Public Health, World Health Organization, World Bank.

Murray, C.J.L. (1994) Quantifying the burden of disease: the technical basis for disability-adjusted life years. *Bulletin of the World Health Organization* **72**(3), 429–445.

Murray, C.J.L. (1996) Rethinking DALYs. In *The Global Burden of Disease and Injury Series. Volume I. The Global Burden of Disease* (eds C.J.L. Murray and A.D. Lopez), pp. 1–98, Boston: Harvard School of Public Health, World Health Organization, World Bank.

Murray, C.J.L. and Lopez, A.D. (1996a) *The Global Burden of Disease and Injury Series. Volume I. The Global Burden of Disease. A comprehensive assessment of mortality and disability from diseases, injuries, and risk factors in 1990 and projected to 2020.* Harvard School of Public Health, World Bank, World Health Organization.

Murray, C.J.L. and Lopez, A.D. (1996b) *The Global Burden of Disease and Injury Series. Volume II. Global Health Statistics. A compendium of incidence, prevalence and mortality estimates for over 200 conditions.* Harvard School of Public Health, World Bank, World Health Organization.

Murray, C.J.L. and Lopez, A.D. (1999) Progress and directions in refining the global burden of disease approach. World Health Organization, GPE Discussion Paper No. 1.

Nord, E. (1995) The person-trade-off approach to valuing health care programs. *Medical Decision Making* **15**, 201–208.

Payment, P. (1997) Epidemiology of endemic gastrointestinal and respiratory diseases: Incidence, fraction attributable to tap water and costs to society. *Water Science and Technology* **35**(11–12), 7–10.

Prüss, A. (1998) Review of epidemiological studies on health effects from exposure to recreational water. *International Journal of Epidemiology* **27**(1), 1–14.

Teunis, P.F.M., Van der Heijden, O.G., Van der Giessen, J.W.B. and Havelaar, A.H. (1996) The dose–response relation in human volunteers for gastro-intestinal pathogens. Report no. 284550002, Bilthoven: National Institute of Public Health and the Environment.

Teunis, P.F.M., Medema, G.J., Kruidenier, L. and Havelaar, A.H. (1997) Assessment of the risk of infection by *Cryptosporidium* or *Giardia* in drinking water from a surface water source. *Water Research* **31**, 1333–1346.

Torrance, G.W. (1986) Measurement of health utilities for economic appraisal. *Journal of Health Economics* **5**, 1–30.

Van der Maas, P.J. and Kramers, P.G.N. (eds) (1997) Gezondheid en levensverwachting gewogen Volksgezondheid Toekomst Verkenning 1997, deel III. Bilthoven; Maarssen: Rijksinstituut voor Volksgezondheid en Milieu; Elsevier/De Tijdstroom.

WHO (1993) Drinking Water Quality Guidelines – Volume 1, Recommendations. World Health Organization, Geneva.

WHO (1998) Guidelines for Safe Recreational-water Environments: Coastal and Freshwaters. Draft for consultation, World Health Organization, Geneva.
WHO (1999) The World Health Report. Making a Difference. World Health Organization, Geneva.
Williams, A. (1999) Calculating the global burden of disease: time for a strategic reappraisal? *Health Economics* **8**, 1–8.

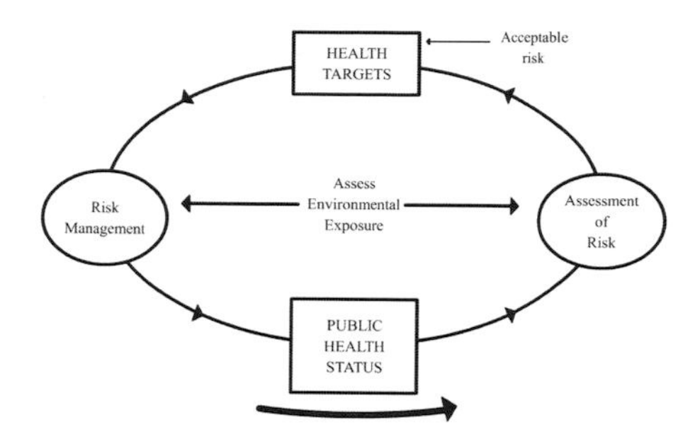

4

Endemic and epidemic infectious intestinal disease and its relationship to drinking water

Pierre Payment and Paul R. Hunter

Guidelines operate from the premise that pathogens do occur in the environment and that there is a susceptible population. This chapter examines this assumption in relation to gastrointestinal pathogens focusing largely on the drinking water environment. It examines the prevalence of diarrhoeal disease in different communities and compares the situation between developing and developed countries.

4.1 INTRODUCTION

It was only in the early part of the twentieth century that waterborne pathogenic micro-organisms and their diseases were finally controlled to an acceptable level in the then rapidly industrialising countries. The control of these diseases was due to

 Water Quality: Guidelines, Standards and Health. Edited by Lorna Fewtrell and Jamie Bartram. Published by IWA Publishing, London, UK. ISBN: 1 900222 28 0

various factors, not least of which were the introduction of adequate water treatment, including filtration and chlorination. Improved wastewater disposal, pasteurisation of milk, and improved food preparation and storage also contributed to a general enhancement of hygiene. It should be remembered that water is usually only one of several possible routes of transmission of 'waterborne diseases'. While it is assumed that a significant proportion of these gastrointestinal illnesses are waterborne, we have very little data to estimate the proportion of the overall burden of disease that is due to transmission by drinking water. Furthermore, considered as part of a holistic approach, a reduction in water resources contamination would result in measurable reductions in gastrointestinal illnesses associated with other transmission routes.

At the onset of the twenty-first century, what do we know of the health effects of drinking water and their impact on our so-called modern societies? There are many reports on the impact of waterborne diseases in countries worldwide revealing thousands of outbreaks due to bacterial, viral, and parasitic micro-organisms associated with the consumption of untreated or improperly treated drinking water (Ford and Colwell 1996; Hunter 1997; WHO 1993; WHO 1996; see also Chapter 6 of this volume).

The World Health Organization (WHO) estimated in 1996 that every eight seconds a child died from a water-related disease and that each year more that five million people died from illnesses linked to unsafe drinking water or inadequate sanitation (Anon 1996). WHO also suggest that if sustainable safe drinking water and sanitation services were provided to all, each year there would be 200 million fewer diarrhoeal episodes, 2.1 million fewer deaths caused by diarrhoea, 76,000 fewer dracunculiasis cases, 150 million fewer schistosomiasis cases and 75 million fewer trachoma cases.

These statistics serve to shock societies complacent about their water supplies, but do little to give a true estimate of the prevalence of waterborne disease in individual communities. In this chapter, we shall attempt to gain an insight into the prevalence of diarrhoeal disease in different communities and then to estimate the proportion of such disease that may be linked to drinking water.

4.2 AETIOLOGY OF GASTROINTESTINAL ILLNESSES

Many infectious causes of acute gastrointestinal symptoms have been described in the literature (Branski 1984; Bryan 1985; Ellner 1984; Goodman and Segreti 1999; Hunter 1997). These include parasitic agents such as *Cryptosporidium parvum*, *Giardia lamblia*, *Cyclospora* and *Entamoeba histolytica*; bacterial pathogens such as *Salmonella*, *Shigella*, *Campylobacter*, *Vibrio cholera*, enterovirulent *Escherichia coli*, *Aeromonas*, *Yersinia* and *Clostridium perfringens*; and viruses such as the enteroviruses, rotaviruses, parvoviruses, Norwalk and Hawaii agents, adenoviruses, caliciviruses, and astroviruses. Many of these pathogens have been transmitted by the water route, in addition to person-to-person, animal-to-human, food-borne and aerosol routes.

However, acute infection is not the only cause of acute diarrhoeal disease. Milk or soy protein intolerance, food abuses or diet changes, side-effects of prescription drugs (especially antibiotics) as well as fungal, algal or shellfish toxins may all cause diarrhoea. A number of chemicals such as monosodium glutamate, organic mercury, antimony, and copper (Branski 1984; Ellner 1984) can also induce gastrointestinal symptoms.

Hodges *et al.* (1956) presented data on the surveillance of infectious diseases in Cleveland (US) and offered a table of circumstances explaining the gastrointestinal symptoms observed. From their data, 116 out of 362 cases were due to acute infectious diseases, 63 were due to dietary indiscretion, 59 to coughing/gagging, 45 to medication, 18 to emotional causes and 61 were of unknown origin.

4.3 INCIDENCE OF GASTROINTESTINAL ILLNESS IN INDUSTRIALISED COUNTRIES

It is generally very difficult to derive a good estimate of the incidence of endemic acute gastrointestinal disease in a community. Only a proportion of people infected develop symptoms, and only a proportion of these will seek medical attention. Even if patients with diarrhoea present to their doctor, the doctor may not report the illness or even take samples. Even if samples are taken, subsequent laboratory analysis may not detect a pathogen because the specific pathogen was not screened for or because the laboratory tests were not sufficiently sensitive. Even when the laboratory correctly identifies a pathogen, this may not be reported to appropriate surveillance systems. In many communities, it is still not clear what proportion of acute episodes reach appropriate surveillance systems (see Chapter 6). The consequence is that existing surveillance systems are likely to dramatically underestimate the real burden of acute gastrointestinal disease. It is only through appropriately designed epidemiological studies that the real incidence of acute diarrhoeal disease in a population can be estimated. There are very few such studies reported in the literature.

Few studies have investigated the incidence of gastrointestinal disease at the community level. Three such studies have described the health status of individuals in the US over a long period of time, the Cleveland study (Dingle *et al.* 1953; Hodges *et al.* 1956) the Tecumseh study (Monto and Koopman 1980, Monto *et al.* 1983) and the Virus Watch Program (Fox *et al.* 1966). These studies have provided information on the illness rates in the northern part of the US. All these studies have reported gastrointestinal illness rates in the range of 0.5–2 episodes/year/person and incidence of 5–100 episodes/1000/ week according to seasons and age. The number of episodes of gastrointestinal illnesses is similar in all these studies despite the fact that more than 40 years have elapsed between some of them.

In the Cleveland study (Hodges *et al.* 1956) the mean incidence of gastrointestinal illness was 1.6 episode/person/year with a maximum of 2.6 observed in children aged four and an incidence of 0.9 in adults. A seasonal pattern was observed, with the lowest incidence in July and the highest in November in 1948/49/50. The Tecumseh study (Monto and Koopman 1980) reported a maximum incidence in children and a mean incidence of 1.2. However, their definitions of gastrointestinal illness were stricter than those of the earlier Cleveland study. Mean rates of enteric illness syndromes were 0.35 episodes/person/year for vomiting, 0.40 for diarrhoea, 0.23 for both at the same time and 0.22 for nausea and/or upset stomach for a mean total of 1.20 (std. dev: 1.5). They also report variation in the seasonal incidence of gastrointestinal illness according to age, with adults (over 20 years old) being least susceptible, and children under five years old being the most susceptible. Peaks of up to 90 episodes/1000/week were observed in children in November, while values as low as five episodes/1000 persons/week were reported in July.

Studies in the UK have suggested a lower level of diarrhoeal disease than those in the US. Feldman and Banatvala (1994) added questions about diarrhoeal disease to the monthly OPCS (Office of Population and Census Survey) Omnibus survey. This is a government survey that interviews about 2000 adults each month on a range of issues. Some 8143 adults were asked whether they had had diarrhoea in the previous month. From the responses, the authors calculated an annual attack rate of 0.95 episodes per person per year. In a study of two general practice populations in South Wales, another group conducted a postal questionnaire survey (Palmer *et al.* 1996). This group estimated an attack rate of 0.89 episodes per person per year.

Also in the UK, a study was carried out to establish the incidence and aetiology of infectious intestinal disease in the community and presenting to general practitioners in comparison with incidence and aetiology of cases reaching the national laboratory surveillance scheme (Wheeler *et al.* 1999). An incidence of 0.194 episodes/person/year was observed in the community-based study. Based on 8770 cases presenting to their general practitioner the incidence was only 0.033 episodes/person/year. One case was reported to national surveillance for every 1.4 laboratory identifications, 6.2 stools sent for laboratory investigation, 23 cases presenting to general practice, and 136 community cases. The ratio of cases in the community to cases reaching national surveillance was lower for bacterial pathogens (*Salmonella* 3.2:1, *Campylobacter* 7.6:1) than for viruses (rotavirus 35:1, small round structured viruses 1562:1). There were many cases for which no organism was identified. The authors concluded that infectious intestinal disease occurs in one in five people each year, of whom one in six presents to a general practitioner. The proportion of cases not recorded by national laboratory surveillance is large and varies widely by micro-organism. The attack rate of only 0.194 episodes per person per year is well under usually reported values. The large discrepancy between this and earlier studies probably reflects the case definition and the different study design, a prospective longitudinal study. Instead of asking whether respondents had had diarrhoea in the previous month, the study asked participants to send in postcards on a weekly basis for six months declaring the absence of symptoms (those with symptoms sent a stool sample). Interestingly, at the start of the prospective community study, participants were asked whether they had had diarrhoea in the past month. Based on this figure the estimated attack rate would be 0.55 episodes per person per year, with the difference thought to be due to 'telescoping' of recent events.

In many developed countries, food-borne infections are under surveillance and data on the occurrence of cases can be found in reports produced by various agencies. While they provide an interesting list of the micro-organisms implicated, the reported cases are often only the tip of the iceberg as very few individuals are severely affected. Reports to physicians and samples sent to laboratories for analysis do not reflect the true level of food-borne disease. The United States have increased their level of surveillance through a network called 'FoodNet' (www.cdc.gov/ncidod/dbmd/foodnet) (CDC 1997, 1998, 1999). Their data provides information on several pathogens by age, sex, site and pathogen on an ongoing basis. Cases reported through active surveillance represent a fraction of the number of cases in the community. To better estimate the number of cases of food-borne disease in the community, FoodNet conducted surveys of laboratories, physicians, and the general population in the FoodNet sites. Of the 10,000 residents covered by the FoodNet survey, 11% reported a diarrhoeal illness during the previous month or 1.4 episodes of diarrhoea per person per year. Of those who were ill, only 8% sought medical care. Of those seeking medical care, 20% reported submitting a stool specimen for culture. Through active surveillance and additional studies, FoodNet is providing better estimates of the burden of food-borne illness and is tracking trends in these diseases over time. In 1997, surveillance of the seven pathogens studied showed that 50 cases of these infections were diagnosed per 100,000 population, representing a total of 130,000 culture-confirmed cases in the entire US population. Additional FoodNet surveys showed that these cases represent a fraction of the burden of food-borne illness. Based on these surveys, at least 60 more of these infections may have occurred for each one that was diagnosed, suggesting that approximately eight million cases of these bacterial infections occurred in 1997 in the US. However, the only community-based study carried out as part of the FoodNet programme was a retrospective study which has been shown in the UK to overestimate the incidence of diarrhoeal disease by about a factor of three (Wheeler *et al.* 1999).

Even taking into consideration differences in methodology between some of the US and UK studies, it would appear that the incidence of acute diarrhoeal disease in the US is about twice that in the UK. It is not clear to the authors of this chapter why this should be, although the more extreme differences of the US climate and a higher level of convenience food consumption may be factors.

4.3.1 Waterborne disease

While the most often reported disease associated with drinking water remains gastroenteritis, this is probably due to the very apparent nature of the symptoms and the fact that the attack rates for these infections can reach over 50% of the exposed population. Even infectious disease specialists often forget that enteric micro-organisms are associated with a wide range of symptoms and diseases. Protozoan parasites such as amoebae can cause severe liver or brain infections and contact-lens wearers are warned of the dangers of eye infections. Bacteria can cause pneumonia (*Legionella*) and some are suggesting the possibility that *Helicobacter pylori*, which has been associated with gastric ulcers, could be transmitted by the water route (Hulten *et al.* 1998).

In all industrialised countries, a steady decline in gastrointestinal disease was evidenced by the virtual elimination of cholera and the reduction of waterborne outbreaks to very low levels. Most bacterial waterborne pathogens have been eliminated by the simple use of chlorine disinfection. However, we are finding strains of *Vibrio cholerae* that are more resistant to disinfection, *Legionella* has been found in water heaters and the *Mycobacterium avium* complex (MAC) is now on the list of potential pathogens.

The micro-organisms implicated in waterborne diseases have been well described (Hunter 1997; Hurst *et al.* 1997; Murray *et al.* 1995). Waterborne diseases are usually described in terms of outbreak reporting in the various countries. Two countries, the US (Craun 1992) and the UK (Hunter 1997) have produced most of the available data (see Chapter 6). In other countries, data gathering is often very poorly performed because of lack of resources to identify the water-related events as well as the lack of centralised official data-gathering authorities. Several methods for the detection and investigation of waterborne outbreaks have been described, but are still not widely used (Craun 1990) as resources and funds are critically lacking even in industrialised countries. An enormous effort is needed to educate populations on the importance of water in the dissemination of disease. All levels of society, from consumer to politicians, must be educated about the benefits of improving water quality (Ford and Colwell 1996) as a major step in improving quality of life and health.

Since the 1950s, with the development of methods to detect and identify viruses, many outbreaks of waterborne gastrointestinal illness that would have been simply classified as of non-bacterial origin have been attributed to enteric viruses. Hepatitis A and E, Norwalk, small round structured viruses (SRSV), astroviruses, caliciviruses and many others are now well-known names in the water industry.

Parasites are being identified as pathogens of importance even in industrialised countries. Numerous waterborne outbreaks of giardiasis have been reported in the US (Craun 1986). During the last twenty years there have also been a number of outbreaks of cryptosporidiosis in the UK (Badenoch 1990a,b). The continued problem with parasitic infection in drinking water is largely related to parasites' resistance to the water disinfection process. Dozens of outbreaks of cryptosporidiosis have now been reported worldwide, but most are small compared to the explosive outbreak experienced in Milwaukee (US) in the spring of 1993 (Edwards 1993; Mackenzie *et al.* 1994). Following what appears to be a combination of storm-washed faecal contamination from a compromised catchment and failure in treatment, water that met US microbiological water quality guidelines caused gastrointestinal illnesses in an estimated 400,000+ people, or one-third of the population of this city, over a period of one month. Most of these illnesses were cryptosporidiosis but many were probably of viral origin. The most surprising aspect of this event is that cryptosporidiosis was probably occurring even before it was detected following a report from an inquisitive pharmacist (Morris *et al.* 1998). This fact suggests that unless an effort is made to identify waterborne diseases they will remain undetected, buried in the endemic level of illness in the population.

Enteric viruses are also excreted through faeces into the environment by infected individuals with or without clinical illness. There are over 100 types of enteric virus, including enterovirus (poliovirus, coxsackievirus, echovirus, hepatitis A), reovirus, rotavirus, adenoviruses, coronavirus, calicivirus, astrovirus, Norwalk-like agents, and so on. While viruses are excreted in large numbers in the faeces of infected individuals, the low incidence of infection in a population, the dilution factor after their release in the water, and difficulty in detection accounts for low numbers in contaminated surface waters (Bitton 1980; Bitton et al. 1985; Rao and Melnick 1986). Reported numbers range from absence of enteric virus in uncontaminated waters to several thousand or millions of viruses per litre of untreated wastewater (Bitton 1980; Bitton et al. 1985, Rao and Melnick 1986). Cultivable enteric viruses are however quite prevalent and can serve as an indicator of the overall viral population (Payment 1993a).

The Lubbock health effect study has also provided a very comprehensive data set on the prevalence of antibodies to several enteric viruses in an North American population (Camaan *et al.* 1985). Data on the seroprevalence of several enteric viruses in the Montreal area were obtained during an epidemiological survey on water-related illnesses (Payment 1993a,b). The seroprevalence of antibodies to several enteric viruses including hepatitis A virus (Payment 1991) and Norwalk virus (Payment *et al.* 1994) were reported (Table 4.1). Results indicated that the rates of hepatitis A viral infections are slightly lower than those reported for other countries. The data indicate that the hepatitis A virus is an infection progressively acquired in life and that in the Montreal area relatively few children have antibodies to this virus. This observation is in contrast with many countries where, due to a low level of hygiene, these infections are acquired early in life (Brüssow *et al.* 1990; Brüssow *et al.* 1988; Morag *et al.* 1984; Nikolaev 1966; Papaevangelou 1980).

Table 4.1. Seroprevalence (expressed as a percentage) to selected enteric pathogens in a French-Canadian population (Payment 1991; Payment *et al.* 1994; unpublished data for *Cryptosporidium*)

	Age groups (years)				
Micro-organism	9–19	20–39	40–49	50–59	60+
Hepatitis A	1	10	49	60	82
Cryptosporidium	21	55	59	52	20
Norwalk virus	36	67	80	70	64
Echovirus 9	40	69	70	51	60
Coxsackievirus B-2	51	60	67	66	60
Coxsackievirus B-3	51	64	63	55	60
Coxsackievirus B-4	44	80	77	74	80
Coxsackievirus B-5	58	74	61	62	20
Echovirus 11	78	84	91	83	80
Echovirus 30	96	98	92	96	100
Rotavirus	100	100	100	100	100

4.4 ENDEMIC WATERBORNE DISEASE IN INDUSTRIALISED COUNTRIES

While many micro-organisms have been implicated in outbreaks of various diseases, there is little epidemiological data on the endemic level of waterborne diseases. Those studies that have attempted to define the burden of waterborne disease have generally concentrated on gastrointestinal illness.

Batik *et al.* (1979) using hepatitis A virus cases as an indicator could not establish a correlation with water quality or find a correlation between current indicators and the risk of waterborne outbreaks (Batik *et al.* 1983).

In France, Collin *et al.* (1981) prospectively studied gastrointestinal illnesses associated with the consumption of tap water using reports from physicians, pharmacists and teachers. Their results were based on more than 200 distribution systems or treated or untreated water and they reported five epidemics (more than 1000 cases) associated with poor quality water. This study is typical of most studies which rely on the detection of epidemics to assess the level of water quality: they do not address the endemic level of gastrointestinal illnesses which may be due to low-level contamination of the water. The same group, in a prospective follow-up study on 48 villages for 64 weeks, evaluated untreated groundwater and found a relationship between faecal streptococci and acute gastrointestinal disease (Ferley *et al.* 1986; Zmirou *et al.* 1987). Faecal coliforms did not appear to be independently related to acute disease. Total coliforms and total bacteria showed no correlation with disease.

In a more recent study carried out in France, Zmirou *et al.* (1995) investigated the effect of chlorination alone on water that did not otherwise satisfy microbiological criteria. They prospectively followed up some 2033 schoolchildren aged between 7 and 11 years from 24 villages. In 13 villages the water had no treatment as it met current microbiological criteria in its raw state. In the other 11 villages, the raw water had evidence of faecal pollution and was chlorinated before it met current standards. The gastrointestinal morbidity of the children was recorded daily under the supervision of the schoolteachers. The crude incidence of diarrhoea was 1.4 times more frequent in the children from villages whose water supplies had evidence of faecal pollution, even after chlorination. These results strongly suggest that there are some pathogens in faecally polluted drinking water which are not adequately treated by chlorination alone.

In Israel, Fattal *et al.* (1988) addressed the health effects of both drinking water and aerosols. Their studies on kibbutz water quality and morbidity were performed in an area with relatively high endemicity of gastrointestinal disease and did not show a relationship between health effects and total or faecal coliforms. This study was, however, based only on morbidity reported to physicians, data that is considered to represent only 1% of the actual cases in a population. In Windhoek (Namibia), Isaäcson and Sayed (1988) conducted a similar study over several years on thousands of individuals served by recycled waste water as well as normal drinking water. They did not observe an increased risk of reported acute gastrointestinal illness associated with the consumption of recycled waters. The populations compared had higher incidence rates than those observed in the US and they were subjected to a high endemicity level due to other causes, thus masking low levels of illnesses.

4.4.1 Intervention studies

Two major epidemiological studies have been conducted in Canada to evaluate the level of waterborne disease. The results of these studies suggest that a very high proportion of gastrointestinal illnesses could still be attributable to tap water consumption, even when water met the current water quality guidelines (Payment *et al.* 1991a,b, 1997). The first study was carried out from September 1988 to June 1989. It was a randomised intervention trial carried out on 299 randomly selected eligible households which were supplied with domestic water filters (reverse-osmosis (RO)) which eliminated microbial and most chemical contaminants from their tap water and on 307 randomly selected households which were left with their usual untreated tap water. The gastrointestinal symptomatology was evaluated by means of a family health diary maintained prospectively by all study families. The estimated annual incidence of gastrointestinal illness was 0.76 among tap water drinkers as compared with 0.50 among RO-filtered water drinkers ($p<0.01$). Because participants in the RO-filter group were still exposed to tap water (i.e. about 40% of their water intake was tap water), it was estimated that about 50% of the illnesses were probably tap-water-related and thus preventable. The remaining illnesses were probably attributable to the other possible causes such as endemic infectious illnesses, food-related infections, allergies, etc.

Attempts were also made to determine the aetiology of the observed illnesses. Sera were collected on four occasions from volunteers and they were tested for antibodies to various pathogens. There was no indication, by serology, of water-related infections caused by enteroviruses, hepatitis A virus and rotavirus or Norwalk virus infections (Payment *et al.* 1994).

The second Canadian study (Payment *et al.* 1997) was more complex: its objective was not only to re-evaluate the level of waterborne illness, but also to identify the source of the pathogens responsible for them. It was conducted from September 1993 until December 1994 and compared the levels of gastrointestinal illness in four randomly selected groups of 250 families, who were served water from one of the following sources:

- normal tap water
- tap water with a valve on the cold water line (to examine the effect of home plumbing)
- plant effluent water as it leaves the plant and bottled (not influenced by the distribution system) ('plant')

- plant effluent water further treated and bottled (to remove any contaminants) ('purified')

The treatment plant was selected for the poor raw water quality (i.e. high microbial contamination) and for its treatment performance. Raw water entering the plant was contaminated with parasites, viruses and bacteria at levels typically found in faecally contaminated waters. The end product met or exceeded current Canadian and US regulations for drinking water quality. The distribution system was in compliance for coliforms but residual chlorine was not detectable at all times in all parts of the distribution system.

The rates of highly credible gastrointestinal illnesses (HCGI) were within the expected range for this population at 0.66 episodes/person/year for all subjects and 0.84 for children aged 2 to 12. The rate of illness was highest in autumn/winter and lowest in summer. Overall, there were more illnesses among tap water consumers than among subjects in the 'purified' water group, suggesting a potential adverse effect originating from the plant or the distribution system. Children were consistently more affected than adults and up to 40% of their gastrointestinal illnesses were attributable to water. The rates of gastrointestinal illness among consumers of water obtained directly at the treatment plant were similar to the rate of illness among consumers of purified water. Two periods of increased tap-water-attributable illnesses were observed in November 1993 and in March 1994.

Subjects in the two bottled water groups (i.e. 'purified' and 'plant') still consumed about one-third of their drinking water as tap water. They were thus exposed to some tap water and its contaminants: as a result the risks due to tap water may be underestimated.

Consumers of water from a continuously running tap had a higher rate of illness than any other group during most of the observation period. This was completely unexpected, since the continuously running tap was thought to minimise the effects of regrowth in home plumbing. Although there are several theories as to the cause of this effect, they remain unsubstantiated.

The data collected during those two epidemiological studies suggest that there are measurable gastrointestinal health effects associated with tap water meeting current standards and that contaminants originating from the water treatment plant or the distribution system could be the source of these illnesses. Short-term turbidity breakthrough from individual filters at the water treatment plant might explain the observed health effects. Potential follow-up research should further examine the relationship of turbidity breakthrough and should investigate the role of the continuously running tap in the occurrence of gastrointestinal illness.

In the Canadian studies, it was not possible to assign a single cause (or aetiological agent) to the observed effects although the authors suggested three explanations: low level or sporadic breakthrough of pathogens at the water treatment plant, intrusions in the distribution system (repairs, breaks, etc.) and finally bacterial regrowth in the mains or in the household plumbing.

4.4.2 Health significance of bacterial regrowth

Bacterial regrowth is common in water and has been observed even in distilled water. In water distribution systems, the heterotrophic plate count can occasionally be elevated and there have been concerns that this flora could contain opportunistic pathogens. Data from epidemiological studies involving reverse-osmosis units suggested that there was a correlation between gastrointestinal illnesses and heterotrophic plate counts at 35°C (Payment *et al.* 1991b). However, a few outliers in the data set drove the correlation and the study would have to be repeated in order to confirm the relationship. Furthermore, this observation could be limited to certain water purification devices such as those in which a rubber bladder is used to accumulate the purified water.

4.4.3 Health significance of turbidity

The Milwaukee outbreak with an estimated 400,000+ cases of gastrointestinal illness occurring in the spring of 1993 is a good illustration of turbidity-related health effects (Mackenzie *et al.* 1994). The outbreak occurred at the beginning of April and was linked to inadequate water treatment as well as to a decrease in river water quality. Turbidity data from the water treatment plant revealed that one of the two water treatment plants was distributing finished water with a turbidity of more than 1.5 NTU. The lag between the turbidity increases and reported illness was seven days in children and eight days in adults. This lag time may reflect the incubation period of *Cryptosporidium* which was identified as the aetiological agent in many of the cases of gastroenteritis. Subsequent to the investigation of the main Milwaukee outbreak, Morris *et al.* (1996) carried out an analysis of hospital records and water turbidity readings over a period of 16 months before the recognised outbreak. They found that attendance of children with gastrointestinal illness at hospital emergency departments showed a strong correlation with rises and falls in turbidity, but did not describe any specific time-lag relationships.

Beaudeau *et al.* (1999) reported data from Le Havre (France) which operates two water treatment plants and distributes water to 200,000 people. The karstic resources used are subject to episodic microbiological quality degradations. One plant only chlorinates, whereas the second plant normally uses slow sand filtration before chlorination but can also implement coagulation-settling when turbidity of the raw water exceeds 3 NTU. During the study period there were several occasions when the chlorine residual was not maintained and there were also significant variations in turbidity. Despite these occurrences the treated water still met all microbiological criteria for potable water in France and the study was undertaken to determine if public health was adequately protected. An ecological time series study was carried out on data collected between April 1993 and September 1996. Records of sales of prescribed and off-the-shelf gastroenteritis medication were provided by the pharmacists participating in the epidemiological surveillance network of Le Havre. Sales data, residual chlorine and turbidity measurements were analysed. Interruption of chlorination of the unfiltered water resulted in a significant increase of medication sales three to eight days later. Raw water turbidity increases resulted in increases of medication sales during the following three weeks. The data analysed suggest that about 10% of the annual cases of gastrointestinal illnesses could be due to the consumption of tap water. This annual average does not reflect the proportional attributable risk occurring during specific periods and underscores the fact that current regulations still do not provide complete protection of public health. Furthermore, such failures have the potential of causing major outbreaks if raw water microbiological quality degrades significantly after rainfall events.

Similar data from the city of Philadelphia has also been studied (Schwartz *et al.* 1997). In this study the researchers examined the association between daily measures of drinking water turbidity and both emergency visits and admissions to the Children's Hospital of Philadelphia for gastrointestinal illness, controlling for time trends, seasonal patterns, and temperature. The data suggested a relationship between hospital admissions for gastrointestinal illnesses and increases in turbidity at the Philadelphia water treatment plant. At all times, the water produced met federal regulations, perhaps suggesting that the standards need to be re-evaluated. The study has been heavily criticised and several potential confounding factors as well as methodological errors have been raised. The turbidity levels examined were in the range of 0.14–0.22 NTU – considerably below the levels of many water supplies in the US and other developed nations. The turbidity meters used at the time of the study were calibrated at approximately four month intervals using standards from 0.2–1.0 NTU: most of the readings used in the analysis were below the calibration range and could be considered unreliable. One of the three water treatment plants routinely rounded turbidity readings to the nearest 0.05 NTU, while the others reported to the nearest 0.01 NTU. No consideration was given to the effect of chemical corrosion inhibitors on turbidity. Although the authors were supplied with minimum and maximum turbidity readings as well as mean readings, these were apparently not analysed. If the basic hypothesis is correct, one would expect stronger effects to be seen with maximum readings or turbidity spikes.

The findings from the US and France complement the Canadian studies, which concluded that a fraction of gastrointestinal illness attributable to drinking water arises from microbiological events in the distribution system, but did not discount the treatment plant as a source of pathogens. Beaudeau *et al.* (1999); Morris *et al.* (1996) and Schwartz *et al.* (1997) suggested that variations in rates of illness were due to changes in the numbers of pathogens (carried in or on small suspended particles) coming through the distribution system from the treatment plant. Given that water treatment is a continuous process, constantly responding to changes in demand, that rapid sand filters are not homogeneous and their performance may vary considerably, and that the distribution system is subjected to numerous challenges, the assertion that pathogens may sometimes be present in treated water seems a reasonable one. At most water treatment plants, even if the water produced by the plant always achieves an average turbidity of less than 1 NTU, individual filters may produce water with significantly higher values for short periods. Such a burst of turbidity could be sufficient to introduce pathogens into the treated water at a level sufficient to explain observations such as those in Canada (Payment *et al.* 1991a,b, 1997) and Philadelphia (Schwartz *et al.* 1997).

Turbidity has been suggested by several groups as a potential indicator of waterborne disease. Much remains to be studied on the value of this easy-to-measure parameter, but it is one indicator that promises to better protect public health and one of the rare indicators that could be used in real time.

4.5 WATERBORNE GASTROINTESTINAL DISEASE IN OTHER COUNTRIES

4.5.1 Incidence of endemic gastrointestinal disease

It is even more difficult to get a clear understanding of the incidence of diarrhoea in developing countries than it is for industrialised countries. Nevertheless, several community-based studies have been reported in the literature. These have generally employed regular visits from a health-care worker. It is, however, difficult to use the results of studies to produce an overall estimate of diarrhoeal disease in developing countries for a number of reasons:

- Levels of diarrhoeal disease may differ markedly between relatively close communities due to different socio-economic factors, such as the availability of a clean water supply and hygiene behaviour.
- The information that is available comes from a variety of sources and has been collected and analysed in different ways.
- The definition of diarrhoea often differs between studies.
- Quite frequently data collection has been part of a prospective epidemiological study designed to investigate the role of some factor such as water supply on health.

In Table 4.2 (see p. 76) we list the estimates of diarrhoeal disease incidence from a variety of prospective epidemiological studies. Where possible the overall incidence of diarrhoea, broken down into broad age groups, has been given. If the study compared two groups during an active intervention such as a health education campaign or improvement in water supply, then we have given the data only for the control group.

It can be seen from Table 4.2 that the incidence of diarrhoeal illness varies markedly between studies. While one should not draw too many conclusions from the data as presented here, it seems clear that incidence is higher in rural rather than urban environments and also in poorer communities.

The age distribution of diarrhoeal disease seems to be similar in all regions where reported. Disease incidence is relatively low in the first few months of life, then peaks at about 24 months before declining towards adulthood. Figure 4.1 shows the incidence of disease in various cohorts during the first five years of life in one study (Schorling *et al.* 1990).

One of the issues that has been raised in recent papers has been the role of population immunity on the epidemiology of water-related disease in general and cryptosporidiosis in particular (Craun *et al.* 1998; Hunter and Quigley 1998; Hunter 1999). The implication of this work is that exposure to diarrhoeal pathogens is far more common than observable disease, the difference being due to pre-existing immunity. Evidence for this in developed countries comes from the investigation of outbreaks of waterborne disease which have shown lower attack rates in residents compared to visitors (Hunter 1999).

Table 4.2. Estimates of diarrhoeal incidence in developing countries from various prospective epidemiological studies

Location	Type of area	Study type	Definition of diarrhoea	Age groups	Inc.*
South-eastern China[1]	Rural villages	Prospective community-based	Passing of 2 or more loose or watery stools in 24 h and lasting for less than 5 days	All age groups Children <5 Adult men Adult women	0.730 2.254 0.750 0.627
North-eastern Brazil[2]	Urban slum	Prospective cohort	Increase in stool frequency or a decrease in consistency, as noted by the caretaker, lasting at least 1 day	Children <5	11.29
Bangladesh[3]	Periurban village	Control group from soap hand washing trial	2 or more watery stools or 4 or more loose stools in 24 hours	All ages	1.114
Nicaragua[4]	City	Prospective cohort	Increased frequency to $\geq$3 liquid stools in 24 h or presence of blood/mucus	Children <2	1.88
Zaire[5]	City	Prospective cohort of control group of HIV negative infants	Change in normal stool pattern with at least one day of increased frequency, blood or mucus	Children <16 m	1.0
Guatemala[6]	Poor rural village	Control group in double blind randomised trial of zinc administration	Mother's definition	Children 6-18 m	23.0
Haiti[7]	Rural	Control group in double blind randomised trial of vitamin A administration	Watery stools four or more times in one day	Children <7	3.29
Columbia[8]	City	Prospective cohort study of diarrhoeal attack rate in those not at day care	3 or more loose stools in 24 hours	Children <24 m 24-35 m 36-60 m	6.8 2.2 1.2

Table 4.2 (cont'd)

Location	Type of area	Study type	Definition of diarrhoea	Age groups	Inc.*
Zambia[9]	City	Retrospective community survey, control group for study of diarrhoea in HIV+ patients	Respondent-defined	Adults	1.74
Bangladesh[10]	Urban	Cohort of children with a non-improved water supply	3 or more loose or watery motions in 24 hours	Children 1–6	3.2
Peru[11]	Poor peri-urban	Prospective cohort	3 or more liquid or semi-liquid stools in 24 hours	0–11 m 12–23 m 24–35 m	8.74 10.18 6.32
Nigeria[12]	Rural	Community survey	Not stated	Children <5	2.12

* Incidence as episodes/person/year (m = months)
[1] Chen *et al.* 1991 [2] Schorling *et al.* 1990 [3] Shahid *et al.* 1996 [4] Paniagua *et al.* 1997
[5] Thea *et al.* 1993 [6] Ruel *et al.* 1997 [7] Stansfield *et al.* 1993 [8] Hills *et al.* 1992
[9] Kelly *et al.* 1996 [10] Henry and Rahim 1990 [11] Yeager *et al.* 1991 [12] Jinadu *et al.* 1991

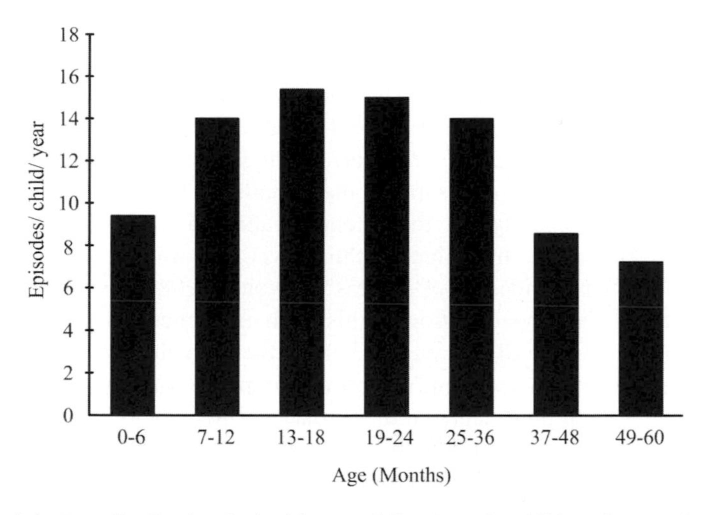

Figure 4.1. Age distribution in incidence of diarrhoea in children in an urban slum in north-eastern Brazil (Schorling *et al.* 1990).

For those interested in the impact of prior immunity on the epidemiology of diarrhoeal disease in developing countries, we are fortunate in having access to a considerable literature on travellers' diarrhoea (Table 4.3).

Table 4.3. Estimated incidence of travellers' diarrhoea from various studies[a]

Home country	Destination country	Study population	Inc.*	Reference
US	Mexico	Adult students	15.64	Johnson *et al.* 1984
US	Mexico	Adult students	9.21	Ericsson *et al.* 1985
US	Mexico	Adult students	6.95	DuPont *et al.* 1987
US	Thailand	Peace core volunteers	4.95	Taylor *et al.* 1985
Sweden	Various	Various age 10+	4.68	Ahlm *et al.* 1994
Various	Jamaica	>16 years	11.3	Steffen *et al.* 1999
Switzerland	Various	0–2	5.00	Pitzinger *et al.* 1991
		3–6	1.45	
		7–14	4.86	
		15–20	5.57	
Netherlands	Various	Adults	11.5[b]	Coeblens *et al.* 1998

* Incidence in episodes/person/year.
[a] Figures usually represent number of affected individuals and so do not count repeated episodes in the same individual and consequently underestimate incidence.
[b] Includes multiple episodes.

Although the studies of travellers' diarrhoea are not directly comparable with studies of local people, it clear that the attack rate in travellers is several-fold higher. This difference is even more notable considering that travellers usually live in rather more hygienic surroundings then do locals. Thus the evidence presented here would support the hypothesis that local people build up a substantial immunity to those enteropathogens circulating in their communities. However, to achieve this level of immunity there is a substantially higher incidence of illness in young children in developing countries than in developed nations. This high incidence of gastrointestinal disease in children is one of the reasons behind the high childhood mortality in developing countries. Consequently there is no place for any argument that allows less than the highest achievable standards of hygiene or water quality in order to build up population immunity. Such arguments would, if implemented, inevitably lead to a rise in childhood morbidity and mortality.

4.5.2 Waterborne disease

Clearly, determining the proportion of diarrhoeal disease in developing countries that is due to contaminated water is problematic. As with the determination of incidence rates, the proportion of diarrhoeal disease due to water consumption varies substantially between communities because of varying water quality and other behavioural and socio-economic factors. An estimate of the proportion of diarrhoeal disease due to water consumption comes from those studies that have compared illness rates between two communities with different water supply or in the same system before and after improvements in water supply.

Esrey and colleagues (1991) published a review of studies that investigated the impact of improved water supply and sanitation on various waterborne diseases. They were able to identify 16 studies that examined the health impacts of pure water over contaminated water. Of these studies, ten reported a positive effect. In only seven studies was it possible to calculate the percentage reduction, the median being a 17% reduction.

Despite the importance of sanitation and hygiene behaviours a significant proportion of diarrhoeal disease due to waterborne transmission will be related to water quality. Perhaps the source with the poorest quality is river water. Two relatively recent studies have examined diarrhoeal illness in people taking river water. In a study in south-eastern China, the incidence of diarrhoea was related to the source of drinking water (Chen *et al.* 1991). The attack rate was 0.575 per person per year in those drinking piped water, 0.846 in those drinking well water and 4.580 in those drinking river water. Thus, other things being equal (which they most likely were not) about 87% of illnesses in people drinking river water were waterborne. This figure of 87% is similar to the reduction in diarrhoea in a more recent study of families in Uzbekistan (Semenza *et al.* 1998). In those families without a piped mains supply the incidence was 2.15 episodes per person per year, and it was 0.91 in those with a piped supply. If those without a piped supply were taught to chlorinate their water, the incidence fell to 0.35, a reduction of 85%.

Moe *et al.* (1991) reported on a particularly elegant study done in Cebu, the Philippines. They looked at the relationship between the microbiological quality of drinking water and the prevalence of diarrhoeal disease in 690 children under two years old. Faecal pollution, as measured by microbiological indicator organisms, was common. The authors reported that: 21% of 123 spring waters, 21% of 131 open dug wells, 14% of 52 wells with pumps, 6% of 751 boreholes, and 60% of 5 non-municipal piped water supplies all yielded water containing more than 1000 faecal coliforms/100 ml. By contrast, only 5% of

138 municipal piped water samples yielded a count of >1000 faecal coliforms/100 ml. The prevalence of diarrhoea ranged from 5.2–10.0% over the six subsequent two-month periods. It appeared that there was little change in the prevalence of diarrhoea if indicator counts rose to 100/100ml. There was a significant association between diarrhoea and >1000 *E. coli*/100ml (Odds Ratio (OR) 1.92, Confidence Interval (CI) 1.27–2.91), enterococci (OR 1.94, CI 1.20-3.16) and faecal streptococci (OR 1.81, CI 1.10–3.00). The association with faecal coliforms was borderline significant (OR 1.49, CI 1.00–2.22). The probability of diarrhoea in a child during a 24-hour period was 0.09 in those exposed to <1000 *E. coli* and 0.15 in those exposed to >1000. The respective probabilities for enterococci were 0.09 and 0.16.

Also from Cebu, VanDerslice and Briscoe (1995) reported that in areas with poor environmental sanitation, improved drinking water would have little or no effect. However, in areas with good community sanitation, reducing faecal coliform counts by two orders of magnitude would reduce the incidence of diarrhoea by 40%, eliminating excreta from around the house by 30% and providing private excreta disposal by 42%.

In conclusion, it is not really possible to give definitive estimates of the burden of diarrhoeal disease due to water consumption in developing communities as this varies substantially depending upon water source and quality as well as other socio-economic and behavioural factors.

4.6 WATERBORNE OUTBREAKS (DEVELOPED COUNTRIES)

An outbreak can be defined as the occurrence of two or more related cases of infection. Usually family outbreaks, where all cases occur in the same family group, are distinguished from general outbreaks. The reasons for this separation are that person-to-person spread within a family is more likely and members of a family are more likely to be exposed to the same risk factors. Both of these reasons make epidemiological investigation of family outbreaks very difficult.

Unfortunately the definitions given in the previous paragraph do not help greatly when identifying potential outbreaks of waterborne disease. Early in a waterborne outbreak, obviously related cases are rare. The exception to this observation is with small supplies providing water for a few homes, or an institution such as a hotel or hospital. The detection of waterborne outbreaks is further hampered by the fact that the most common waterborne infections are also endemic in the community. Consequently most waterborne outbreaks are first identified by noting a general increase in cases over what would be expected for the time of year.

A more useful definition is an increase of cases of a particular infection above what would be normally expected. The detection of a potentially waterborne outbreak now becomes a question of identifying an increase in cases as early as possible when still only a few cases have occurred. One approach that has recently been suggested is to define check and alert values based on the usual weekly rate within a population (DETR and DoH 1998). The check and alert values have been calculated to give only a 1/20 and 1/100 probability of occurring by chance in any week. To have cases that exceed the alert value on two consecutive weeks is very strong evidence of an outbreak. Clearly, if numbers are large early on in the outbreak then such statistical tests are not needed. In this case it is usually obvious that there is an outbreak.

4.6.1 Factors leading to waterborne outbreaks

The causes behind the occurrence of outbreaks are numerous and have been well described by Craun (1986) and in subsequent reports from the US EPA and US Centers for Disease Control (CDC) (Herwaldt *et al.* 1991, 1992), and are described more fully in Chapter 6. Because of an increasingly contaminated global water resource, there has been a rise in waterborne disease worldwide (Ford and Colwell 1996). In developing countries, treatment of water and wastes is often non-existent or grossly inadequate and until sanitation is improved it will be impossible to impact greatly on the level of waterborne disease. In developed countries, deficiencies in treatment and delivery systems, anthropogenic impacts on source water, and the emergence of resistant and more virulent micro-organisms pose serious threats to human health. In industrialised countries, an increase in waterborne disease is expected because of a number of factors, including:

- Newly recognised agents (*Cryptosporidium*, *Giardia*, *Cyclospora*) that have a high resistance to chemicals used in water treatment and development of antibiotic resistant strains of pathogens.
- Less immunity to pathogens (because of better sanitary conditions and a higher population of immunocompromised individuals) and the resulting higher susceptibility and risk of disease during systems failures.
- Anthropogenic alterations of water systems that have stimulated eutrophication, changes in food chain structure, and unrestricted growth of 'nuisance species', creating breeding sites for vector-borne diseases.

- Changes in agricultural production methods, including high-density animal operations carried out in proximity to urban development, leading to an increase in transmission of animal pathogens to humans.
- Ageing and deteriorating environmental infrastructure, particularly in inner cities.

4.7 COSTS TO SOCIETY

In developed countries, such as the US, diarrhoeal disease is common but generally not severe. Sufferers frequently downplay its significance and doctors often do not trace the causes of individual cases. Therefore, many illnesses caused by waterborne agents go unreported. Moreover, few physicians are on the look-out for rare or emerging organisms and laboratory analyses that might alert communities to outbreaks of waterborne disease are infrequently done.

The societal cost of the so-called 'mild gastrointestinal illnesses' is several orders of magnitude higher than the costs associated with acute hospitalised cases (Payment 1997). In the US, the annual cost to society of gastrointestinal infectious illnesses was estimated in 1985 as $19,500 million for cases with no consultation by physician, $2,750 million for those with consultations, and only $760 million for those requiring hospitalisation (Garthright *et al.* 1988; Roberts and Foegeding 1991). These estimates, however, do not address the deaths associated with these illnesses, particularly in children and older adults.

From the data collected during the Payment studies the economic costs of endemic waterborne diseases were calculated based on reported symptom and behaviour rates between unexposed and tap water exposed groups (Payment 1997). These estimates were then combined with published figures for the cost of gastrointestinal infectious diseases in the US (Garthright *et al.* 1988; Roberts and Foegeding 1991). Assuming a population of 300 million individuals, the estimate of the cost of waterborne illness ranges from US$269–806 million for medical costs, and US$40–107 million for absences from work. Such figures can only underscore the enormous economic cost of endemic gastrointestinal illnesses, even in societies where they are not perceived to be a problem.

4.8 CONCLUSIONS

In this chapter we have presented evidence examining the levels of infectious intestinal disease in both developed and developing countries. It is clear, however, that deriving such estimates from routinely available data is difficult because of problems in ascertainment. Nevertheless, we can be certain that the incidence of disease is high in all countries. The proportion of endemic disease due to the water route varies substantially from community to community. The water route seems to increase in importance as general levels of hygiene increase in a community. Indeed, in many poorer tropical countries, the priority is not to improve quality of drinking water supplies but to provide adequate water close to the home, and supply or maintain adequate sanitation.

All developed civilisations depend on an adequate supply of safe water for their continuation. We cannot afford to become complacent about the safety and reliability of our water supplies, nor can we afford not to invest in and maintain our infectious disease surveillance systems.

4.9 IMPLICATIONS FOR INTERNATIONAL GUIDELINES AND NATIONAL REGULATIONS

In terms of the framework and guidelines development this chapter clearly shows that a hazard exists and that there can be no room for complacency even in developed countries. It is also clear that there is very weak understanding of risk with regard to endemic rates of illness attributable to drinking water in developed and developing countries.

4.10 REFERENCES

Ahlm, C., Lundberg, S., Fessé, K. and Wiström, J. (1994) Health problems and self-medication among Swedish travellers. *Scand. J. Infect. Dis.* **26**, 711–717.

Anon (1996) Water and sanitation: WHO fact sheet no. 112, World Health Organization, Geneva.

Badenoch, J. (1990a) *Cryptosporidium* – a waterborne hazard. *Letters Appl. Microbiol.* **11**, 269–270.

Badenoch, J. (1990b) *Cryptosporidium* in water supplies. Dept Environment and Dept of Health, HMSO, London.

Batik, O., Craun, G.F., Tuthil, R.W. and Kroemer, D.F. (1979) An epidemiologic study of the relationship between hepatitis A and water supply characteristics and treatment. *Amer. J. Publ. Health* **70**, 167–169.

Batik, O., Craun, G.F. and Pipes, W.O. (1983) Routine coliform monitoring and waterborne disease outbreaks. *J. Env. Health* **45**, 227–230.

Beaudeau, P., Payment, P., Bourderont, D., Mansotte, F., Boudhabay, O., Laubiès, B. and Verdière, J. (1999) A time series study of anti-diarrheal drug sales and tap-water quality. *International J. Environ. Hlth Res.* **9**(4), 293–312.

Bitton, G. (1980) *Introduction to Environmental Virology*, Wiley, New York.

Bitton, G., Farrah, S.R., Montague, C., Binford, M.W., Scheuerman, P.R. and Watson, A. (1985) Survey of virus isolation data from environmental samples. Project report for contract 68-03-3196, US EPA, Cincinnati, OH.

Branski, D. (1984) Specific etiologies of chronic diarrhoea in infancy. *Nestle Nutrition Workshop Series* **6**, 107–145.

Brüssow, H., Werchau, H., Liedtke, W., Lerner, L., Mietens, C., Sidoti, J. and Sotek, J. (1988) Prevalence of antibodies to rotavirus in different age-groups of infants in Bochum, West Germany. *J. Infect. Dis.* **157**, 1014–1022.

Brüssow, H., Sidoti, J., Barclay, D., Sotek, J., Dirren, H. and Freire, W.B. (1990) Prevalence and serotype specificity of rotavirus antibodies in different age groups of Ecuadorian infants. *J. Inf. Dis.* **162**, 615–620.

Bryan, J.P. (1985) Procedures to use during outbreaks of food-borne disease. In *Manual of Clinical Microbiology*, 4th edn, American Society of Microbiology, Washington DC.

Camann, D.E., Graham, P.J., Guentzel, M.N., Harding, H.J., Kimball, K.T., Moore, B.E., Northrop, R.L., Altman, N.L., Harrist, R.B., Holguin, A.H., Mason, R.L., Becker Popescu, C. and Sorber, C.A. (1985) Health effects study for the Lubbock land treatment project. Lubbock Infection surveillance study (LISS), US EPA Report, Cincinnati, OH.

CDC (1997) The food-borne diseases active surveillance network, 1996. *Morbidity and Mortality Weekly Report* **46**(12), 258–261.

CDC (1998) Incidence of food-borne illness – FoodNet, 1997. *Morbidity and Mortality Weekly Report* **47**(37), 782–786.

CDC (1999) Incidence of food-borne illness – FoodNet, 1998. *Morbidity and Mortality Weekly Report* **48**(9), 189–194.

Chen, K., Lin, C., Qiao, Q., Zen, N., Zhen, G., Gongli, C., Xie, Y., Lin, Y. and Zhuang, S. (1991) The epidemiology of diarrhoeal diseases in south-eastern China. *J. Diarrhoeal Dis. Res.* **9**, 94–99.

Coeblens, F.G.J., Leentvaar-Kuijpers, A., Kleijnen, J. and Countinho, R.A. (1998) Incidence and risk factors of diarrhoea in Dutch travellers: consequences for priorities in pre-travel health advice. *Tropical Medicine International Health* **3**, 896–903.

Collin, J.F., Milet, J.J., Morlot, M. and Foliguet, J.M. (1981) Eau d'adduction et gastroentérites en Meurthe-et-Moselle, *J. Franc. Hydrologie* **12**, 155–174. (In French.)

Craun, G.F. (1986) Waterborne diseases in the United States. CRC Press, Boca Raton, FL.

Craun, G.F. (1990) Methods for the investigation and prevention of waterborne disease outbreaks. EPA/600/1-90/005a. US EPA, Washington DC.

Craun, G.F. (1992) Waterborne disease outbreaks in the United States of America: causes and prevention. *World Health Stat. Q.* **45**, 192–199.

Craun, G.F., Hubbs, S.A., Frost, F., Calderon, R.L. and Via, S.H. (1998) Waterborne outbreaks of cryptosporidiosis. *J. Am. Water Works Ass.* **90**, 81–91.

DETR and DoH (1988) *Cryptosporidium* in Water Supplies: 3rd report of the Group of Experts. Department of the Environment, Transport and Regions, London.

Dingle, J.H., Badger, G.F., Feller, A.E., Hodges, R.G., Jordan, W.S. and Rammelkamp, C. (1953) A study of illness in a group of Cleveland families. I. Plan of study and certain general observations. *Amer. J. Hyg.* **58**, 16–30.

DuPont, H.L., Ericsson, C.D., Johnson, P.C., Bitsura, J.A.M., DuPont, M.W. and de la Cabada, F.J. (1987) Prevention of travelers' diarrhoea by the tablet formulation of bismuth subsalicylate. *JAMA* **2257**, 1347–1350.

Edwards, D.D. (1993) Troubled water in Milwaukee. *ASM News* **59**, 342–345.

Ellner, P.D. (1984) Infectious diarrhoeal diseases. *Microbiology Series* **12**, 1–175.

Ericsson, C.D., DuPont, H.L., Galindo, E., Mathewson, J.J., Morgan, D.R., Wood, L.V. and Mendiola, J. (1985) Efficacy of bicozamycin in preventing travelers' diarrhoea. *Gastroenterology* **88**, 473–477.

Esrey, S.A., Potash, J.B., Roberts, L. and Shiff, C. (1991) Effects of improved water supply and sanitation on ascariasis, diarrhoea, dracunculiasis, hookworm infection, schistosomiasis, and trachoma. *Bull. World Health Org.* **69**, 609–621.

Fattal, B., Guttman-Bass, N., Agursky, T. and Shuval, H.I. (1988) Evaluation of health risk associated with drinking water quality in agricultural communities. *Water Sci. Technol.* **20**, 409–415.

Feldman, R.A. and Banatvala, N. (1994) The frequency of culturing stools from adults with diarrhoea in Great Britain. *Epidemiol. Infect.* **113**, 41–44.

Ferley, J.P., Zmirou, D., Collin, J.F. and Charrel, M. (1986) Etude longitudinale des risques liés à la consommation d' eaux non conformes aux normes bactériologiques. *Rev. Epidemiol. Sante Publique* **34**, 89–99. (In French.)

Ford, T.E. and Colwell, R.R. (1996) A global decline in microbiological safety of water: A call for action. American Academy of Microbiology, Washington DC.

Fox, J.P.,Elveback, L.R., Wassermann, F.E., Ketler, A., Brandt, C.D. and Kogon, A. (1966) The virus watch program. *Amer. J. Epi.* **83**, 389–412.

Garthright, W.E., Archer, D.L. and Kvenberg, J.E. (1988) Estimates of incidence and costs of intestinal infectious diseases. *Public Health Reports* **103**, 107–116.

Goodman, L. and Segreti, J. (1999) Infectious Diarrhoea, *Disease-a-Month*, 268–299.

Henry, J.H. and Rahim, Z. (1990) Transmission of diarrhoea in two crowded areas with different sanitary facilities in Dhaka, Bangladesh. *J. Trop. Med. Hyg.* **93**, 121–126.

Herwaldt, B.L., Craun, G.F., Stokes, S.L. and Juranek, D.D. (1991) Waterborne disease outbreaks, 1989–1990. CDC Surveillance Summaries. December 1991. *MMWR* **40**, 1–22.

Herwaldt, B.L., Craun, G.F., Stokes, S.L. and Juranek, D.D. (1992) Outbreaks of waterborne diseases in the United States: 1989–1990. *J. AWWA* **83**, 129.

Hills, S.D., Miranda, C.M., McCann, M., Bender, D. and Weigle. K. (1992) Day care attendance and diarrheal morbidity in Columbia. *Pediatrics* **90**, 582–588.

Hodges, R.G., McCorkle, L.P., Badger, G.F., Curtiss, C., Dingle, J.H. and Jordan, W.S. (1956) A study of illness in a group of Cleveland families. XI. The occurrence of gastrointestinal symptoms. *Amer. J. Hyg.* **64**, 349–356.

Hulten, K., Enroth, H., Nystrom, T. and Engstrand, L. (1998) Presence of *Helicobacter* species DNA in Swedish water. *J. Appl. Microbiol.* **85**(2), 282–286.

Hunter, P.R. (1997) *Waterborne Disease: Epidemiology and Ecology.* Wiley, Chichester, UK.

Hunter, P.R. (1999) Modelling the impact of prior immunity on the epidemiology of outbreaks of cryptosporidiosis. Proceedings of the International Symposium on Waterborne Pathogens, Milwaukee, WI.

Hunter, P.R. and Quigley, C. (1998) Investigation of an outbreak of cryptosporidiosis associated with treated surface water finds limits to the value of case control studies. *Comm. Dis. Public Health* **1**, 234–238.

Hurst, J.H., Knudsen, G.R., Melnerney, M.J., Stetzenbach, L.D. and Walter, M.V. (1997) Manual of environmental microbiology. ASM Press, American Society for Microbiology, Washington DC.

Isaäcson, M. and Sayed, A.R. (1988) Health aspects of the use of recycled water in Windhoek, SWA/Namibia, 1974–1983. Diarrhoeal diseases and the consumption of reclaimed water. *South African Medical Journal* **7**, 596–599.

Jinadu, M.K., Olusi, S.O., Agun, J.I. and Fabiyi, A.K. (1991) Childhood diarrhoea in rural Nigeria. I. Studies on prevalence, mortality and socio-economic factors. *J. Diarrhoeal Dis. Res.* **9**, 323–327.

Johnson, P.C., Ericsson, C.D., Morgan, D.R. and DuPont, H.L. (1984) Prophylactic norfloxacin for acute travelers' diarrhoea. *Clin. Res.* **32**, 870A.

Kelly, P., Baboo, K.S., Wolff, M., Ngwenya, B., Luo, N. and Farthing, M.J. (1996) The prevalence and aetiology of persistent diarrhoea in adults in urban Zambia. *Acta Tropica* **61**, 183–190.

Mackenzie, W.R., Hoxie, N.J., Proctor, M.E., Gradus, M.S., Blair, K.A., Peterson, D.E., Kazmierczak, J.J., Addiss, D.G., Fox, K.R., Rose, J.B. and Davis, J.P. (1994) A massive outbreak in Milwaukee of *Cryptosporidium* infection transmitted through the public water supply. *New Engl. J. Med.* **331**, 161–167.

Moe, C.L., Sobsey, M.D., Samsa, G.P. and Mesolo, V. (1991) Bacterial indicators of risk of diarrhoeal disease from drinking-water in the Philippines. *Bull. World Health Organ.* **69**, 305–317.

Monto, A.S. and Koopman, J.S. (1980) The Tecumseh Study: XI. Occurrence of acute enteric illness in the community. *Amer. J. Epidemiol.* **112**, 323–333.

Monto, A.S., Koopman, J.S., Longini, I.M. and Isaacson, R.E. (1983) The Tecumseh Study: XII. Enteric agents in the community, 1976–1981. *J. Infect. Dis.* **148**, 284–291.

Morag, A., Margalith, M., Shuval, H.I. and Fattal, B. (1984) Acquisition of antibodies to various Coxsackie and Echo viruses and Hepatitis A virus in agricultural settlements in Israel. *J. Med. Virol.* **14**, 39–47.

Morris, R.D., Naumova, E.N., Levin, R. and Munasinghe, R.L. (1996) Temporal variation in drinking water turbidity and diagnosed gastroenteritis in Milwaukee. *American Journal of Public Health* **86**, 237–239.

Morris, R.D., Naumova, E.N. and Griffiths, J.K. (1998) Did Milwaukee experience waterborne cryptosporidiosis before the large documented outbreak of 1993? *Epidemiology* **9**, 264–270.

Murray, P.R., Baron, E.J., Pfaller, M.A., Tenover, F.C. and Yolken, R.H. (1995) *Manual of Clinical Microbiology*, 6th edn, ASM Press, Washington DC.

Nikolaev, V.P. (1966) A study of neutralizing antibodies against various enteroviruses in various age groups of the population of Leningrad. *Voprosy Virusology* 11, 307–311.

Palmer, S., Houston, H., Lervy, B., Ribeiro, D. and Thomas, P. (1996) Problems in the diagnosis of food-borne infection in general practice. *Epidemiol. Infect.* 117, 497–484.

Paniagua, M., Espinoza, F., Ringman, M., Reizenstein, E., Svennerholm, A.M. and Hallander, H. (1997) Analysis of incidence of infection with enterotoxigenic *Escherichia coli* in a prospective cohort study of infect diarrhoea in Nicaragua. *J. Clin. Microbiol.* **35**, 1404–1410.

Papaevangelou, G.J. (1980) Global epidemiology of Hepatitis A. In *Hepatitis A* (ed. R.J. Gerety), pp. 101–132, Academic Press, Orlando, FL.

Payment, P. (1991) Antibody levels to selected enteric viruses in a normal randomly selected Canadian population. *Immunology and Infectious Diseases* **1**, 317–322.

Payment, P. (1993a) Viruses: Prevalence of disease levels and sources. In *Safety of Water Disinfection: Balancing Chemical and Microbial Risks* (ed. G. Craun), pp. 99–113, ILSI Press, Washington DC.

Payment, P. (1993b) Viruses in water: an underestimated health risk for a variety of diseases. In *Disinfection Dilemma: Microbiological Control versus By-products* (eds W. Robertson, R. Tobin and K. Kjartanson), pp. 157–164, American Water Works Association, Denver, CO.

Payment, P. (1997) Epidemiology of endemic gastrointestinal and respiratory diseases – incidence, fraction attributable to tap water and costs to society. *Water Science and Technology* **35**, 7–10.

Payment, P., Richardson, L., Siemiatycki, J., Dewar, R., Edwardes, M. and Franco, E. (1991a) A randomized trial to evaluate the risk of gastrointestinal disease due to the consumption of drinking water meeting currently accepted microbiological standards. *Amer. J. Public Health* **81**, 703–708.

Payment, P., Franco, E., Richardson, L. and Siemiatycki, J. (1991b) Gastrointestinal health effects associated with the consumption of drinking water produced by point-of-use domestic reverse-osmosis filtration units. *Appl. Env. Microbiol.* **57**, 945–948.

Payment, P., Franco, E. and Fout, G.S. (1994) Incidence of Norwalk virus infections during a prospective epidemiological study of drinking-water-related gastrointestinal illness. *Can. J. Microbiol.* **40**, 805–809.

Payment, P., Siemiatycki, J., Richardson, L., Renaud, G., Franco, E. and Prévost, M. (1997) A prospective epidemiological study of gastrointestinal health effects due to the consumption of drinking water. *Int. J. Environ. Health Research* **7**, 5–31.

Pitzinger, B., Steffen, R. and Tschopp, A. (1991) Incidence and clinical features of traveler's diarrhoea in infants and children. *Pediatr. Infect. Dis. J.* **10**, 719–723.

Rao, V.C. and Melnick, J.L. (1986) *Environmental Virology*, American Society of Microbiology, Washington DC.

Roberts, T. and Foegeding, P.M. (1991) Risk assessment for estimating the economic costs of food-borne diseases caused by micro-organisms. In *Economics of Food Safety* (ed. J.A. Caswell), pp. 103–130, Elsevier, New York.

Ruel, M.T., Rivera, J.A., Santizo, M.C., Lonnerdal, B. and Brown, K.H. (1997) Impact of zinc supplementation on morbidity from diarrhoea and respiratory infections among rural Guatemalan children. *Pediatrics* **99**, 808–813.

Schorling, J.B., Wanke, C.A., Schorling, S.K., McAuliffe, J.F., de Souza, M.A. and Guerrant, R.L. (1990) A prospective study of persistent diarrhoea among children in an urban Brazilian slum. *Am. J. Epidemiol.* **132**, 144–156.

Schwartz, J., Levin, R. and Hodge, K. (1997) Drinking water turbidity and pediatric hospital use for gastrointestinal illness in Philadelphia. *Epidemiology* **8**, 615–620.

Semenza, J.C., Roberts, L., Henderson, A., Bogan, J. and Rubin, C.H. (1998) Water distribution system and diarrhoeal disease transmission: a case study in Uzbekistan. *Am. J. Trop. Med. Hyg.* **59**, 941–946.

Shahid, N.S., Greenough 3rd, W.B., Samadi, A.R., Huq, M.I. and Rahman, N. (1996) Hand washing with soap reduces diarrhoea and spread of bacterial pathogens in Bangladesh village. *J. Diarrhoeal Dis. Res.* **14**, 85–89.

Stansfield, S.K., Pierre-Louis, M., Lerebours, G. and Augustin, A. (1993) Vitamin A supplementation and increased prevalence of childhood diarrhoea and acute respiratory infections. *Lancet* **342**, 578–582.

Steffen, R., Collard, F., Tornieporth, N., Campbell-Forrester, S., Ashley, D., Thompson, S., Mathewson, J.J., Maes, E., Stephenson, B., DuPont, H.L. and von Sonnenburg, F. (1999) Epidemiology, etiology, and impact of traveler's diarrhoea in Jamaica. *JAMA* **281**, 811–817.

Taylor, D.N., Echeverria, P., Blaser, M.J., Pitangsi, C., Blacklow, N., Cross, J. and Weniger, B.G. (1985) Polymicrobial aetiology of traveller's diarrhoea. *Lancet* **i**, 381–383.

Thea, D.M., St. Louis, M.E., Atido, U., Kanjinga, K., Kembo, B., Matondo, M., Tshiamala, T., Kamenga, C., Davachi, F., Brown, C., Rand, W.M. and Keusch, G.T. (1993) A prospective study of diarrhoea and HIV-1 infection among 429 Zairian infants. *New England Journal of Medicine* **329**, 1696–1702.

VanDerslice, J. and Briscoe, J. (1995) Environmental interventions in developing countries, and their implications. *Am. J. Epidemiol.* **141**, 135–144.

Wheeler, J.G., Sethi, D., Cowden, J.M., Wall, P.G,. Rodrigues, L.C,. Tomkins, D.S., Hudson, M.J. and Roderck, P.J. (1999) Study of infectious intestinal disease in England: rates in the community, presenting to general practice, and reported in national surveillance. *Brit. Med. J.* **318**, 1046–1050.

WHO (1993) Guidelines for Drinking-water Quality. Volume 1: Recommendations. Second edition. World Health Organization, Geneva.

WHO (1996) Guidelines for Drinking-water Quality. Volume 2: Health Criteria and Other Supporting Information Second edition, World Health Organization, Geneva.

Yeager, B.A.C., Lanata, C.F., Lazo, F., Verastegui, H. and Black, R.E. (1991) Transmission factors and socio-economic status as determinants of diarrhoeal incidence in Lima, Peru. *J. Diarrhoeal Dis. Res.* **9**, 186–193.

Zmirou, D., Ferley, J.P., Collin, J.F., Charrel, M. and Berlin, J. (1987) A follow-up study of gastro-intestinal diseases related to bacteriologically substandard drinking water. *Am. J. Public Health* **77**, 582–584.

Zmirou, D., Rey, S., Courtois, X., Ferley, J.P., Blatier, J.F., Chevallier, P., Boudot, J., Potelon, J.L. and Mounir, R. (1995) Residual microbiological risk after simple chlorine treatment of drinking ground water in small community systems. *European Journal of Public Health* **5**, 75–81.

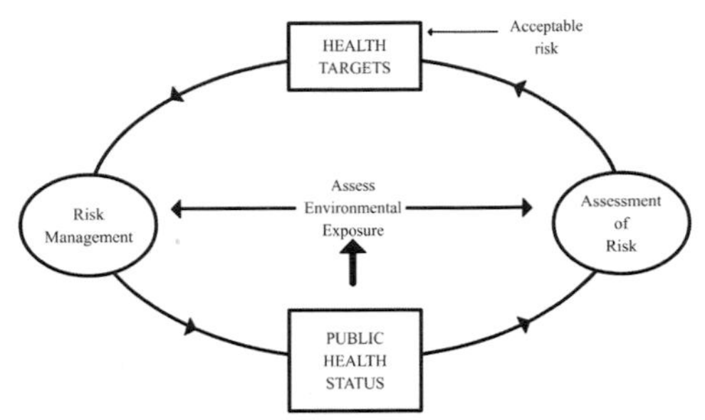

5

Excreta-related infections and the role of sanitation in the control of transmission

Richard Carr[1]

This chapter examines the role sanitation (in its widest sense) plays in preventing the transmission of excreta-related diseases. The proper management of excreta acts as the primary barrier to prevent the spread of pathogens in the environment. It, thus, directly impacts disease transmission through person-to-person contact, water and the food chain. This chapter focuses on the health dimensions and relative importance of sanitation measures, and discusses technical options for the containment and treatment of excreta. It highlights the need to consider water-related guidelines and standards in terms of the 'greater picture', utilising an integrated approach rather than proceeding on a case by case basis.

[1] With contributions from Martin Strauss

 Water Quality: Guidelines, Standards and Health. Edited by Lorna Fewtrell and Jamie Bartram. Published by IWA Publishing, London, UK. ISBN: 1 900222 28 0

5.1 INTRODUCTION

Human excreta and the lack of adequate personal and domestic hygiene have been implicated in the transmission of many infectious diseases including cholera, typhoid, hepatitis, polio, cryptosporidiosis, ascariasis, and schistosomiasis. The World Health Organization (WHO) estimates that 2.2 million people die annually from diarrhoeal diseases and that 10% of the population of the developing world are severely infected with intestinal worms related to improper waste and excreta management (Murray and Lopez 1996; WHO 2000a). Human excreta-transmitted diseases predominantly affect children and the poor. Most of the deaths due to diarrhoea occur in children and in developing countries (WHO 1999).

Proper excreta disposal and minimum levels of personal and domestic hygiene are essential for protecting public health. Safe excreta disposal and handling act as the primary barrier for preventing excreted pathogens from entering the environment. Once pathogens have been introduced into the environment they can be transmitted via either the mouth (e.g. through drinking contaminated water or eating contaminated vegetables/food) or the skin (as in the case of the hookworms and schistosomes), although in many cases adequate personal and domestic hygiene can reduce such transmission. Excreta and wastewater generally contain high concentrations of excreted pathogens, especially in countries where diarrhoeal diseases and intestinal parasites are particularly prevalent. Therefore for maximum health protection, it is important to treat and contain human excreta as close to the source as possible before it gets introduced into the environment.

Although the principal focus of the guideline documents examined in this book is water, in many settings other disease transmission pathways are at least as important. In microbiological terms, the traditional approach of examining each guideline area in isolation ignores the inter-related pathways and also the root of the problem, namely excreta and inadequate hygiene.

5.2 TRANSMISSION ROUTES

Human excreta may contain many types of pathogens. When these pathogens are introduced into the environment some can remain infectious for long periods of time (Table 5.1) and, under certain conditions, they may be able to replicate in the environment. The presence of pathogens presents a potential threat to human health. However, for an actual risk of disease an infectious dose of the excreted pathogen must reach a human host.

Table 5.1. Pathogen and indicator survival in different environmental media

	Pathogen survival (time in days unless otherwise indicated)			
Organism	Freshwater	Saltwater	Soil	Crops
Viruses	11–304	11–871	6–180	0.4–25
Salmonellae	<10	<10	15–100	5–50
Cholera	30	+285	<20	<5
Faecal coliforms	<10	<6	<100	<50
Protozoan cysts	176	1yr	+75	ND
Ascaris eggs	1.5yr*	2*	1–2 yr	<60
Tapeworm eggs	63*	168*	7 months	<60
Trematodes	30-180	<2	<1*	130**

ND No data; *Not considered an important transmission pathway; **Aquatic macrophytes

Sources: Feachem *et al.* 1983; Mara and Cairncross 1989; National Research Council 1998; Robertson *et al.* 1992; Rose and Slifko 1999; Schwartzbrod 2000; Tamburrini and Pozio 1999.

Note: Differing survival times for each organism (or group of organisms) may be related to temperature.

Disease transmission is determined by several pathogen-related factors including:

- An organism's ability to survive or multiply in the environment (some pathogens require the presence of specific intermediate hosts to complete their lifecycles).
- Latent periods (many pathogens are immediately infectious, others may require a period of time before they become infective).
- An organism's ability to infect the host (some pathogens can cause infections when present in small numbers e.g. *Ascaris*, others may require a million or more organisms to cause infection; Feachem *et al.* 1983).

Disease transmission is also affected by host characteristics and behaviour, including:

- immunity (natural or as a result of prior infection or vaccination)
- nutritional status
- health status
- age
- sex
- personal hygiene
- food hygiene.

In general, pathogenic micro-organisms may be transmitted from source to new victim in a number of ways including direct person-to-person spread and indirect routes including inanimate objects (fomites), food, water or insect vectors.

Table 5.2 details a selection of faecal-oral pathogens and their transmission routes. As the table shows, multiple transmission routes are the norm, rather than the exception, for many pathogenic organisms.

Table 5.2. Selected faecal-oral pathogens and selected transmission routes (adapted from Adams and Moss 1995)

Pathogen	Important reservoir/carrier	Transmission			X in food
		water	food	p-to-p	
Campylobacter jejuni	Variety of animals	+	+	+	+
Enterotoxigenic *E. coli*	Man	+	+	+	+
Enteropathogenic *E. coli*	Man	+	+	+	+
Enteroinvasive *E. coli*	Man	+	+	Ni	+
Enterohaemorrhagic *E. coli*	Man	+	+	+	+
Salmonella typhi	Man	+	+	±	+
Salmonella (non-*typhi*)	Man and animals	±	+	±	+
Shigella	Man	+	+	+	+
Vibrio cholerae O1	Man, marine life?	+	+	±	+
Vibrio cholerae, non O1	Man and animals	+	+	±	
Hepatitis A	Man	+	+	+	–
Norwalk agents	Man	+	+	Ni	–
Rotavirus	Man	+	ni	+	–
Cryptosporidium parvum	Man, animals	+	+	+	–
Entamoeba histolytica	Man	+	+	+	–
Giardia lamblia	Man, animals	+	±	+	–
Ascaris lumbricoides	Man	–	+	–	–

X in food - multiplication in food p-to-p – person-to-person
+ yes ± rare - no ni - no information

Figure 5.1 outlines the routes of transmission, important pathogen and host-related transmission factors and also possible barriers to transmission for excreted pathogens. As Figure 5.1 illustrates, sanitation is the primary barrier for preventing faecal-oral disease transmission. If excreta disposal is ineffective or non-existent (or other animals serve as sources of excreted pathogens) other measures must be taken to avoid disease transmission. Removing or destroying infectious agents by disinfecting drinking water prior to consumption or preparation of food; cleaning hands, utensils, and surfaces before food preparation and consumption; and cooking food thoroughly are interventions that will reduce disease transmission (WHO 1993). For example, the simple act of washing one's hands with soap can reduce diarrhoea by a third (WHO 2000a).

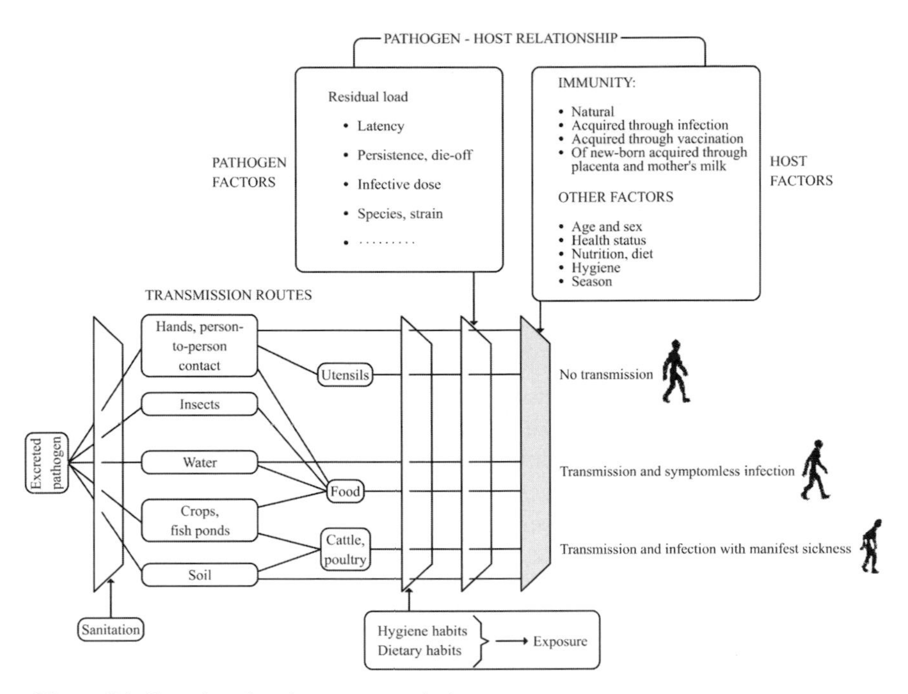

Figure 5.1. Faecal-oral pathogen transmission routes.

Faecally contaminated water, both marine and fresh, is a frequent cause of food-borne illness. For example, some shellfish (such as mussels, oysters, and clams) obtain their food by filtering large quantities of water and are therefore particularly likely to accumulate contamination. Excreta-related human pathogens, heavy metals and other chemical contaminants are taken in with the food particles and can be concentrated in the tissues. Shellfish are also frequently eaten raw or partially cooked. Fish, and non-filter feeding shellfish (crabs, lobsters, prawns, shrimps) grown in faecally contaminated water containing high levels of human pathogens can also concentrate pathogens in their intestinal tracts and on their skin surfaces. When concentrations of faecally derived bacteria exceed a certain level they can be found in muscle tissues (WHO 1989). Infection may occur when the contaminated fish is consumed raw or lightly cooked. Food handlers may also be at risk during preparation of the contaminated product.

When untreated or inadequately treated wastewater or excreta (faecal sludge) is applied to soil and crops, disease transmission can occur. The persons at risk are the farmers, farm workers and their families as well as consumers of crops produced in such a way. The use of inadequately treated wastewater in irrigation and of faecal sludges in soil amendment and fertilisation is especially associated with elevated prevalence of intestinal helminth infection. For example, in a study in Mexico, irrigation with untreated or partially treated wastewater was directly responsible for 80% of all *Ascaris* infections and 30% of diarrhoeal disease in farm workers and their families (Cifuentes *et al.* 2000). Trematode infections are caused by parasitic flatworms (also known as flukes) that infect humans and animals. Infected individuals transmit trematode larvae in their faeces. Infections with trematode parasites can cause mild symptoms such as diarrhoea and abdominal pain or, more rarely, debilitating cerebral lesions, splenomegaly and death, depending on the parasite load. In many areas of Asia where trematode infections are endemic, untreated or partially treated excreta and nightsoil are directly added to fishponds. The trematodes complete their lifecycles in intermediate hosts and subsequently infect fish, shellfish, or encyst on aquatic plants. Humans become infected when they consume the fish, shellfish, or plants raw or partially cooked. It has been estimated that more than 40 million people throughout the world are infected with trematodes and that over 10% of the global population is at risk of trematode infection (WHO 1995).

5.3 THE ROLE OF IMPROVED EXCRETA MANAGEMENT

Numerous studies have shown that the incidence of many diseases is reduced when people have access to, and make regular use of, effective basic sanitary installations. As previously illustrated in Table 5.1 and Figure 5.1, it is particularly important to keep pathogens out of the environment in the first place because many of these organisms are capable of surviving for long periods of time under different conditions. Therefore, effective excreta management at the household and community levels produces far ranging societal benefits by helping to protect water resources and the food supply from faecal contamination. The following sections describe the health benefits of improved excreta management and provide an overview of the current state of coverage worldwide.

5.3.1 The health dimension of poor sanitation

In the Global Burden of Disease (GBD) study (outlined in detail in Chapter 3) disability adjusted life years (DALYs) were ascribed to 10 selected risk factors. Water, sanitation (i.e. excreta disposal) and hygiene accounted for the second biggest percentage of DALYs behind malnutrition. Worldwide, it is estimated that there are approximately 4 billion cases of diarrhoea per year (resulting in

2.2 million deaths), 200 million people with schistosomiasis and as many as 400 million people infected with intestinal worms (Murray and Lopez 1996; UN 1998; WHO 2000a,b). All of these diseases are largely excreta-related. In less developed countries, poor nutritional status and poverty exacerbate morbidity and mortality associated with excreta-related diseases. For example, most deaths attributed to diarrhoea occur in children below the age of five (WHO 2000b). Rice *et al.* (2000) reviewed 21 studies on infant mortality associated with diarrhoea and found that children with low weight for their age had a much higher risk of mortality. Overall, malnutrition is thought to have a role in about 50% of all deaths among children worldwide.

Two literature reviews assessing the health impact of water and sanitation interventions have been published (Esrey *et al.* 1985, 1991). The first review focused on water and sanitation interventions with one of three outcomes (diarrhoea or a specific pathogen e.g. *Shigella* spp., nutritional status and mortality). The second study expanded the literature on diarrhoea or similar outcomes to include: ascariasis, dracunculiasis, hookworm, schistosomiasis and trachoma as well as diarrhoea. Median values, rather than means, were used to summarise the findings.

In general, impacts measured as reduction in morbidity ranged from low (4% for hookworm) to high (78% for guinea worm). The mean reduction from diarrhoea from the better studies was 26%, ranging from 0–68%. Different levels of impact (summarised in Table 5.3) were found according to which intervention (i.e. improved excreta disposal, water quality, water quantity or hygiene) was examined. The largest effect was seen for interventions focusing on improved excreta disposal, reflecting excreta as being the source of pathogens and the multiple routes of transmission. Moreover, it is important that all members of a community, particularly the children, make use of improved sanitation installations. Children are frequently the victims of diarrhoeal disease and other faecally/orally transmitted illnesses, and thus may act as sources of pathogens. Getting children to use sanitation facilities (or designing child-friendly toilets) and implementing school sanitation programmes are important interventions for reducing the spread of disease associated with waste and excreta (WHO 1993).

Combining the results of the many studies and reviews conducted, it becomes evident that improvements in excreta management, hygiene and water supply may reduce diarrhoeal morbidity, diarrhoea mortality and child mortality by significant amounts (WHO 1993). For example, Esrey *et al.* (1991) found reductions in diarrhoea mortality and overall child mortality of 65% and 55% respectively when improved water and sanitation were introduced. However, the size of the impact is likely to vary according to a wide range of factors, including current sanitary conditions, food supply, breast-feeding habits, education level and uptake of new facilities and behaviours. Clearly, tackling the problem at source assists in reducing transmission via all routes.

Table 5.3. Reduction in diarrhoeal morbidity from specified water and sanitation improvements based on rigorous studies (Esrey *et al.* 1991)

Water and sanitation measure	Percentage reduction in diarrhoea morbidity
Sanitation (improved excreta disposal)	36
Improved hygiene	33
Water and sanitation	30
Water quantity	20
Water quality and quantity	17
Water quality	15

5.3.2 Sanitation coverage

Despite the fact that access to an adequate water supply and sanitation is a fundamental need (and, indeed, arguably a right) for all people, a recent survey shows that almost two and a half billion people do not have access to improved sanitation (WHO 2000a).

As might be expected, sanitation coverage varies dramatically around the world, as illustrated in Figure 5.2. While Figure 5.2 shows the differences between regions, it does not show the fact that the percentage coverage is barely increasing over time. Table 5.4 shows sanitation coverage for Africa and at the global level in 1990 and 2000. It can be seen that increases on a global scale are negligible, while in Africa coverage is standing still or even decreasing. It is also likely that much of the 'improvement' seen may be due to a change in reporting methods.

It can been seen from Figure 5.2 and Table 5.4 that the situation is particularly severe in rural areas, where coverage lags behind that reported from urban areas. However, increasing urbanisation and concentrations of poor in urban slums is likely to be associated (in many cases) with higher risks of transmission, thus posing much greater sanitation challenges.

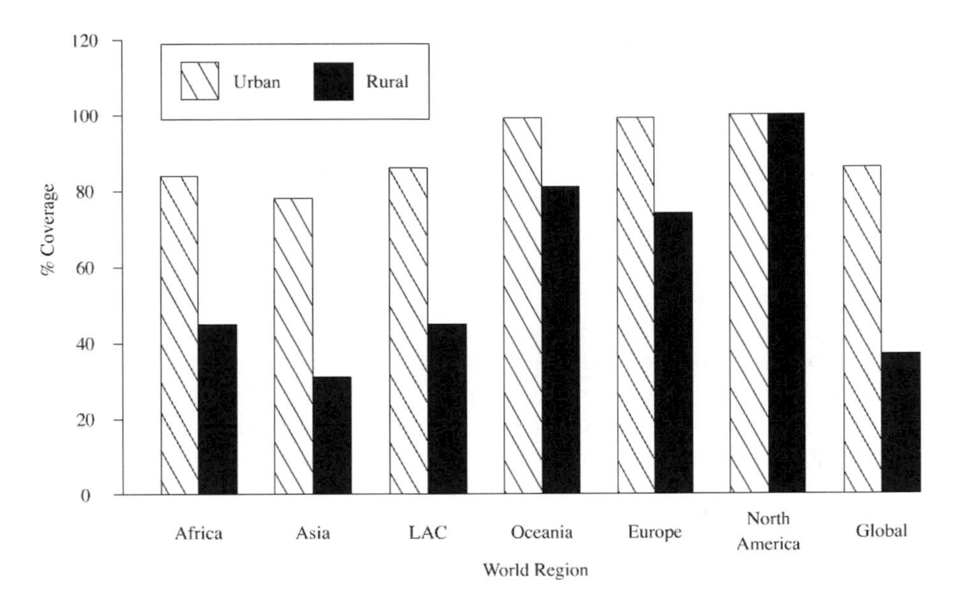

Figure 5.2. Global coverage of sanitation by world region in 2000 (WHO 2000a).

While these figures are disturbing in their own right they do not tell the whole story, as sanitation is not just the presence of available facilities (although that is a start), but for it to be effective it is also the proper use and maintenance of the facilities.

Table 5.4. Sanitation coverage in 1990 and 2000 globally and in Africa (WHO 2000a)

	1990 pop. in millions				2000 pop. in millions			
	Total Pop.	Pop. served	Pop. unserved	% served	Total pop.	Pop. served	Pop. unserved	% served
Global								
Urban	2292	1869	423	82	2844	2435	409	86
Rural	2974	1029	1945	35	3210	1201	2009	37
Total	5266	2898	2368	55	6054	3636	2418	60
Africa								
Urban	197	166	31	84	297	249	48	84
Rural	418	205	213	49	487	217	270	45
Total	615	371	244	60	784	466	318	60

From an industrialised country perspective it is often hard to visualise the sanitation conditions that may be experienced in a developing country. Box 5.1 outlines a scenario detailing some of the conditions experienced by millions of impoverished people in developing countries, and places the sanitation situation in context of other risk factors and general living conditions.

Box 5.1. Low-income peri-urban neighbourhood.

The neighbourhood is located in a city in a tropical coastal area with perennially high temperatures (between 28 and 35°C). A typical family might consist of two adults and four children (two further children having died before the age of two after repeated and heavy attacks of diarrhoea). The family occupies one rented room in a brick-built, tin-roofed, one-storey compound house. The father finds occasional employment as a daily labourer, while the elder children sell goods and gadgets at traffic lights. The family fetches water from a neighbourhood/communal tapstand some 100 metres away from their home. The supply to the tapstand is intermittent (every second day for two hours). The nearest excreta disposal facility is a communal toilet located close to the public tapstand. The user's fees for the toilet, taken by the neighbourhood committee, are improperly used (misappropriated) and so the toilet is not maintained properly, leading to the surrounding area being used for defecation. The family also uses buckets for night-time defecation which are emptied on to an unused piece of land close by. All the children of the compound aged four or under defecate indiscriminately in the lane outside their home.

The children typically experience several episodes of diarrhoea per year. The nationwide introduction of oral rehydration therapy seems to have some effect in preventing death due to diarrhoeal attacks, but chronic malnutrition is a problem. There is a high prevalence of ascariasis and hookworm infection among children aged two or over. The four children all experienced hepatitis A infection in their early childhood and developed immunity. The parents are struck by typhoid fever, amoebic or bacillary dysentery on average about once a year.

5.4 EXCRETA DISPOSAL

The problem of excreta disposal is clearly as old as mankind itself and the need for careful disposal is highlighted in a number of religious books including Hindu, Islamic and Christian texts. The following sub-sections outline a number of excreta management options. Although these are essentially 'technological' answers, albeit of varying complexity, it is important to remember that experience indicates that technology alone is inadequate to secure health gains. Without local interest, involvement and commitment facilities may remain unused or fall out of use. As Samanta and van Wijk (1998) point out 'access to a latrine is not the same as adoption of sanitary practices in dealing with human waste'. Moreover, technical measures for improved sanitary installations, excreta treatment and use or disposal must be complemented by personal and domestic hygiene measures. This section focuses on technical measures for excreta disposal and treatment.

There are numerous technical options for excreta management, many of which, if properly designed, constructed, operated and maintained will provide adequate and safe service as well as health benefits. It is necessary to choose technically, economically and financially feasible options for sustainable excreta management. Equally important is the involvement of all stakeholders playing a role in sanitation development, including users (or customers), community organisations, authorities and entrepreneurs. In particular, it is essential to involve women in the design and selection of domestic sanitation facilities. Research conducted in South Asia demonstrates that involving women in sanitation programmes has resulted in higher coverage, better maintenance of the facilities, increased hygiene awareness, and lower incidence of faecal-oral disease in the community (Neto and Tropp 2000). In addition, for sanitation programmes to be sustainable there must be the political will and institutional capacity to ensure adequate public services and the proper maintenance of sanitation systems (Simpson-Hébert and Wood 1998). Indeed, there are numerous instances where public toilets, in particular, are poorly maintained and the latrine contents inappropriately disposed. Although, happily, this is not always the case and successful schemes (often run on a franchise basis by profit-making organisations or by social organisations) exist in a number of places including Ghana and India (Gear *et al.* 1996; National Institute of Urban Affairs 1990).

It is important to note that there is no single appropriate technology for all circumstances and all socio-economic segments of a community, town or city. The more costly or, apparently, convenient technologies may not provide the greatest health benefit or may be unsustainable from an economic or technological viewpoint.

For practical purposes sanitation can be divided into on-site and off-site technologies. On-site systems (e.g. latrines) store and/or treat excreta at the point of generation. In off-site systems (e.g. sewerage), excreta is transported to another location for treatment, disposal or use. Some on-site systems, particularly in densely populated regions, require off-site treatment components as well. For example, the faecal sludges accumulating in single pit or vault latrines in urban areas and in septic tanks periodically need to be removed and treated off-site for use or disposal.

For sanitary installations to deliver health benefits they need to be able to:

- isolate the user from their own excreta
- prevent nuisance animals (e.g. flies) from contacting the excreta and subsequently transmitting disease to humans
- contain the excreta and/or inactivate the pathogens.

It must be noted that not all excreted components contain pathogens. Urine in most circumstances is sterile (unless it is cross-contaminated by faecal matter caused by the inappropriate design, or use, of the urine-diverting toilet) and contains most of the agriculturally useful nutrients. To reduce required excreta storage volumes, some sanitation facilities promote the separation of urine and faeces. Once it is separated, and diluted, urine can be used immediately as a crop fertiliser with minimal risk to public health (Esrey 2000; SEPA 1995; Wolgast 1993).

5.4.1 Technical sanitation options

This sub-section examines a number of selected technical sanitation options (Franceys *et al.* 1992; Mara 1996b; WHO 1996; WELL/DfID 1998), including both on- and off-site alternatives.

5.4.1.1 On-site installations

On-site installations comprise so-called 'dry' and 'wet' systems. Pit, ventilated improved pit (VIP) and urine-separating latrines are operated without flush water and are designated 'dry'. Pour-flush latrines and septic tanks are 'wet' systems in that they require water, albeit only two or three litres in the case of pour-flush latrines.

Latrine systems may be built with either one or two pits or vaults, depending on affordability, housing density and socio-cultural habits. In the case of two pit/vault systems, only one is in active use at any one time with the other being used to allow pathogen inactivation and decomposition of the excreted material. A storage period of 6–12 months is required in a tropical, year-round warm climate to render the faecal sludges of dry or pour-flush latrines safe for handling and agricultural use (Feachem *et al.* 1983; Peasey 2000; Strauss 1985; WHO 1996). Such pit contents will satisfy the WHO guideline equivalent of 3–8 nematode eggs/g of dry matter.

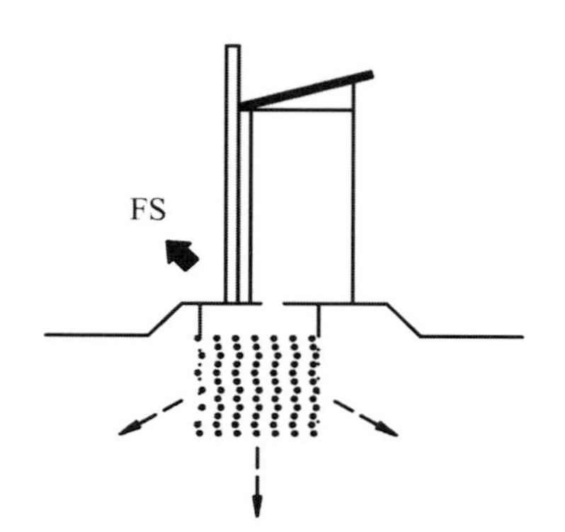

Figure 5.3. VIP latrine (FS = faecal solids).

The VIP latrine (Figure 5.3) is an improvement over a simple pit as the screened vent pipe removes odours from the interior of the toilet superstructure and helps to prevent problems with flies and mosquitoes. Reducing the ability of flies (and other insects) to transmit pathogenic organisms from faeces to food or drink is important in public health terms.

In the no-mix latrine (Figure 5.4), urine and faecal material are collected separately. The diluted urine can be used immediately as a fertiliser. Deposited faeces must be covered with lime, ash, or earth to lower the moisture content, reduce the smell and make the faecal material less attractive to flies.

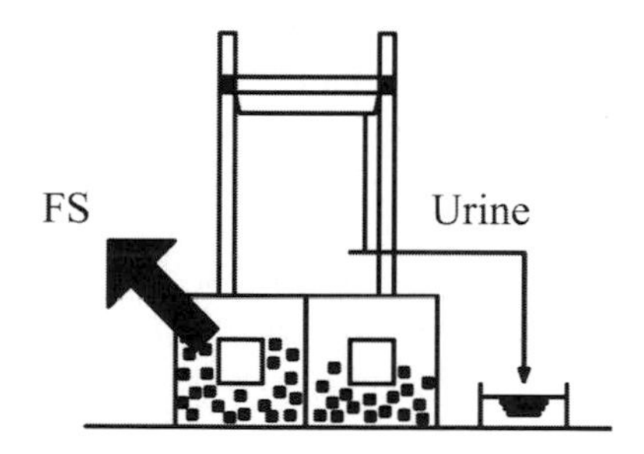

Figure 5.4. Double vault no-mix latrine (FS = faecal solids)

Like 'dry' latrines, pour-flush toilets can be built with one or two pits for excreta disposal. They have a special pan, which is cast into the cover slab and is preferably equipped with a water seal for odour and fly control. Pour-flush latrines require between two and three litres of water per flush and are not suitable for areas with cold climates, impermeable soils or high water tables where the groundwater is a source of drinking water (WHO 1996). They are also inappropriate where the use of solid objects for anal cleansing is the custom, as these may cause siphon blockage.

Septic tanks (Figure 5.5) are watertight chambers sited below ground level that receive excreta and flush water ('blackwater') from flush toilets and also household sullage or 'greywater'. The solids settle out and undergo partial anaerobic degradation in the tank, while the effluent stays in the tank for a short period before, according to conventional design, overflowing into a soakpit or drainfield. Septic tanks should not be used where the soil is impermeable or where the water table is high and the groundwater is a source of drinking water (WHO 1996). Septic tanks may be used in sanitation upgrading, by making them an integral part of low-cost sewerage and enabling a solids-free transportation of wastewater (Mara 1996a).

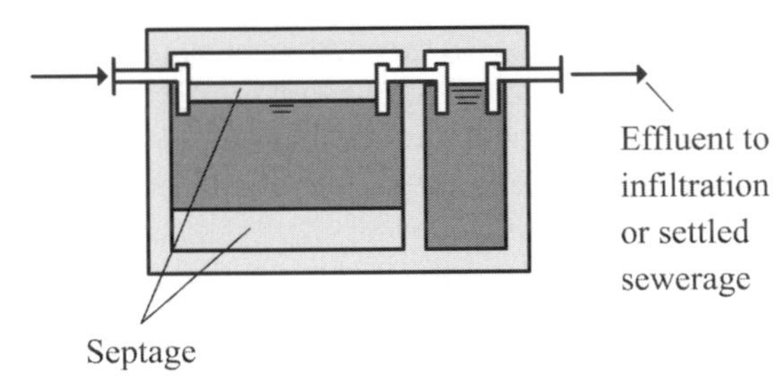

Figure 5.5. A septic tank.

The faecal material or sludges accumulating in septic tanks, single-pit or vault latrines and unsewered public toilets in urban areas must periodically be removed and hauled away. In many developing countries, however, reasonable emptying intervals are rarely observed due either to cost, inefficient emptying services or access difficulties. As a result, in many cities that rely on on-site sanitation systems only a fraction of the faecal sludge generated is collected and accounted for.

5.4.1.2 Groundwater pollution risks from on-site sanitation

Where on-site sanitation systems with unsealed pits for excreta storage or with liquid soakage pits and drainfields are used, there exists a potential risk of microbiological and chemical groundwater pollution. The risk is particularly high for shallow groundwater covered by only a few metres of permeable strata. It is virtually zero for groundwater flowing in deeper aquifers, which are usually protected by impermeable strata. This section focuses on the risk from pathogens.

As early as the 1950s a safety distance of 15–30 metres between latrines and wells was stipulated (California State Water Pollution Control Board 1954;

Wagner and Lanoix 1958). This rule of thumb has persisted and been repeatedly cited since then. It does not, however, take into account the fact that actual groundwater pollution and the concurrent public health implications are dependent on many factors and conditions. It may be overly strict in some and too lenient in other cases. In many cases, such as in densely populated low-income housing areas of cities in developing countries, demanding a distance of even 15 m is impractical. Several factors play a role and must interact for a potential risk of groundwater pollution to turn into actual pollution. The important factors are:

- Characteristics of the strata (soils, rocks) between an infiltration pit or field and the groundwater table.
- Distance between the bottom of a latrine pit and the groundwater table, i.e. the depth of the so-called unsaturated zone.
- Whether the latrine pit leaches into the groundwater (seasonally or permanently).
- The hydraulic gradient and the rate of groundwater flow.
- Hydraulic loading from the sanitary installation; this is related to the type of on-site installation (i.e. 'dry' latrines with minimal water use versus 'wet' installations such as pour-flush latrines or septic tank soak pits receiving both black and greywater).
- Depth of the filter screen below the groundwater surface in a tube or bored well (vertical permeability in unconsolidated soils is much lower than horizontal permeability).
- The temperature in the soil strata and in the aquifer (this is the major factor determining pathogen die-off).

Unsaturated, well-graded and finely divided, so-called unconsolidated soils constitute a very effective defence against the penetration of micro-organisms, helminth eggs and protozoal cysts and their reaching the groundwater table (Lewis *et al.* 1982; Schertenleib 1988). Therefore, the ideal situation is where the groundwater level does not reach latrine pits year-round and an unsaturated soil layer can act as a permanent barrier.

5.4.1.3 Faecal sludge treatment

Few developing countries, to date, have seen investment in faecal sludge treatment as a priority, due to the paucity of treatment options suited to the economic and institutional conditions prevailing in many developing countries. However, several basic options depending on the goal of treatment, the type of faecal sludge collected, and economic and climatic conditions may prove suitable (Heinss *et al.* 1998; Montangero and Strauss 2000; Strauss *et al.* 2000). The sludges may be treated separately, e.g. in pond systems (with or without prior solids separation in settling tanks), unplanted or planted sludge drying beds or drying lagoons. Alternatively, options exist which treat the sludges in combination with wastewater (e.g. in pond systems comprising separate pre-treatment of faecal sludges and combined treatment of faecal sludge liquids with municipal waste), solid organic waste (so-called co-composting) or with sewage treatment plant sludge. The treatment of faecal sludges, whether singly or in combination with other wastes, calls for criteria and procedures that differ from those used for wastewater. Faecal sludges are usually low in chemical contaminants and thus lend themselves well to agricultural use; if they are used in this way nematode egg counts would be the most appropriate criterion to assess suitability.

Faecal sludge collection, haulage and treatment strategies should, ideally, focus on decentralised solutions in order to minimise haulage distances, prevent the uncontrolled dumping of sludges, keep land requirements for individual treatment schemes modest and keep the distance to suitable agricultural areas short.

5.4.1.4 Off-site (sewered) sanitation

Sewerage is the removal of excreta, flushing water and household greywater through a pipe network to a treatment works or a point of disposal or use. In order to minimise environmental pollution and disease transmission it is important that the sewage is properly treated and not allowed to flow untreated into rivers or other water bodies. Estimates suggest that less than 5% of all sewage in developing countries receives any treatment before it is discharged into the environment (World Resources Institute 1998). Industrialised countries also need to improve their sewage, excreta and sludge management practices. In the US, for example, the number of waterborne disease outbreaks and the number of affected individuals per outbreak has increased since 1940 (Hunter 1997). Similarly, water quality monitoring of major European rivers indicates that average coliform levels have been steadily increasing for decades (Meybeck *et al.* 1990).

The cost of a conventional sewerage system (which is in the order of 20–70 times that of dry on-site alternatives; see Table 5.7) and its requirement for a piped water supply preclude its adoption in many communities in developing countries (Franceys *et al.* 1992). Low-cost sewerage (a sewerage alternative whereby the design and construction standards associated with conventional sewerage are greatly relaxed) is increasingly being adopted. Although the costs still exceed those of on-site systems (except septic tanks) by a factor of 5–40 (see Table 5.7) low-cost sewerage might be the option of choice in very densely inhabited areas where a regular and adequate water supply is affordable and available.

Conventional sewerage combined with sewage treatment, which is the predominant mode of excreta management in many industrialised countries, is often considered to be the 'gold standard' in terms of excreta disposal and achieving health benefits. For this reason it has often been uncritically transferred to developing countries. However, in many instances it has proved to be far from ideal, not least because of its high cost and need for in-house water supply. The myth that health benefits accrue only from a 'conventional' sewerage system is gradually fading away as suitable alternative sanitation options have been revitalised, developed and promoted during recent decades. There has also been recognition of the need to reduce serious downstream health impacts associated with waterborne sewage including contamination of recreational waters and shellfish beds. As Cairncross (1989) writes:

> No one can plead ignorance of its [waterborne sewage] disadvantages as a sanitation system for low-income communities. Its excessive cost, its wasteful water consumption, its unreliability in conditions of intermittent water supply and its technical impossibility in the narrow, winding alleys of the slums and shanty towns of the Third World are only the better known arguments against it. (p. 304)

While progress in the implementation of appropriate versus industrialised country options has been made, setbacks are constantly occurring. Recently, following a cholera outbreak, the Deputy Minister of Health in Ghana declared that latrines will be phased out and homeowners will be required to install flush toilets.

5.4.1.5 Wastewater treatment

So-called 'conventional' wastewater treatment options (primary and secondary treatment), as are widely applied in industrialised countries, have traditionally focused on the removal of suspended solids and pollutants that require oxygen in the receiving waters to decompose (biochemical oxygen demanding substances (BOD)) and not on the reduction of pathogens and nutrients. These processes are usually difficult, and costly, to operate due to their high energy, skilled labour, infrastructure, and maintenance requirements. Tertiary treatments must be added to the process to effectively reduce pathogen and nutrient levels. A combination of different tertiary treatments such as filtration and chlorination must be used to reduce pathogen levels to very low or undetectable levels. Addition of such treatment steps, however, significantly increases the cost and complexity of the process. The cited options are, therefore, inappropriate in less industrialised or less economically advanced countries.

Waste stabilisation ponds (WSP) are receiving increasingly wide acceptance in developing countries. They can be designed to provide partial treatment (i.e. the removal of helminth eggs to protect farmers and their families who use the effluent for irrigation) or full treatment, which is equivalent to conventional tertiary treatment and achieves inactivation of viruses and pathogenic bacteria (Mara and Pearson 1998). Such effluents, according to WHO (1989), may safely be used for unrestricted vegetable irrigation (i.e. irrigation of vegetables that may be consumed uncooked). In warm climates, where land is available at low cost, WSP have, thus, become a proven method for treating wastewater. When designed properly, WSP are more effective, reliable and robust at removing pathogens than most 'conventional' treatment options. Moreover, WSP remove pathogens without the addition of costly chemicals such as chlorine, are simple to operate and maintain, and promote the use of the water and nutrient resources in the wastewater (Mara and Cairncross 1989). They do, however, have relatively large land requirements.

In order for any sanitation option to be effective it needs to either contain the pathogens or destroy them. The effectiveness of some of the options has been alluded to above. The following two sections, however, explicitly examine pathogen inactivation in general and also look at the containment and inactivation for a range of sanitation options.

5.4.2 Pathogen inactivation

Survival of pathogens derived from faeces is an important factor in disease transmission. Table 5.5 indicates survival times for different pathogens in faecal sludge under both temperate and tropical conditions.

Table 5.5. Organism survival periods in faecal sludge (Feachem *et al.* 1983; Strauss 1985) * survival periods are much shorter if faecal sludge is exposed to the sun

Organism	Av. survival time (days) in wet faecal sludge at ambient temp*	
	Temperate climate (10–15°C)	Tropical climate (20–30°C)
Viruses	<100	<20
Salmonellae	<100	<30
Cholera	<30	<5
Faecal coliforms	<150	<50
Amoebic cysts	<30	<15
Ascaris eggs	2–3 years	10–12 months
Tapeworm eggs	12 months	6 months
Trematodes	<30	<30

5.4.3 Containment

A number of sanitation options are rated on their containment ability in Table 5.6. It is important to bear in mind that containment can act at different levels, protecting the household, the community and 'society'. In the case of the VIP latrine it is easy to see that the containment acts at a household level. However, poor design or inappropriate location may lead to migration of waste matter and contamination of local water supplies putting the community at risk. In terms of waterborne sewage, the containment may be effective for the individual and possibly also the community, but effects may be seen far downstream of the original source, hence affecting 'society'.

Table 5.6. Sanitation options and their containment efficiency

Sanitation option	Containment		
	Household	Community	'Society'
Pit latrine	±	–	+
VIP latrine	+	±	+
No-mix double vault	±	+	+
Pour-flush latrine	+	±	+
Septic tanks	+	±	±
Sewerage/sewage treatment	+	±	–

+ good protection ± some protection – poor protection

Table 5.7 expands upon some of the points within Table 5.6 in terms of potential risk and effectiveness of the barrier to transmission of illness, and also examines the relative construction costs, affordability and institutional implications of selected sanitation options.

Table 5.7. Characterisation of selected excreta management and treatment options

Man/treat option†	Water*	Disease barrier/potential risks	Relative construct. cost	Affordability[a]
VIP latrines	0	Single pit installations also contain fresh excreta; pit contents thus need to be treated and precautionary measures observed during emptying, collection and transport. Groundwater pollution risk where soils are fissured or groundwater levels rise to the pit during wet seasons. Reduces potential disease transmission from flies. Pit contents of double-pit latrines are hygienically safe after storage periods of 6–12 months (tropical climate) or 18–24 months (sub-tropical climate). Such stored contents may be safely used in agriculture	1–2[b] For single pit latrine in urban or peri-urban areas; mechanically emptied every three years; including off-site treatment of faecal sludge.	6[b]
No-mix double-vault latrines	0	Handling and use of urine does not pose health risks in most situations. Hygienic safety of vault contents (as above). Potential disease transmission from flies can be reduced by covering fresh faecal matter with lime, ashes, or soil.	1[b] Manual emptying; no treatment required for pit contents.	3[b]
Pour-flush double-pit latrines	10–15	Handling and use of pit contents: as for no-mix double-vault latrines. Groundwater contamination possible where water levels are (periodically) high. Water seal must be maintained to prevent flies from contacting faeces.	1[b] Manual emptying; no treatment required for pit contents.	3[b]
Septic tanks	20–30	Septage (the settled and floating solids mixed with interstitial wastewater) require treatment as they contain the bulk of excreted pathogens carried in wastewater; high level of pathogen viability in recently deposited solids. Effluent liquids, unless allowed to infiltrate, require treatment, to minimise the pollution load on receiving waters, as they also contain pathogens. Potential pollution of groundwater if water levels are high and soils not consolidated. Fly problem minimised by water barrier.	15–25 Including infiltration system, emptying and off-site treatment of septage.	30–50

Man/treat option†	Water*	Disease barrier/potential risks	Relative construct. cost	Affordability[a]
Waste stabilisation ponds	20–100	A well designed series of ponds is capable of high pathogen removal rates, particularly in warm climates. Ponds must be designed to increase retention times and prevent short-circuiting. Some precautions may be needed to prevent disease vectors from breeding (e.g. mosquitoes or snails).	5–40 Requires large amounts of land and thus depends on the price and availability of land.	5–15
Simplified sewerage	60–100	Solids retention chambers in settled sewerage schemes contain fresh, highly pathogenic contents that require treatment and hygiene precautions in emptying and haulage. Wastewater collected through low-cost sewerage requires treatment for pathogen removal prior to use and discharge and for organics removal prior to discharge.	5–40 Decreasing with increasing housing density and number of houses connected.	12–15
Conventional sewerage	>100	Primary and secondary sewage treatments are not highly effective at reducing pathogen levels. Disinfection and/or additional tertiary treatments are required to reduce pathogen concentrations to acceptable levels.	20–70 Decreasing with increasing housing density and number of houses connected.	30–50

† Management/treatment options

* Water required for operation (litres/capita/day)

[a] Approximate total annual investment and current cost as a percentage of yearly income (assumed to amount to $180/capita and $900/household) of an average low-income household in 1990.

[b] In urban areas, latrine installations might be shared by several families; investment and annual economic cost would accordingly be lower relative to the other sanitation options (the investment cost of a VIP latrine for example, might be lower by a factor of 3–4 if used by 5–8 families instead of 1).

Sources: Cotton *et al.* 1995; Kalbermatten *et al.* 1980; Mara 1996a,b; WELL/DfID 1998; Whittington *et al.* 1992.

5.5 IMPLICATIONS FOR INTERNATIONAL GUIDELINES AND NATIONAL REGULATIONS

Poor sanitation practices lead to disease transmission through numerous pathways. To manage the risks of excreta-related disease transmission, it is important to apply a multiple barrier approach (similar to the hazard assessment and critical control point (HACCP) type programs discussed in Chapters 1 and 12) to sanitation. The use of safe sanitary installations and the appropriate handling, treatment and use of excreta are important barriers or critical control points in the transmission of faecal-oral disease. Effective excreta management programmes will reduce disease transmission via drinking water, contact with recreational water and via the food chain. As discussed earlier, when such management fails, other interventions are necessary to prevent the spread of disease. Numerous studies have helped to identify additional barriers to the spread of faecal-oral disease. Many of these barriers are related to behaviours such as good personal and domestic hygiene practices, water storage and food preparation. Therefore, behaviour modifications as well as technical sanitation solutions are necessary to reduce the transmission of excreta-related disease.

Although the guidelines under consideration in this book focus on water-related areas, it is clear from a public health perspective that consideration of sanitation provision, under the auspices of the harmonised framework, is vital in terms of both international guidelines and national standards.

5.6 REFERENCES

Adams, M.R. and Moss, M.O. (1995) *Food Microbiology*, Royal Society of Chemistry, Cambridge.

Cairncross, S. (1989) Water supply and sanitation: an agenda for research. *Journal of Tropical Medicine and Hygiene* **92**, 301–314.

California State Water Pollution Control Board (1954) *Report on the Investigation of Travel of Pollution*. Sacramento, California. Publication no. 11.

Cifuentes, E., Blumenthal, U., Ruiz-Palacios, G., Bennett, S. and Quigley, M. (2000) Health risks in agricultural villages practising wastewater irrigation in Central Mexico: perspectives for protection. In *Water Sanitation & Health* (eds I. Chorus, U. Ringelband, G. Schlag and O. Schmoll), pp. 249–256, IWA Publishing, London.

Cotton, A., Franceys, R., Pickford, J. and Saywell, D.F. (1995) *On-Plot Sanitation in Low-Income Urban Communities – a Review of the Literature*. Water, Engineering and Development Centre (WEDC), Loughborough, UK.

Esrey, S.A. (2000) Rethinking sanitation: panacea or Pandora's box. In *Water Sanitation & Health* (eds I. Chorus, U. Ringelband, G. Schlag and O. Schmoll), pp. 7–14, IWA Publishing, London.

Esrey, S.A., Feachem, R.G. and Hughes, J.M. (1985) Interventions for the control of diarrhoeal diseases among young children: improving water supplies and excreta disposal facilities. *Bulletin of the World Health Organization* **63**(4), 757–772.

Esrey, S.A., Potash, J.B., Roberts, L. and Shiff, C. (1991) Effects of improved water supply and sanitation on ascariasis, diarrhoea, dracunculiasis, hookworm infection, schistosomiasis, and trachoma. *Bulletin of the World Health Organization* **69**(5), 609–621.

Feachem, R.G., Bradley, D.J., Garelick, H. and Mara, D.D. (1983) Sanitation and disease: health aspects of excreta and wastewater management. World Bank Studies in Water Supply and Sanitation 3, Wiley, Chichester, UK.

Franceys, R., Pickford, J. and Reed, R. (1992) A Guide to the Development of On-site Sanitation. World Health Organization, Geneva.

Gear, S., Brown, A. and Mathys, A. (1996) *Strategic Sanitation Plan – the Kumasi Experience*. UNDP/World Bank Water Supply and Sanitation Program. Regional Water and Sanitation Group – West Africa.

Heinss, U., Larmie, S.A. and Strauss, M. (1998) *Solids Separation and Pond Systems for the Treatment of Faecal Sludges in the Tropics – Lessons Learnt and Recommendations for Preliminary Design*. SANDEC Report No. 05/98, EAWAG/SANDEC, Duebendorf, Switzerland.

Hunter, P. (1997) *Waterborne Disease: Epidemiology and Ecology*, Wiley, Chichester, UK.

Kalbermatten, J., Julius, D.S. and Gunnerson, C. (1980) Appropriate Technology for Water Supply and Sanitation – Technical and Economic Options. World Bank.

Lewis, W.J., Foster, S.S.D. and Drasar, B.S. (1982) *The Risk of Groundwater Pollution by On-Site Sanitation in Developing Countries – A Literature Review*. IRCWD/SANDEC Report No. 01/82.

Mara, D. and Cairncross, S. (1989) Guidelines for the safe use of wastewater and excreta in agriculture and aquaculture. World Health Organization, Geneva.

Mara, D.D. (ed.) (1996a) *Low-Cost Sewerage*, Wiley, Chichester, UK.

Mara , D.D. (1996b) *Low-Cost Urban Sanitation*, Wiley, Chichester, UK.

Mara, D.D. and Pearson, H. (1998) *Design Manual for Waste Stabilization Ponds in Mediterranean Countries.* European Investment Bank, Mediterranean Environmental Technical Assistance Programme.

Meybeck, M., Chapman, D. and Helmer, R. (1990) *Global Freshwater Quality: A First Assessment*, Blackwell, Oxford.

Montangero, A. and Strauss, M. (2000*) Faecal Sludge Management – Strategic Aspects and Treatment Options*. Proceedings International Workshop on Biosolids Management and Utilisation, Nanjing (China) and Forum on Biosolids Management and Utilisation, Hong Kong, September 2000.

Murray, C.J.L. and Lopez, A.D. (eds) (1996) *The Global Burden of Disease, Vol. II, Global Health Statistics: A compendium of incidence, prevalence and mortality estimates for over 200 conditions*, Harvard School of Public Health on behalf of the World Health Organization and The World Bank, Cambridge, MA.

National Institute of Urban Affairs (1990) *A Revolution in Low Cost Sanitation: Sulabh International New Delhi Case Study*. NIUA, 11 Nyaya Marg, Chanakyapuri, New Dehli-110021, India.

National Research Council (1998) *Issues in Potable Reuse: The Viability of Augmenting Drinking Water Supplies With Reclaimed Water,* National Academy Press, Washington, DC.

Neto, F. and Tropp, H. (2000) Water supply and sanitation services for all: global progress during the 1990s. *Natural Resources Forum* **24**, 225–235.

Peasey, A. (2000) *Health Aspects of Dry Sanitation with Waste Reuse*. WELL, London School of Hygiene and Tropical Medicine, London.

Rice, A.L., Sacco, L., Hyder, A. and Black, R.E. (2000) Malnutrition as an underlying cause of childhood deaths associated with infectious diseases in developing countries. *Bulletin of the World Health Organization* **78**(10), 1207–1221.

Robertson, L.J., Campbell, A.T. and Smith, H.V. (1992) Survival of *Cryptosporidium parvum* oocysts under various environmental pressures. *Applied and Environmental Microbiology* **58**(11), 3494–3500.

Rose, J.B. and Slifko, T.R. (1999) *Giardia*, *Cryptosporidium*, and *Cyclospora* and Their Impact on Foods: A Review. *Journal of Food Protection* **62**(9), 1059–1070.

Samanta, B.B. and van Wijk, C.A. (1998). Criteria for successful sanitation programmes in low income countries. *Health Policy and Planning* **13**(1), 78–86.

Schertenleib, R. (1988) *Risk of Groundwater Pollution by On-Site Sanitation in Developing Countries*. Unpublished report, SANDEC.

Schwartzbrod, L. (2000) Human Viruses and Public Health: Consequences of Use of Wastewater and Sludge in Agriculture and Aquaculture. Unpublished document commissioned by WHO, Nancy, France. (In French.)

SEPA (1995) *Vad innehaller avlopp fran hushall?* (Content of wastewater from households). Report 4425, Stockholm, Swedish Environmental Protection Agency. (In Swedish.)

Simpson-Hébert, M. and Wood, S. (eds) (1998) *Sanitation Promotion*. Unpublished document WHO/EOS/98.5, World Health Organization/Water Supply and Sanitation Collaborative Council (Working Group on Promotion of Sanitation), Geneva.

Strauss, M. (1985). *Health Aspects of Nightsoil and Sludge Use in Agriculture and Aquaculture. Part II: Pathogen Survival*. IRCWD Report No. 04/85, IRCWD (now SANDEC), Duebendorf, Switzerland.

Strauss, M., Heinss, U., Montangero, A. (2000). On-Site Sanitation: When the Pits are Full – Planning for Resource Protection in Faecal Sludge Management. In *Proceedings, Int. Conference, Bad Elster, 20–24 November 1998. Schriftenreihe des Vereins fuer Wasser-, Boden- und Lufthygiene,* **105***: Water, Sanitation & Health – Resolving Conflicts between Drinking-Water Demands and Pressures from Society's Wastes* (eds I. Chorus, U. Ringelband, G. Schlag and O. Schmoll), WHO Water Series, IWA Publishing, London.

Tamburrini, A. and Pozio, E. (1999) Long-term survival of *Cryptosporidium parvum* oocysts in seawater and in experimentally infected mussels (*Mytilus galloprovincialis*). *International Journal for Parasitology* **29**, 711–715.

United Nations Population Division (1998) *World Population Nearing 6 Billion Projected Close to 9 Billion by 2050*. New York, United Nations Population Division, Department of Economic and Social Affairs (Internet communication of 21 September 2000 at www.popin.org/pop1998/1.htm).

Wagner, E.G. and Lanoix, J.N. (1958) *Excreta Disposal for Rural Areas and Small Communities*, World Health Organization, Geneva.

WELL/DfID (1998) DfID Guidance Manual on Water Supply and Sanitation Programmes. Water and Environmental Health at London and Loughborough (WELL) and the Department for International Development (DfID), UK.

Whittington, D., Lauria, D.T., Wright, A., Choe, K., Hughes, J.A. and Swarna, V. (1992) *Household Demand for Improved Sanitation Services: a Case Study of Kumasi, Ghana*. UNDP/World Bank Water and Sanitation Program.

WHO (1989) *Health Guidelines for the Use of Wastewater in Agriculture and Aquaculture*. Report of a WHO Scientific group, Technical Report Series No. 778, World Health Organization, Geneva.

WHO (1993) *Improving Water and Sanitation Hygiene Behaviours for the Reduction of Diarrhoeal Disease*. Report of an Informal Consultation, WHO.CWS 90.7, 18–20 May 1992, Geneva.

WHO (1995) *Control of Food-borne Trematode Infections*. Technical Report Series 849, World Health Organization, Geneva.

WHO (1996) *Cholera and other epidemic diarrhoeal diseases control. Fact sheets on environmental sanitation*. World Health Organization, Geneva.

WHO (1999) *WHO Report on Infectious Diseases – Removing obstacles to healthy development*. World Health Organization, Geneva.

WHO (2000a) *Global Water Supply and Sanitation Assessment*. World Health Organization, Geneva.

WHO (2000b) *The World Health Report 2000 – Health systems: Improving performance*. World Health Organization, Geneva.

Wolgast, M. (1993) *Clean Waters: Thoughts About Recirculation*, Creamon, Uppsala.

World Resources Institute (1998) *A Guide to the Global Environment: Environmental Change and Human Health*, Oxford University Press, New York.

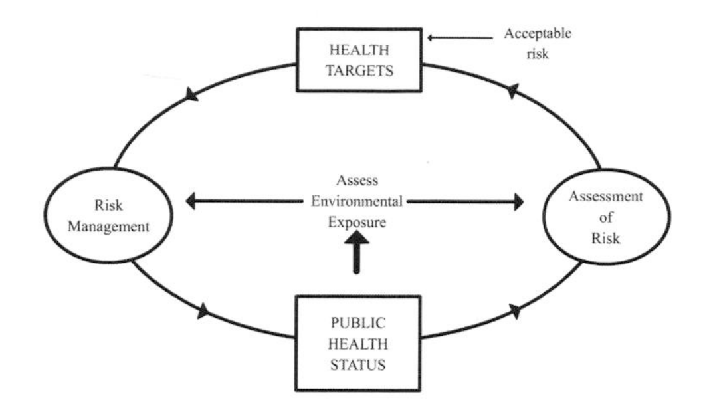

6

Disease surveillance and waterborne outbreaks

Yvonne Andersson and Patrick Bohan

Outbreaks are both a demonstration of a breakdown or failure in the system and, by acting as a 'natural experiment', present an opportunity to provide new insights into disease transmission and, perhaps, improvements to the system. This chapter outlines in detail the surveillance systems in Sweden and the US that are designed to detect waterborne disease outbreaks, and examines the actions taken upon suspecting an outbreak. It also examines some of the outbreaks that have occurred, principally from drinking water, details lessons that can be learnt from well-conducted investigations, and briefly looks at the worldwide situation.

6.1 THE SWEDISH SITUATION

Sweden has a long history of communicable disease awareness, with legislation dating back to 1875. The regulations are based on a selection of disease agents,

 Water Quality: Guidelines, Standards and Health. Edited by Lorna Fewtrell and Jamie Bartram. Published by IWA Publishing, London, UK. ISBN: 1 900222 28 0

their occurrence and the severity of the disease. Under the Communicable Disease Act, County Medical Officers have the main responsibility for dealing with such diseases and they have an overseeing and co-ordinating role for combating communicable diseases in their region.

Local doctors are responsible for any epidemiological investigation relating to a patient and also for giving hygiene advice to people who have contracted communicable diseases. A doctor who identifies a person with a notifiable disease is required to inform the County Medical Officer and the Swedish Institute for Infectious Disease Control (SIIDC) of the case and, in relation to diseases which may have been contracted via food, water and the environment, the local Environmental and Public Health Committee. The reporting of waterborne outbreaks, as such, however, is not mandatory.

Water is included under the Food Act in Sweden, the responsible authority being the National Food Administration. The reporting and investigation system can be quite complex with a large number of different bodies being involved.

6.1.1 Waterborne disease outbreaks in Sweden

Sweden has a long tradition in the reporting and surveillance of communicable diseases, including waterborne diseases. The first reported outbreaks of waterborne disease in Sweden were cholera epidemics between 1834 and 1874 (Arvidsson 1972). Based on historical data a retrospective summary of incidents and outbreaks has been made, dating back to 1880. The number of outbreaks and aetiological agents has varied over the years, according to prevailing knowledge, the interest of local authorities and diagnostic capabilities. Over a period of 100 years (between 1880 and 1979) 77 waterborne outbreaks, with 26,867 reported cases and 789 deaths were known (Andersson 1992). Most of the outbreaks (88%) during that period were due to known agents. At the start of the twentieth century the most commonly reported diseases (possibly of waterborne origin) were typhoid fever and shigellosis. The picture of hepatitis and polio reporting has changed in line with the improving general standards of hygiene in society.

6.1.1.1 Waterborne outbreaks since 1980

An improved reporting system, which includes the results of epidemiological investigations, has existed since 1980. Improvements have included the systematic investigation of possible waterborne outbreaks with a standardised questionnaire for large outbreaks, as well as clinical and environmental sampling. The enhanced system has resulted in an increase in the number of detected waterborne outbreaks. During the period 1980–99, 116 outbreaks of waterborne diseases were reported from both large and small water supply systems, the majority affecting systems with less than 15,000 consumers. In total, about 57,500 people were affected, but only two deaths were recorded. These numbers are based on epidemiological follow-ups and sometimes local authority reports. More than 70% of the outbreaks were due to unknown agents, and are termed Acute Gastrointestinal Illness (AGI). The most commonly identified agents were *Campylobacter* sp. and *Giardia lamblia*. A few outbreaks also involved *Entamoeba histolytica*, enterotoxigenic *E. coli* (ETEC) and *Cryptosporidium*. During this period, *Salmonella* spp. and *Shigella* spp. were only isolated from outbreaks associated with private wells. Over the last few years, the number of reported waterborne outbreaks involving caliciviruses has increased, owing to the use of better laboratory methods for clinical samples. The numbers of outbreaks and cases are shown in Table 6.1.

Table 6.1. Waterborne disease outbreaks in Sweden (1980–99)

Years	No. of outbreaks	No. of cases
1980	3	4030
1981	3	105
1982	3	622
1983	3	1266
1984	9	1149
1985	12	5256
1986	12	5575
1987	8	900
1988	5	13,144
1989	4	223
1990	4	100
1991	4	935
1992	4	588
1993	5	297
1994	8	4070
1995	10	13,574
1996	7	3135
1997	6	209
1998	4	2310
1999	2	180

In Sweden, surface water is used for approximately half of community water supplies; the remainder being supplied by groundwater or artificially recharged groundwater. The number of outbreaks attributable to surface water since 1980 is relatively small: however, as a source it has been responsible for the largest reported outbreaks, affecting thousands of people. Problems often occur in early spring, when the surface of the water is still frozen and the final water receives little or no chlorination.

The largest outbreak between 1980 and 1999 occurred in early 1988 (Andersson 1991) and affected approximately 11,000 people (with an attack rate of 41%). Investigation revealed that the water treatment plant was undergoing refurbishment and as a consequence there was a chlorination failure. During the short period of chlorination failure the raw surface water received only filtration and pH adjustment.

The other large outbreak due to surface water (affecting 10,000 people) was due to a change in pipeline (Wahren 1996). A pipeline, containing stagnant raw water, was brought into use without being flushed first.

Groundwater was most commonly associated with the outbreaks outlined in Table 6.1. Generally, however, the problem was not the quality of the groundwater *per se*, but technical difficulties or communication breakdowns leading to cross-connections with sources of contamination. In a Swedish ski resort 3600 people became ill (*Giardia* and *Entamoeba histolytica*) when a drinking water reservoir was contaminated with sewage through a pipeline connected to a spillway overflow (Andersson and de Jong 1989; Ljungstrom and Castor 1992). A damaged septic tank led to contamination of a drinking water well which supplied water to a restaurant resulting in at least ten customers reporting campylobacteriosis. An illegal cross-connection to a creek to serve as a private source of irrigation led to contaminated creek water being pumped into the community water supply and approximately 600 people falling ill with a variety of infections including campylobacteriosis, giardiasis and cryptosporidiosis (Thulin 1991).

6.1.1.2 Recognition of waterborne outbreaks

An outbreak or epidemic normally means that more cases are clustered than the anticipated, endemic, background level. The World Health Organization (WHO) definition of a food- or waterborne outbreak is when two or more persons experience a similar illness after ingestion of the same type of food or water from the same source and when the epidemiological evidence implicates the food or the water as the source of the illness (Schmidt 1995).

The probability of detecting an outbreak depends on both knowledge and resources (both microbiological and personnel). Rapid recognition of the possibility of an outbreak and a timely start to the investigation greatly increase the likelihood of determining cause.

There are a number of different possibilities that could suggest a waterborne outbreak:

- non-potable water found by routine sampling
- complaints about water quality
- an increase of AGI in the community, in general practices, or in hospitals (clinical surveillance)
- an increase of positive laboratory results indicating possible waterborne agents (laboratory surveillance).

6.1.1.3 Water sampling

The routine monitoring of drinking-water quality cannot prevent an outbreak but can detect that contamination has occurred, thus it plays an important role as it reveals basic water quality and the likely risk of an outbreak.

Communication can play a vital role in the detection, and prevention, of outbreaks. In the investigations of some Swedish surface-water-related outbreaks it was found that the raw-water quality deteriorated every spring with high levels of faecal coliforms and/or coliforms. Although this information was collected each year, it was not interpreted and as a result no action was taken. If this type of water-quality monitoring had been used as intended, an appropriate action might have been to increase disinfection levels each spring, possibly averting an outbreak.

Outbreaks may start with complaints about water quality (Thulin 1991). A rapid collection of water samples and technical investigation may confirm deficient water quality. Taking prompt control measures may prevent a waterborne outbreak or at least reduce the number of cases.

6.1.1.4 Clinical and laboratory surveillance

In Sweden, there are two mandatory surveillance systems: the reporting of notifiable diseases and reporting from the laboratories. Diseases that should be reported by doctors which may be of interest in waterborne outbreaks are hepatitis A, cholera, typhoid fever, paratyphoid fever, salmonellosis, shigellosis, campylobacteriosis, yersiniosis, enterohaemorrhagic *E. coli* O157, giardiasis and amoebiasis.

The diseases reported only by laboratories (voluntarily) are enterohaemorrhagic *E. coli* (other than serotype O157), caliciviruses, rotaviral enteritis, cryptosporidiosis and diarrhoea caused by *Cyclospora* sp.

To recognise an increase in illness from the reporting system is a slow way of discovering a waterborne outbreak. Normally, it will take about one to two weeks before the surveillance system recognises an increase. It also suffers from a lack of sensitivity, as outlined below.

One major problem with outbreak detection is that a significant number of people may not consult a doctor. There have been waterborne outbreaks with several hundred or a thousand people affected, which were discovered more or less accidentally. Therefore, even with a surveillance system waterborne outbreak detection can be down to luck (Figure 6.1).

Gastrointestinal symptoms → will see a doctor
→ will not see a doctor

→ the person will be sampled
→ the person will not be sampled

→ negative result
→ positive result

Figure 6.1. Conditions for a pathogenic micro-organism to be diagnosed.

There have been very few examples of outbreaks in Sweden in which the surveillance system first revealed that a waterborne outbreak existed. Two such examples are:

> One small outbreak of *Giardia lamblia* in which a private well at a 'holiday village' was suspected as the source of the cases.
> A laboratory reported seven patients with campylobacteriosis at the hospital to the County Medical Office. All of them came from the same small town. It was later revealed that a large, waterborne, *Campylobacter* outbreak had occurred with about 2500 people falling ill (Andersson *et al.* 1994).

Investigations, based on interviews and standardised questionnaires, often reveal many more cases of illness, as shown in Table 6.2. The attack rate is unexpectedly high, confirming the underestimation of cases.

Table 6.2. Initially reported cases and actual numbers of cases in selected outbreaks in Sweden

Causative agent	Initially reported sick	Sick identified by lab	Estimated no. of sick	No. at risk	Attack rate (%)
Campylobacter	380	221	2000	15,000	13
Unknown	45	–	2000	2500	82
Unknown & *Giardia*	Several ill	56	550	750	73
Unknown	Several ill	Unknown	1000	1200	85
Unknown, *Giardia* & *Entamoeba*	Several ill	*Giardia*: 1480 *Entamoeba:* 106	3600	4000	90
Unknown	700	Unknown	11,000	26,000	41
Campylobacter	200	7 initially	2500	10,000	25

6.1.1.5 Common causes of outbreaks

A thorough investigation is vital to determine the outbreak cause (and an example of outbreak management is given later). In Nordic countries outbreak investigation analysis has revealed a number of common causes of outbreaks (Stenstrom *et al.* 1994). From community systems supplied with surface water the following occurrences were highlighted:

- Wastewater contamination of raw water source in combination with disinfection deficiencies
- No disinfection
- Cross-connections
- Regrowth in the distribution system.

Similar occurrences were identified from outbreaks involving groundwater, with the most common problem being source water contamination through wastewater infiltration. These problems and deficiencies are not confined to Nordic countries as similar causes have been reported elsewhere (e.g. Tulchinsky *et al.* 1988).

Realising some of the common causes of outbreaks led to a Swedish survey and inventory of all community supplies in the country and an examination of some of the larger private water supplies. Over 4000 supplies were subject to survey, of which 2281 were community supplies. Table 6.3 shows the risk factors that were identified as a result of the survey.

Table 6.3. Risk factors identified from a water supply survey (community supplies and larger private supplies (adapted from Hult 1991)

Factor	Percentage of total number
Safety area for source not established	74
Risk due to wastewater pipes close to source	13
Pollution risk at groundwater source	9
Pollution risk at low reservoir from drain gutter	4
Pollution risk at low reservoir from overflow pipe	8
No disinfection	79*
Unsatisfactory control of disinfection	69
Unsatisfactory water treatment (other than disinfection)	5
Unsatisfactory control programme for distribution system	62

* mainly small groundwater systems

Although sanitary inspection is a sensible step in developed countries (Prescott and Winslow 1931), a literature search suggests that it receives very little attention (Bartram 1996). In a number of developing countries, however, it is used extensively as a primary monitoring tool, in line with recommendations by WHO (1997).

6.2 THE SITUATION IN THE US

The surveillance system for Waterborne Disease Outbreaks (WBDO) in the US (while voluntary in nature) has much in common with that in Sweden, and suffers many of the same problems. In line with worldwide definitions, the unit of analysis for the WBDO surveillance system in the US is an outbreak rather than an individual case of a particular disease. Two criteria must be met for an event to be defined as a WBDO. First, two or more people must have experienced a similar illness after either ingestion of drinking water or exposure to water used for recreational purposes (this stipulation is waived for single cases of laboratory-confirmed primary amoebic meningoencephalitis). Second, epidemiologic evidence must implicate water as the probable source of the illness. Outbreaks caused by contamination of water or ice at the point of use are not classified as WBDOs.

6.2.1 Overview

Since 1971, the Centers for Disease Control (CDC) and the US Environmental Protection Agency (EPA) have maintained a collaborative surveillance system for collecting and periodically reporting data that relate to occurrences and causes of waterborne disease outbreaks. The surveillance system includes data about outbreaks associated with drinking water and recreational water, and these data are published in Morbidity and Mortality Weekly Reports (MMWR) approximately every two years (CDC 1990, 1991, 1993; Kramer *et al.* 1996; Levy *et al.* 1998; Louis 1988).

State, territorial, and local public health departments are primarily responsible for detecting and investigating WBDOs and for voluntarily reporting them to CDC on a standard form. CDC annually requests reports from state and territorial epidemiologists or from persons designated as the WBDO surveillance co-ordinators. When necessary, additional information about water quality and treatment is obtained from the state's drinking-water agency. There is no national surveillance system in place for waterborne disease outbreaks and all the data gathered is voluntarily reported to CDC.

6.2.1.1 Considerations

The waterborne disease surveillance data, which identify the types of water systems, their deficiencies, and the respective aetiologic agents associated with the outbreaks, are useful for evaluating the adequacy of current technologies for providing safe drinking and recreational water. However, the data presented here have at least one important limitation: they almost certainly do not reflect the true incidence of WBDOs or the relative incidence of outbreaks caused by various aetiologic agents. Not all WBDOs are recognised, investigated, and reported to CDC or EPA; and clearly, the extent to which WBDOs are unrecognised and under-reported is unknown.

The likelihood that individual cases of illness will be detected, epidemiologically linked, and associated with water varies considerably depending on locale, and is dependent upon a number of factors, including:

- public awareness
- the likelihood that several ill people consult the same rather than different health-care providers
- the interest of health-care providers
- availability of laboratory testing facilities
- local requirements for reporting cases of particular diseases
- surveillance and investigative activities and capacities of state and local health and environmental agencies.

Therefore, the states that report the most outbreaks might not be those in which the most outbreaks occur, but those with the most rigorous investigation procedures. Recognition of WBDOs is also dependent on certain outbreak characteristics:

- Those involving serious illness are most likely to receive the attention of health authorities.
- Outbreaks of acute diseases, particularly those characterised by a short incubation period, are more readily identified than those associated with disease from chronic, low-level exposure to an agent such as a chemical.
- Outbreaks associated with community water systems are more likely to be recognised than those associated with non-community systems because the latter serve mostly non-residential areas and transient populations.
- Outbreaks associated with individual systems are the most likely to be under-reported because they generally involve relatively few people.

The identification of the aetiologic agent of a WBDO is dependent on the timely recognition of the outbreak so that appropriate clinical and environment samples can be obtained. The interests and expertise of investigators and the routine practices of local laboratories can also influence whether the aetiologic agent is identified. Diarrhoeal stool specimens, for example, are generally examined for bacterial pathogens, but not for viruses. In most laboratories, testing for *Cryptosporidium* is carried out only if requested and is not included in routine stool examinations for ova and parasites. The water quality data that are collected vary widely among outbreak investigations, depending on such factors as available fiscal, investigative, and laboratory resources. Furthermore, a few large outbreaks can substantially alter the relative proportion of cases of waterborne disease attributed to a particular agent. Finally, the number of reported cases is generally an approximate figure, and the method and accuracy of the approximation vary among outbreaks.

6.2.2 Waterborne outbreaks between 1995–6

During the two-year period between January 1995 and December 1996, 13 states reported a total of 22 outbreaks associated with drinking water, of which 15 were attributed to infectious agents. A total of 36 outbreaks were attributed to recreational water affecting an estimated 9129 people, including 8449 people in two large outbreaks of cryptosporidiosis. Twenty-two of the recreational water incidents were outbreaks of gastroenteritis.

6.2.2.1 Drinking water

Of the 15, non-chemically-related, drinking-water outbreaks the aetiological agent was identified in 7 cases. The outbreaks are summarised in Table 6.4.

Table 6.4. Waterborne disease outbreaks associated with drinking water, by aetiological agent and water system type.

	Type of water system							
	Community		Non-com.		Individual		Total	
Agent	O	C	O	C	O	C	O	C
AGI	1	18	6	658	1	8	8	684
Giardia lamblia	1	1449	0	0	1	10	2	1459
Shigella sonnei	0	0	2	93	0	0	2	93
SRSV	1	148	0	0	0	0	1	148
P. shigelloides	0	0	1	60	0	0	1	60
E. coli O157:H7	0	0	1	33	0	0	1	33
Total	3	1615	10	844	2	18	15	2477

AGI = acute gastrointestinal illness of unknown aetiology; SRSV = small round structured virus; Non-com. = non-community; O = outbreaks; C = cases.

Both outbreaks of giardiasis were associated with surface water. The small outbreak occurred in Alaska and was caused by untreated surface water, and the second outbreak occurred in New York affecting an estimated 1449 people, and was associated with surface water that was both chlorinated and filtered. A dose–response relation was found between consumption of municipal water and illness. No interruptions in chlorination were identified at the water plant; however, post-filter water turbidity readings exceeded the regulated limit before and during the outbreak.

One outbreak of shigellosis occurred in Idaho and affected 83 people. This outbreak was at a resort supplied by untreated well water, which became contaminated with sewage from a poorly-draining line (CDC 1996). The other outbreak of shigellosis was in Oklahoma and affected 10 people. It was associated with tap water in a convenience store that was supplied by chlorinated well water. Although the factors contributing to contamination of the water were not determined, the water was thought to have been inadequately chlorinated.

The outbreak of E. coli O157:H7 infection occurred at a summer camp in Minnesota that was supplied by chlorinated spring water. Several of the 33 affected persons had stool samples that also were positive for Campylobacter jejuni and Salmonella serotype London. Water samples from the spring and distribution system were positive for coliforms and E. coli. The contamination was attributed to flooding from heavy rains and to an improperly protected spring.

A non-community water system supplying a New York restaurant was responsible for the outbreak of Plesiomonas shigelloides infection. This outbreak affected 60 people and is thought to be the largest outbreak of Plesiomonas infection reported in the US (CDC 1998a). Chlorinated spring water that supplied a kitchen tap in the restaurant had a high coliform count (including E. coli) and the disinfectant residual was zero. The chlorinator was found to be depleted of disinfectant, and cultures of water from the river adjacent to the uncovered reservoir where treated water was stored grew Plesiomonas.

One outbreak in 1995 was thought to have been caused by a Small Round Structure Virus (SRSV). It occurred at a high school in Wisconsin and affected 148 people. The school received its drinking water from a community water supply. Contamination is thought to have occurred from back-siphonage of water through hoses submerged in a flooded football field. The source of the virus was not determined.

Eight of the WBDOs associated with drinking water had no identified aetiologic agent. Of these, three outbreaks were associated with untreated well water, three with inadequate chlorination of unfiltered well water and one with possible short-term cross-connection and back-siphonage problems in the distribution system. The other outbreak was associated with water from an outside tap at a wastewater treatment plant that was not marked as non-potable.

6.3 OUTBREAK MANAGEMENT

Once a potential waterborne outbreak has been identified, the public health authorities have the responsibility of conducting further investigations. The objectives of these investigations are to determine the size and nature of the outbreak and its cause. This is important in order to implement control measures to reduce the number of cases and to ensure that the outbreak does not happen again. A more detailed description of the general approach to outbreak investigation is given elsewhere (Hunter 1997). This chapter presents a brief outline based upon UK procedure.

Even before the outbreak is detected, good outbreak management depends on prior planning. This planning will have identified the agencies that need to be involved and will have obtained agreement with them over their roles. The prior planning will also have led to the setting up of appropriate surveillance systems (as already outlined), without which outbreaks are unlikely to be identified.

Once a possible outbreak is identified, the next step is outbreak confirmation. This is essentially a quick look at possible alternative explanations for the apparent increase in illness, such as laboratory false positives or changes in notification behaviour. Should there be no alternative explanation, an outbreak control team is formed.

The first action of the outbreak control team is to agree an explicit statement of the case definition. This is essential to know whether individual illnesses should be included in the outbreak. Case definitions may include a range of possible onset dates, clinical symptoms, geographical locations and microbiological results. Case definitions can be very broad or very narrow to either include many possible cases or few. The broader the definition, the more cases will be identified, although many of these additional cases may not be related to the main outbreak. Case definitions can and should change as new information becomes available.

Once a case definition has been agreed, case finding is the next step. For case definitions that include a microbiological diagnosis the easiest way of identifying cases is to review microbiology laboratory results. A positive microbiological result will be very specific. However, relying on such results will exclude those patients who have not had microbiological investigation samples taken. It may be necessary to encourage doctors to increase their sampling rate or to report all episodes of particular clinical syndromes. A common alternative is to develop more than one case-definition, one that includes microbiology data and one that relies exclusively on clinical features. These can be called confirmed cases and presumptive cases.

The next stage of the investigation is outbreak description. Outbreak description requires that a basic set of data is collected on every individual who satisfies the case definitions. As a minimum, these data will include name, address, age, sex, date of onset, the results of microbiological examination and sufficient clinical information to prove that the individual satisfies the case definition. It is also usual to record place of work or schooling, a basic food or contact history and any travel history. This type of data may be collected by a trawling questionnaire that asks a series of open questions covering activities during the period before the onset of illness. The results of these early investigations are usually presented in tabular and graphical form.

At this stage it may be possible to develop a hypothesis as to the cause of the outbreak. The hypothesis generated at this point may then indicate possible control measures. One of the more difficult decisions in any outbreak investigation is when control measures should be implemented. For control measures to be effective, they have to be implemented early in the outbreak at a time when the working hypothesis is still far from proven. The damage to a water company's image and financial position may be great if it has to make major changes to its treatment processes or issue a notice for its customers to boil their water. If the outbreak is eventually proven to be due to another cause, this will have been for no purpose.

Once a hypothesis as to the cause of the outbreak has been generated, the next step is to prove it. This may involve further epidemiological investigations such as case-control studies, more microbiological investigations such as typing any isolates or environmental investigations into the treatment plant and its records. If the hypothesis is proven by the further investigations, then more definitive control measures may be put in place to prevent a recurrence.

The final phase in any outbreak investigation is the dissemination of lessons learnt. It is usual for a detailed report to be prepared for local stakeholders. This report may be used by legal staff in possible civil and criminal proceedings. As we have seen earlier, in many outbreaks more general lessons are learnt and these should be published more widely in the medical or scientific literature.

6.4 UNDER-REPORTING

The previous sections have touched upon the problems and reasons for under-reporting. Estimates of the level of under-reporting vary, reflecting differences in surveillance systems and access to medical care as well as true differences in disease incidence. Ford (1999) cites an analysis recently conducted in India, where it was estimated that hospital incidence data from Hyderabad underestimated the incidence of waterborne disease by a factor of approximately 200 (Mohanty 1997). In their study of food-related illness, Mead and colleagues (1999) used adjustment factors ranging from 20 to 38, depending upon the pathogen concerned, to account for under-reporting of gastrointestinal symptoms. Table 6.5 (adapted from WHO (1999)) illustrates the number of waterborne outbreaks in Europe following a survey conducted in 1997. Of the 52 European countries asked for information on waterborne disease outbreaks, 26 returned information and 19 provided information specifically on outbreaks.

Table 6.5 probably sheds considerably more light on the enthusiasm for surveillance and outbreak detection than it does on the actual level of outbreaks. Interestingly, the figures reported for Sweden are considerably lower than those reported in Table 6.1! The survey response in general would seem to suggest that a degree of confusion exists, since in many cases countries reported fewer cases of gastrointestinal disease linked with drinking water than cases of gastrointestinal illness linked with waterborne outbreaks.

Table 6.5. Reported waterborne disease outbreaks associated with drinking and recreational water in 19 European countries, 1986–96 (adapted from WHO 1999)

Country	Agent or disease (no. of outbreaks)	Total no. of outbreaks	No. of cases (with details)
Albania	Amoebic dysentery (5), typhoid fever (5), cholera (4)	14	59 (3)
Croatia	Bacterial dysentery (14), gastroenteritis (6), hepatitis A (4), typhoid (4), cryptosporidiosis (1)	29[1]	1931 (31[1])
Czech Republic	Gastroenteritis (15), bacterial dysentery (2), hepatitis A (1)	18[2]	76 (3)
England & Wales	Cryptosporidiosis (13), gastroenteritis (6), giardiasis (1)	20	2810 (14)
Estonia	Bacterial dysentery (7), hepatitis A (5)	12	1,010 (12)
Germany	No outbreaks reported	0	0
Greece	Bacterial dysentery (1), typhoid (1)	2	16 (1)
Hungary	Bacterial dysentery (17, gastroenteritis (6), salmonellosis (4)	27[3]	4884 (27)
Iceland	Bacterial dysentery (1)	1	10 (1)
Latvia	Hepatitis A (1)	1	863 (1)
Lithuania	No outbreaks reported	0[4]	0
Malta	Gastroenteritis (152), bacterial dysentery (4), hepatitis A (4), giardiasis (1), typhoid (1)	162	19 (6)
Norway	No outbreaks reported	0	0
Romania	Bacterial dysentery (36), gastroenteritis (8), hepatitis A (8), cholera (3), typhoid (1), methaemoglobinaemia (1)	57	745 (1)
Slovak Republic	Bacterial dysentery (30), gastroenteritis (21), hepatitis A (8), typhoid (2)	61	5173 (61)
Slovenia	Gastroenteritis (33), bacterial dysentery (8), hepatitis A (2), amoebic dysentery (1), giardiasis (1)	45	n.a.
Spain	Gastroenteritis (97), bacterial dysentery (47), hepatitis A (28), typhoid (27), giardiasis (7), cryptosporidiosis (1), unspecified (1)	208	n.a.
Sweden	Gastroenteritis (36), campylobacteriosis (8), Norwalk like virus (4), giardiasis (4), cryptosporidiosis (1), amoebic dysentery (1), *Aeromonas* sp. (1)	53[5]	27,074 (47)

[1] Discrepancies in data were noted in different sections of the questionnaire
[2] One year of reporting only
[3] Outbreaks associated with drinking water (n = 12) and recreational water (n = 15)
[4] Ten years of reporting only
[5] In one outbreak *Campylobacter* sp., *Cryptosporidium* sp. and *Giardia lamblia* were identified as aetiologic agents (all three are listed in the relevant column)

Water may play an additional role in disease outbreaks through the use of contaminated water in food irrigation or food processing. Such a route was suspected in an outbreak of shigellosis that affected several countries in North West Europe during 1994. The source of the pathogen was identified as lettuce imported from Spain, and irrigation with contaminated water was strongly suspected (Frost *et al.* 1995; Kapperud *et al.* 1995). In North America outbreaks of cyclosporiasis have been associated with raspberries imported from Guatemala; again wastewater irrigation was noted as a possible source of contamination (CDC 1998b). A case-control study in Fuerteventura during an outbreak of vero cytotoxin-producing *E. coli* O157 showed an association with the consumption of raw vegetables (odds ratio 8.4, 95% CI 1.5–48.2) which were believed to have been washed in water from a contaminated private well (Peasbody *et al.* 1999).

6.5 CONCLUSIONS

A good surveillance system requires strong epidemiological and laboratory inputs as well as consideration of environmental factors. Outbreak investigation will only be as strong as the weakest link and it is not enough to only make the connection between the host and agent. The ability to identify the environmental antecedents of an outbreak will enable a move to be made towards developing relevant interventions.

Outbreaks point to a failure in the public health system. However, they are an important source of information, especially on contributory factors, which are often inadequately used to inform disease prevention measures. Suggested additional surveillance tools include monitoring issuances of boil-water advisories and keeping track of pharmacy dispensing.

Lessons have been learned as a result of outbreak intervention and new regulations introduced. In the US, the outbreak of cryptosporidiosis in Milwaukee, for example, led to more stringent EPA standards for acceptable turbidity values. These have become effective in all states and may have contributed to the fact that no outbreaks of drinking water associated with *Cryptosporidium* were reported in 1995–6.

6.6 IMPLICATIONS FOR INTERNATIONAL GUIDELINES AND NATIONAL REGULATIONS

Surveillance of infectious illness and good outbreak investigation does not give an exposure assessment but it does provide important insights into risk factors and major public health events, and can usefully inform the international guideline-setting process. Additionally, lessons learned from outbreaks and routine monitoring should help to define priority microbiological hazards on a country by country basis and drive the setting of location specific health targets. Such systems are also likely to play an important role in deciding upon appropriate management techniques and testing management interventions. This is important at both international guideline and national standards level.

6.7 ACKNOWLEDGEMENTS

The data and information presented in 'The Situation in the US' was principally taken from Levy *et al.* 1998.

6.8 REFERENCES

Andersson, Y. (1991) A waterborne disease outbreak. *Water Science and Technology* **24**(2), 13–15.

Andersson, Y. (1992) Outbreaks of waterborne disease in Sweden from a historical, hygienic and technical perspective. Master of Public Health, Nordic School of Public Health, Gothenburg, Sweden. (In Swedish.)

Andersson, Y. and de Jong, B. (1989) An outbreak of giardiasis and amoebiasis at a ski resort in Sweden. *Water Science and Technology* **3**, 143–146.

Andersson, Y., Bast, S., Gustavsson, O., Jonsson, S. and Nillsson, T. (1994) Outbreak of *Campylobacter*. *Epid. Aktuellt* **17**(6), 9. (In Swedish.)

Arvidsson, S.O. (1972) The Swedish epidemics of cholera, a study. Stockholm Diss. (In Swedish.)

Bartram, J.K. (1996) Optimising the monitoring and assessment of rural water supplies. PhD thesis, University of Surrey, UK.

CDC (1990) Waterborne disease outbreaks, 1986–88. *MMWR* **39**(SS–2), 1–13.

CDC (1991) Waterborne disease outbreaks, 1989–90. *MMWR* **40**(SS–3), 1–21.

CDC (1993) Surveillance for waterborne-disease outbreaks – United States, 1991–2. *MMWR* **42**(SS–5), 1–22.

CDC (1996) *Shigella sonnei* outbreak associated with contaminated drinking water - Island Park, Idaho, August 1995. *MMWR* **45**, 229–231.

CDC (1998a) *Plesiomonas shigelloides* and *Salmonella* serotype Hartford infections associated with a contaminated water supply – Livingston County, New York, 1996. *MMWR* **47**, 394–396.

CDC (1998b) Outbreak of cyclosporiasis – Ontario, Canada, May 1998. *MMWR* **47**, 806–809.

Ford, T.E. (1999) Microbiological safety of drinking water: United States and global perspectives. *Environmental Health Perspectives* **107**(S1), 191–206.

Frost, J.A., McEvoy, M.B., Bentley, C.A., Andersson, Y. and Rowe, B. (1995) An outbreak of *Shigella sonnei* infection associated with consumption of iceberg lettuce. *Emerging Infectious Diseases* **1**(1), 26–28.

Hult, A. (1991) Risk factors and controls at premises for drinking water. *SLV Rapport* 1991, 1. (In Swedish.)

Hunter, P.R. (1997) *Waterborne Disease: Epidemiology and Ecology*, Wiley, Chichester, UK.

Kapperud, G., Rorvik, L.M., Hasseltvedt, V., Hoiby, E.A., Iversen, B.G., Staveland, K., Johnsen, G., Leitao, J., Herikstad, H., Andersson, Y., Langeland, G., Gondrosen, B. and Lassen, J. (1995) Outbreak of *Shigella sonnei* infection traced to imported iceberg lettuce. *Journal of Clinical Microbiology* **33**(3), 609–614.

Kramer, M.H., Herwaldt, B.L., Craun, G.F., Calderon, R. and Juranek, D.D. (1996) Surveillance for waterborne-disease outbreaks – United States, 1993–4. *MMWR* **45**, 1–15.

Levy, D.A., Bens, M.S., Craun, G.F., Calderon, R.L. and Herwaldt, B.L. (1998) Surveillance for waterborne-disease outbreaks – United States, 1995–6. *MMWR* **47**(SS-5), 1–34.

Louis, M.E. (1988) Water-related disease outbreaks, 1985. *MMWR* **37**(SS–2), 15–24.

Ljungstrom, I. and Castor, B. (1992) Immune response to *Giardia lamblia* in a water-borne outbreak of giardiasis in Sweden. *Med. Microbil.* **36**, 347–352.

Mead, P.S., Slutsker, L., Dietz, V., McCraig, L.F., Bresee, J.S., Shapiro, C., Griffen, P.M. and Tauxe, R.V. (1999) Food-related illness and death in the United States. *Emerging Infectious Diseases* **5**(5), 607–625.

Mohanty, F.C. (1997) Environmental health risk analysis of drinking water and lead in Hyderabad city, India. PhD thesis, Harvard University, Cambridge, MA.

Peasbody, R.G., Furtado, C., Rojas, A., McCarthy, N., Nylen, G., Ruutu, R., Leino, T., Chalmers, R., deJong, B., Donnelly, M., Fisher, I., Gilham, C., Graverson, L., Cheasty, T., Willshaw, G., Navarro, M., Salmon, R., Leinikki, P., Wall, P. and Bartlett, C. (1999) An international outbreak of vero cytotoxin-producing *Escherichia coli* O157 infection among tourists: a challenge for the European infectious disease surveillance network. *Epidemiology and Infection* **123**, 217–223.

Prescott, S.C. and Winslow, C.E.A. (1931) Elements of Water Bacteriology with Special Reference to Sanitary Water Analysis, 5th edn, Wiley, New York and Chapman & Hall, London.

Schmidt, K. (1995) WHO surveillance programme for control of foodborne infections and intoxications in Europe. Sixth report, 1990–2. BgVV, Berlin, p. 14.

Stenstrom, T.A., Boisen, F., Georgsen, F., Lahti, K., Lund, V., Andersson, Y. and Omerod, K. (1994) Waterborne infections in the Nordic countries. *Tema Nord* **1994**, 585. (In Swedish.)

Thulin, R. (1991) Contamination of tap water in Jonkoping in summer 1991. Rapport Jonkopings community, Sweden. (In Swedish.)

Tulchinsky, T.H., Levine, I., Abrookin, R. and Halperin, R. (1988) Waterborne enteric disease outbreaks in Israel, 1976–85. *Israel Journal of Medical Sciences* **24**, 644–651.

Wahren, H. (1996) A large waterborne disease outbreak in Skane 1995: an evaluation. National Food Administration Rapport 3/96. (In Swedish.)

WHO (1997) Guidelines for Drinking Water Quality. Volume 3: Surveillance and control of community supplies. World Health Organization, Geneva.

WHO (1999) Water and Health in Europe. World Health Organization Regional Office for Europe.

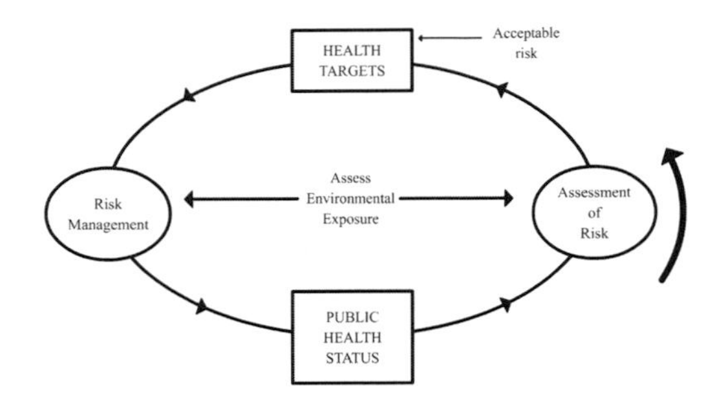

7

Epidemiology: a tool for the assessment of risk

Ursula J. Blumenthal, Jay M. Fleisher, Steve A. Esrey and Anne Peasey

The purpose of this chapter is to introduce and demonstrate the use of a key tool for the assessment of risk. The word epidemiology is derived from Greek and its literal interpretation is 'studies upon people'. A more usual definition, however, is the scientific study of disease patterns among populations in time and space. This chapter introduces some of the techniques used in epidemiological studies and illustrates their uses in the evaluation or setting of microbiological guidelines for recreational water, wastewater reuse and drinking water.

7.1 INTRODUCTION

Modern epidemiological techniques developed largely as a result of outbreak investigations of infectious disease during the nineteenth century.

Environmental epidemiology, however, has a long history dating back to Roman and Greek times when early physicians perceived links between certain environmental features and ill health.

John Snow's study of cholera in London and its relationship to water supply (Snow 1855) is widely considered to be the first epidemiological study (Baker *et al.* 1999). Mapping cases of cholera, Snow was able to establish that cases of illness were clustered in the streets close to the Broad Street pump, with comparatively few cases occurring in the vicinity of other local pumps.

Epidemiological investigations can provide strong evidence linking exposure to the incidence of infection or disease in a population. They can provide estimates of the magnitude of risk related to a particular level of exposure or *dose* and so can be used in the evaluation of appropriate microbiological quality guideline levels or standards. Epidemiological methods can quantify the probability that observed relationships occurred by *chance* factors and they also have the potential to control for other risk factors and/or confounders of the outcome illness being studied. Epidemiological studies used for the evaluation or setting of guidelines must be of high quality, so that there is confidence in the validity of the results.

The following sections outline the basic elements of epidemiological studies (including comments on features that are important for high quality studies), the different types of epidemiological study, and the use of epidemiology in guideline setting, with case studies of the use of epidemiology in recreational water, drinking water and wastewater reuse settings.

7.2 BASIC ELEMENTS OF EPIDEMIOLOGICAL STUDIES

The basic elements of an epidemiological study can be characterised as follows:

- formulation of the study question or hypothesis
- selection of study populations and study samples
- selection of indicators of exposure
- measurement of exposure and disease
- analysis of the relationship between exposure and disease
- evaluation of the role of bias
- evaluation of the role of chance.

These elements will be considered here in a simplified format. Readers are referred to epidemiology textbooks for consideration of the factors in more detail (Beaglehole *et al.* 1993; Friis and Sellers 1996; Hennekens and Buring

1987; Rothman and Greenland 1998). The case studies include examples of the elements described here.

7.2.1 Formulation of the study question or hypothesis

The study question must be formulated so that it can be tested using statistical methods. For example:

- Exposure to wastewater (meeting the WHO guidelines) compared with no exposure to wastewater does not increase the rate of *Ascaris* infection.

The *null hypothesis* (which implies there is no relationship between postulated cause and effect) states that observed differences are due to sampling errors (i.e. to chance). Stated in the null form, the propositions are refutable and can be assessed using statistical tests (see section 7.2.6).

7.2.2 Selection of study populations

A study population exposed (to the factor of interest) and a control population (not exposed to the factor of interest) need to be selected (except in a prospective cohort study where a single cohort is studied and analysis is on exposure status). A sample from the exposed and control populations needs to be selected to be as similar as possible in all factors other than the factor of interest e.g. socio-economic status, and other risk factors for the disease outcome of interest. Since samples are never totally similar, we need to record possible confounding factors and control for them in the analysis (see below). For enteric infections arising from exposure to contaminated water, such factors would include sanitation, personal hygiene, drinking-water supply, food hygiene, and travel. It is important that both exposure and disease can be measured as accurately as possible in the chosen populations. For example, in studies on drinking water, the drinking water source (and therefore the quality) for each household needs to be known accurately. In most studies, a sample will be selected from a larger population exposed to the factor of interest, using a *sampling frame*. This needs to be done so that it is representative of the larger population – difficulties here can arise due to *selection bias* and inadequate sample size (see also sections 7.2.6. and 7.2.7). The choices of study population will depend on the type of epidemiological study selected (see section 7.3).

7.2.3 Selection of indicators of exposure

The quality of the water to which the population is exposed needs to be measured. The use of indicators of contamination are preferred to measurements of pathogenic organisms in the water due to the low numbers of pathogenic organisms present, the difficulties in detecting them and the expense involved (see Chapter 13). Indicators should be selected that are appropriate to the water being studied e.g. thermotolerant coliforms or *E.coli* are used in assessing the quality of drinking water whereas these are less suitable for assessing the quality of coastal recreational waters where enterococci and faecal streptococci are generally preferred. Where the density of an indicator does not accurately reflect the relative density of the underlying pathogenic organism, then it is not a valid indicator organism. This is a particular concern when bacterial indicators are used to indicate the presence of both bacterial and viral pathogens, as treatment methods are often less effective against viruses. This has led to concern about the adequacy of the zero faecal coliform guideline for drinking water quality (Payment *et al.* 1991).

7.2.4 Measurements of exposure and disease status

In the study population measurements of exposure and disease status need to be made while minimising the various types of error that can occur. Where errors occur, this is called *information bias* and results in *misclassification* (see below). For *exposure* to occur, an individual must have contact with water of a given quality. It is preferable to measure exposure at an individual level, but in many studies, exposure status is measured at a group level, which can give rise to misclassification of exposure for the individual. For example, in studies of the effects of aerosol exposure from wastewater irrigation in Israel, exposure status was assigned at the kibbutz level and no differences in individual exposure status were measured. However, the effect of exposure was assessed separately for children and agricultural workers and for the general population, so allowing for some differences in exposure between sub-groups (Fattal *et al.* 1986; Shuval *et al.* 1989). Where the misclassification does not depend on disease status, then this is called *non-differential* misclassification, and the bias would be towards the null, making it more difficult to detect true associations between exposure and disease. This is important in studies assessing the validity of specific microbiological quality guideline levels, as a study may fail to show an effect of exposure to the guideline level whereas a true effect may exist. Recent studies of recreational water exposure and wastewater reuse have put a lot of effort into avoiding misclassification of exposure (see section 7.5). Differential misclassification can either overestimate or underestimate the effect of exposure on disease. One source of misclassification of exposure results from the limited precision of current techniques for the enumeration of indicator organisms (Fleisher and McFadden 1980). This has not been taken into account in most epidemiological and experimental studies of the health impact of contaminated recreational water, drinking water or treated wastewater.

7.2.5 Analysis of the relationship between exposure and disease

The basic measures of disease frequency in each population are described by using the prevalence rate (which is the proportion of the population that has the disease at a specific point in time) or the incidence rate (the number of new cases of disease per unit of person-time). Measuring the difference between disease frequencies in the exposed and control populations is usually done using a relative measure. The relative risk (RR) estimates the magnitude of an association between exposure and disease. It indicates the likelihood of developing the disease in the exposed group relative to those who are not exposed. If the disease is rare the odds ratio will approximate to the relative risk. The odds ratio (OR) is the ratio of the odds of exposure among the cases (numbers exposed divided by numbers not exposed) to the odds in favour of exposure among the controls. Where multivariate analysis is carried out (a technique that allows an assessment of the association between exposure and disease, while taking account of other risk factors that may be confounding factors) the odds ratios is the relative measure normally calculated. In many studies, the effect of different levels or *doses* of exposure will be calculated in order to see if there is a *dose–response* relationship. Response is defined as the proportion of the exposed group that develops a specific effect in comparison to the control group. Such information is very important in the setting of guideline levels where the guideline can be set at the level at which a response first occurs, or can be set at a level that is deemed 'acceptable' (see Chapter 10).

7.2.6 Evaluation of the role of chance

This involves two components. The first is hypothesis testing, or performing a test of statistical significance to determine the probability that chance can explain the observed results. The role of chance is assessed by calculating the P-value – if this is low, it is unlikely that the observed results would have been caused by chance alone, and if it is high, it is more likely that they are due to chance. Although arbitrary in nature, it is usual to choose either 0.05 (5%) or

0.01 (1%) as significance values for testing the null hypothesis. The P-value reflects both the size of the sample and the magnitude of the effect, e.g., P-values can be above the level of significance where the sample is too small to detect a significant effect. The second component is the estimation of the confidence interval. This indicates the range within which the true estimate of effect is likely to lie (with a certain degree of assurance) thus reflecting the precision of the point estimate of effect. This will be calculated for the chosen measure of effect, and is normally presented as, for example, the relative risk and the 95% confidence intervals.

7.2.7 Evaluation of the role of bias

Bias is any systematic error that results in an incorrect estimate of the association between exposure and disease. The main types of bias include *selection bias, information bias, recall bias,* and *confounding*. The case studies (outlined in Section 7.5) give examples of studies where particular attention has been paid to reducing bias.

Selection bias occurs when inclusion of study subjects on the basis of either exposure or disease is somehow related to the disease or exposure being studied. In a recent study of the risks of enteric disease from consumption of vegetables irrigated with partially treated wastewater (Blumenthal *et al.* 1996) problems were faced in determining a suitable control population. This was due to selection bias, as the other strong risk factors for enteric disease were more prevalent in the only nearby area where fresh water was used for irrigation of vegetables. In this case, the exposed population alone was studied, and individuals with low exposure (infrequent consumption of raw vegetables) compared with individual with higher exposure levels: tests were also done for a dose–response relationship.

Information bias occurs when there are systematic differences in the way data on exposure or outcome are obtained from the different study groups. Recall bias occurs when the reporting of disease status is different depending on the exposure status (or vice versa, in a case-control study). There was potential for recall bias in the cross-sectional study of the effect of wastewater reuse on diarrhoeal disease in Mexico (Blumenthal *et al.* 2001a), where individuals exposed to untreated wastewater may have recalled episodes of diarrhoea more accurately than individuals exposed to partially-treated wastewater. Interviewer bias occurs where interviewers are aware of the exposure status of individuals and may probe for answers on disease status differentially between exposure groups. In cohort studies, where individuals leave the study or are otherwise *lost to follow-up*, there can be bias if those lost are different in status to those who remain. These types of bias can generally be dealt with by careful design and conduct of a study.

Confounding occurs when the relationship between the exposure and disease is attributable (partly or wholly) to the effect of another risk factor, i.e. the confounder. It happens when the other risk factor is an independent risk factor for the disease and is also associated with the exposure. It can result in an over- or underestimate of the relationship between exposure and disease. For example, personal hygiene is a potential confounder of the association between drinking water quality and gastro-intestinal disease status. Risk factors that could potentially act as confounders must be measured during the study and controlled for using statistical analysis (e.g. logistic regression analysis can be used to adjust the measure of association between exposure and disease for the effect of the other risks factors). Many epidemiological studies of water-related infections before the mid-1980s did not adequately control for confounding.

7.3 TYPES OF STUDY

Essentially there are three broad types of epidemiological study design:

- descriptive studies
- analytical or observational studies
- experimental or intervention studies.

These will be outlined, in turn, in the following sections.

7.3.1 Descriptive studies

These examine the distribution of disease and possible determinants of disease in a defined population, and can often lead to suggestions of important risk or protective factors. They aim to identify changes in morbidity and/or mortality in time or to compare the incidence or prevalence of disease in different geographical areas or between groups of individuals with different characteristics. Descriptive studies generally use routinely collected health data, such as infectious disease notifications, and are cheap and quick to carry out. A series of descriptive studies of *Ascaris lumbricoides* infection in Jerusalem have shed light on the role of wastewater irrigation of vegetable and salad crops in the transmission of *Ascaris* infection (Shuval *et al.* 1985, 1986). Analysis of stool samples taken in a hospital in western Jerusalem between 1935 and 1947 showed that 35% were positive for *Ascaris* infection, whereas analysis of samples taken between 1949 and 1960 indicated that only 1% were positive –

the decrease was related by the authors to the partitioning of the city and the cessation in the supply of wastewater irrigated vegetables from valleys to the east of Jerusalem. Further descriptive studies indicated that the prevalence of *Ascaris* increased again when the city was reunited and the supply of wastewater-irrigated vegetables reintroduced, and decreased again when wastewater irrigation of vegetables was stopped. Descriptive studies are useful in generating hypotheses about the causes of certain disease patterns, but are not useful for testing hypotheses concerning the effect of particular exposures on particular disease outcomes.

7.3.2 Analytical studies

These are planned investigations designed to test specific hypotheses, and can be categorised into four groups:

- ecological
- cross-sectional studies
- cohort studies
- case-control studies.

7.3.2.1 Ecological (or correlational) studies

These examine associations between exposures and health outcomes using groups of people, rather than individuals, and often use surrogate measures of exposure, e.g. place and time of residence. Such a study would compare an aggregate measure of exposure (such as average exposure or the proportion of the population exposed) with an aggregate measure of health outcome in the same population. They are sometimes included under descriptive studies (e.g. in the US). In Thailand, for example, the seasonal variation in the reported incidence of acute diarrhoea in selected areas was examined in relation to rainfall and temperature records for the same areas (Pinfold *et al.* 1995). The authors found that the incidence of diarrhoea appeared to be inversely related to a sharp seasonal decrease in temperature. Rainfall did not appear to have a direct effect on the relative incidence of acute diarrhoea. The lack of ability to link individual exposure to individual disease risk and to control for possible confounders are major disadvantages of this approach and severely limit its usefulness in many settings, especially where the exposure changes over time and space and where there are many risk factors for the disease outcome of interest.

7.3.2.2 Cross-sectional studies

In a cross-sectional study exposure and health status are ascertained simultaneously on one occasion, and prevalence rates (or incidence over a limited recent time) in groups varying in exposure are compared. Careful measurement and statistical control of confounding variables is important to assess the effect of other risk factors for the outcome on observed prevalence. This approach has been used to assess the effects of wastewater reuse for irrigation. In India, the prevalence of intestinal parasitic infections was assessed in agricultural workers working on farms which were flood-irrigated with wastewater and compared with a control population where agricultural workers practised irrigation with clean water (Krishnamoorthi *et al.* 1973 cited in Shuval *et al.* 1986). Stool samples were examined for *Ancylostoma duodenale* (hookworm), *Ascaris lumbricoides* (roundworm) and *Trichuris trichiura* (whipworm). The exposed population had at least a two-fold excess of hookworm and *Ascaris* infection as compared to the control population. The usefulness of this study and other past cross-sectional studies has been limited by its failure to control for confounding variables and to document the type and extent of exposure of potentially exposed persons (Blum and Feachem 1985). A cross-sectional study can only provide information on the association between an exposure and disease, and the temporal relationship between exposure and disease cannot be established. Other problems include the need for large sample sizes (for infections where prevalence is low), and potential bias due to exposure and disease misclassification. However, the advantages are that such studies are relatively cheap and can provide meaningful results where exposure and confounding factors are measured carefully.

7.3.2.3 Cohort studies

In a cohort study the population under investigation consists of individuals who are at risk of developing a specific disease or health outcome. These individuals will then be observed for a period of time in order to measure the frequency of occurrence of the disease among those exposed to the suspected causal agent as compared to those not exposed. This type of approach has been used to examine the health effects of recreational water use (Balarajan *et al.* 1991; Cabelli *et al.* 1983). Typically, individuals are recruited immediately before or after participation in some form of recreational water exposure, with controls drawn from a population at the same location not participating in the water-based activity. During the follow-up period, data are acquired on the symptoms experienced by the two cohorts using questionnaire interviews. The quality of the recreational water is defined through sampling on the day of exposure. The exposure data are often combined to produce a daily mean value for the full group of bathers using a particular water on any one day. The problem with this approach is that the aggregation of exposure and subsequent assignment of the same exposure to many people produces a large degree of non-differential misclassification bias, which biases the measure of association. Cohort studies are useful for the study of relatively common outcomes and for the study of relatively rare exposures e.g. risks from occupational exposure to wastewater (Shuval *et al.* 1989). Careful classification of exposures and outcomes is needed, as is the measurement and control for confounding factors. The disadvantages are that the studies are often complex and difficult to manage, the time span is often at least a year (to take into account seasonality of disease incidence) and the studies can therefore be expensive. A wastewater reuse cohort study is outlined in Section 7.5.2.

7.3.2.4 Case-control studies

Case-control studies examine the association between exposure and a health outcome by comparing individuals already ill with the disease of interest (i.e. cases) and a control group who are a sample of the same population from which the cases were identified. Gorter *et al.* (1991) used a case-control study design to examine the effects of water supply and sanitation on diarrhoeal disease in Nicaragua. They compared over 1200 children with diarrhoea with a similar number of controls (children of a similar age with illnesses other than diarrhoea). They found a statistically significant association between water availability and diarrhoea morbidity. Children from homes with water supplies over 500 metres from the house had incidence rates of diarrhoea 34% higher than those of children from houses with their own water supply. This relationship remained significant after controlling for confounding factors. The advantages of case-control studies are that they require smaller sample sizes, fewer resources, require less time and less money, and sometimes are the only way to study rare diseases. The difficulties are in appropriate study design to minimise bias, including the selection of appropriate controls and the control of confounding variables and minimising recall bias. Regarding wastewater reuse and recreational water reuse, the potential for misclassification of exposure is higher within a case-control design than in other types of study due to recall bias. They are therefore of less value than other designs in evaluating risks related to exposure to water of varying qualities.

7.3.3 Experimental or intervention studies

These differ from the observational techniques outlined above in that the investigators determine who will be exposed. A key part of the experimental design consists of randomising a single cohort into two groups. The process of randomisation attempts to ensure the same distribution of various intra-individual traits and potential confounders between study groups so that they are as comparable as possible. One group is then assigned to exposure to the factor under study; the other group is the control and the health outcomes for the groups are compared. Randomisation of subjects is important to minimise the potential for confounding or selection bias. In terms of determining causality this type of study is generally considered to be the most powerful. It is equivalent to the randomised controlled trial used in testing the impact of drugs and other medical interventions. Its use in examining environmental exposures has been limited because of ethical concerns, since many exposures of interest are potentially detrimental. A notable exception is provided by the first case study in this chapter (section 7.5.1), which presents the study design and results of four randomised trials assessing the risk of bathing in marine waters contaminated with domestic sewage (Fleisher *et al.* 1996; Kay *et al.* 1994). In the third case study (in section 7.5.3), intervention trials are described which have recently been used in evaluating the current guidelines for drinking water quality. These have compared persons drinking ordinary tap water with those drinking water that has been 'treated' in the home, using reverse-osmosis filters or UV light (Hellard *et al.* 2000; Payment *et al.* 1991). This type of design is not applicable in the study of wastewater treatment and reuse where the intervention is at a community not an individual level, and it is not possible to assign wastewater treatment plants randomly to a number of different communities (due to costs and practical issues).

7.4 USES OF EPIDEMIOLOGY IN THE SETTING OR EVALUATING OF MICROBIOLOGICAL GUIDELINES

There are several different approaches that can be taken to the use of epidemiological studies in the setting or evaluation of microbiological guidelines for drinking water, recreational water or wastewater:

- Measure the relationship between exposure and disease for a range of levels of indicator organisms to get a dose–response curve. Set an acceptable level of risk and then find the microbiological level related to that level of risk (using the dose–response curve). This method has been used for proposing recreational water guidelines (see section 7.5.1 and Chapter 2).

- Measure the relationship between exposure and disease for water at the current guideline level, and possibly for water above or below the guideline level. Examples of this approach can be provided by both drinking water and wastewater reuse studies. The studies in the drinking-water case study (section 7.5.3) assessed the relationship between exposure and disease for water that met the current drinking-water guideline limits. The studies outlined in the wastewater case study section (section 7.5.2) assessed the relationship between exposure and disease for wastewater meeting the WHO guideline levels (WHO 1989).
- Use the results of several studies where the relationship between exposure and disease has been assessed for water of different qualities, and estimate the level at which no effect would be found. This method was used informally to propose a new faecal coliform guideline to protect agricultural workers involved in wastewater reuse (Blumenthal *et al.* 2000b). Ideally a meta-analysis, such as that conducted by Esrey *et al.* (1985, 1991) would be conducted to combine the results of several studies.

7.5 CASE STUDIES

Three case studies, using different approaches and epidemiological methods, are outlined in the following sections. The recreational water studies have been used to inform standards development, while the wastewater reuse and drinking-water studies are likely to inform future development.

7.5.1 Recreational water case study

Four separate study locations around England and Wales (UK) were used (Fleisher *et al.* 1996; Kay *et al.* 1994). The study locations were sufficiently distant from one another so that site-specific differences in the risk of bathing-associated illness could be assessed. All the study locations met European Community (EC) mandatory bacteriological marine bathing-water quality criteria as well as US EPA bathing-water criteria for marine waters. A randomised controlled trial design was used in order to minimise selection bias and control for intra-individual differences in susceptibility, immune status and so on between study groups. Equally importantly, the risk of non-differential misclassification of exposure was minimised by assigning precise measures of exposure to each individual bather (studies by Cabelli *et al.* (1993) were seriously affected by bias of this type). Healthy volunteers aged 18 or over were randomised into two groups:

- an exposed group where volunteers actually entered the water, and
- an unexposed group where volunteers spent an equal amount of time on the beach but did not enter the water.

All volunteers were blinded to the specific outcome illnesses being studied in order to control for or minimise bias in the reporting of symptoms. Volunteers also did not know which group they would be assigned to until the day of the trial.

Since the mix of underlying pathogens that could possibly be present in the bathing waters remained unknown, five indicator organisms or groups of organisms were used to assess exposure among the bather group:

- total coliforms
- faecal coliforms
- faecal streptococci
- total staphylococci
- *Pseudomonas aeruginosa*

This was done to maximise the chance of finding an indicator organism that directly correlated with the underlying pathogen or pathogens, thus reducing misclassification of exposure.

Duration and precise location of individual bather exposure was rigorously controlled. This is important because of the large spatial and temporal variations in concentration of indicator organisms that are seen at bathing water locations caused by environmental factors. Indicator organism concentration was measured every 30 minutes. Exposure was assigned to each individual bather within 15 minutes of the actual exposure and within a maximum of 10 metres of the actual point of exposure. These measures minimised misclassification of exposure among bathers.

All five indicator organisms used were assessed using the Membrane Filtration method of enumeration. In addition, three replicate determinations were made on each sample taken. Using the most precise method of indicator organism enumeration, coupled with taking three replicate determinations per sample, maximised the precision of each estimate and minimised the bias due to lack of precision in estimation.

In order to control for competing risk factors and/or confounders for the outcome illnesses under study, four separate interviews were held with each study participant. These interviews were conducted two to three days prior to each trial, on the trial day, at seven days post-trial, and at three weeks post-trial. In this manner, information about exposure to competing non-water-related risk factors and/or confounders was recorded for each participant prior to the trial, at the time of the trial, and upon completion of the trial (allowing for a suitable incubation period). These exposures to non-water-related risk factors were then controlled for in the analysis.

The outcome illnesses used were gastroenteritis, acute febrile respiratory illness, and skin, ear, and eye infection. All study participants reporting symptoms of any of these five outcome illnesses during the pre-trial interview or at the interview conducted on the actual trial day were excluded from the study. The same interview was used 7 and 21 days post-trial. Since gastroenteritis is often used as the 'index' illness for assessing waterborne illness, the results presented here are for gastroenteritis. Table 7.1 shows a partial list of the confounders or competing risk factors that were recorded.

Table 7.1. Non-exposure-related risk factors for gastroenteritis

Non-exposure related risk factor
Age – grouped by 10-year intervals
Gender
History of migraine headaches
History of stress or anxiety
Frequency of diarrhoea (often, sometimes, rarely or never)
Current use of prescription drugs
Illness within 4 weeks prior to the trial day (lasting more than 24 hours)
Use of prescription drugs within 4 weeks prior to the trial day
Consumption of the following foods in the period from 3 days prior to 7 days after the trial day:
mayonnaise
purchased sandwiches
chicken
eggs
hamburgers
hot dogs
raw milk
cold meat pies
seafood
Illness in the household within 3 weeks after the trial day
Alcohol consumption within the 7 day period after the trial
Frequency of usual alcohol consumption
Taking of laxatives within 4 weeks of the trial day
Taking of other stomach remedies within 4 weeks of the trial day
Additional bathing within 3 days prior and 3 weeks after the trial day (this was included in order to control for possible confounding due to multiple exposures among bathers and exposure among non-bathers prior to or after the trial day)

Faecal streptococci (FS) was the only indicator organism that predicted gastroenteritis among bathers. Crude rates of illness among bathers versus non-bathers were 14.8% versus 9.7% ($P = 0.01$). Crude rates do not, however, reflect the effects of variation in exposure to differences in indicator organism densities among individual bathers, and should be viewed with caution. Faecal streptococci densities ranged from 0–158 per 100 ml of water. Therefore, the crude difference in rates dampens out this variability in exposure of individual bathers to differing levels of sewage (and thus risk). However, the rates of illness among those exposed to the highest quartile of exposure (50–158 FS) shows the rates of illness to be 24.6% among bathers versus 9.7% for non-bathers. The stratification of rates of illness over increasing levels of indicator organism exposure is an important feature of the analysis. This becomes especially important in the construction of mathematical models used to quantify microbial risk. Using crude rates of illness would invariably lead to an underestimate of risk produced by the model, and possibly question the validity of the model itself.

Using multiple logistic regression modelling, a dose–response curve was produced relating the probability of a bather acquiring gastroenteritis relative to individual bather faecal streptococci exposure while adjusting for the non-water-related risk factors and/or confounders. Using this technique, the probability of competing risk factors for the same illness can be quantified. Such information on competing risk factors can be important in setting water quality criteria.

The results of the randomised trials discussed above are based on a total sample size of only 1216 participants. This illustrates that the use of an appropriate epidemiological study design (randomised trial) can yield extremely informative and precise information regarding quantitative microbiological risk assessment without the need for large sample sizes. In addition, randomised trials can be conducted at multiple sites over wide geographical areas within a region while assessing for any site-specific differences. Such an epidemiological design contains no assumptions, relies solely on data collected during the study, and yields more valid and precise estimates of risk than mathematical risk assessment models.

The implications of the studies for the setting of microbiological guidelines for recreational water are discussed in Chapter 2.

7.5.2 Wastewater reuse case study

A series of epidemiological studies were conducted in Mexico to assess the occupational and recreational risks associated with exposure to wastewater of different microbiological qualities. Observational study methods were used to assess the risks associated with existing practices, as there was no possibility of introducing a wastewater treatment facility and assessing its impact on health through an intervention study or randomised trial. Infections (from helminths, protozoa and diarrhoeal disease) in persons from farming families in direct contact with effluent from storage reservoirs or raw wastewater were compared with infections in a control group of farming families engaged in rain-fed agriculture (Blumenthal *et al.* 1996; Blumenthal *et al.* 2001a; Cifuentes 1998; Peasey 2000). The storage reservoirs fulfilled a 'partial treatment' function and produced water of differing microbiological qualities. The effects of wastewater exposure were assessed after adjustment for many other potential confounding factors (including socio-economic factors, water supply, sanitation and hygiene practices).

Raw wastewater coming from Mexico City to the Mezquital valley, Hidalgo, is used to irrigate a restricted range of crops, mainly cereal and fodder crops, through flood irrigation techniques. Some of the wastewater passes through storage reservoirs and the quality of the wastewater is improved before use. The effluent from the first reservoir (retention time 1–7 months, depending on the time of year) met the WHO guidelines for restricted irrigation (Category B, ≤ 1 nematode eggs/litre), even though a small amount of raw wastewater enters the effluent prior to irrigation. Some effluent from the first reservoir passes into a second reservoir where it is retained for an additional 2–6 months, and the quality improved further. Local farming populations are exposed to the wastewater and effluent through activities associated with irrigation, domestic use (for cleaning, not for drinking) and play.

The untreated wastewater contained a high concentration of faecal coliforms (10^6–10^8/100ml) and nematode eggs (90–135 eggs/l). Retention in a single reservoir reduced the number of nematode eggs substantially, to a mean of <1 eggs/l whereas faecal coliform levels were reduced to 10^5/100 ml (average over the irrigation period) or 10^4/100ml, with annual variations depending on factors such as rainfall. The concentration of nematode eggs remained below 1 egg/l (monthly monitoring) even after a small amount of raw wastewater entered the effluent downstream of the reservoir. Retention in the second reservoir further reduced the faecal coliform concentration (mean 4×10^3/100ml) and no nematode eggs were detected. Faecal coliform levels varied over the year depending on the retention time in each reservoir, which varied according to demand for irrigation water. Three studies were carried out in this study area. The first used a cross-sectional methodology to study the prevalence of a range of parasitic infections and diarrhoeal disease (and included two surveys); the second used a prospective cohort methodology to study the intensity of *Ascaris lumbricoides* infection; and the third used a cross-sectional methodology to study prevalence of diarrhoeal disease. Use of a cross-sectional methodology was recommended by Blum and Feachem (1985) as a cost-effective way to study the association between wastewater exposure and a range of infections.

In the first study (Blumenthal *et al.* 2001a; Cifuentes 1995, 1998) a census was conducted to locate households where one or more persons were actively involved in agriculture. Exposure groups included agricultural households using untreated wastewater for irrigation, households using effluent from a reservoir and households practising rain-fed agriculture (control group). In the first cross-sectional study (rainy season), the reservoir group was exposed to wastewater retained in two reservoirs in series and in the second survey, the reservoir group was exposed to wastewater retained in the single reservoir. Measures were taken to reduce the misclassification of exposure. Data on the siting of agricultural plot(s) worked by the farming families, the irrigation canals feeding them and the source (and therefore quality) of water in the canals was used in an algorithm to define the exposure status of the farming family (Cifuentes 1995). Inclusion criteria for households were: location in an agricultural community, one or more adults with tenure of a farm plot and occupational contact with wastewater of a defined quality (raw wastewater, effluent from the reservoir) or farming of a rain-fed plot (control group). Farmers were excluded if they had contact with an unknown or unclassified source of irrigation water, if they had plots in more than one area or contact with more than one type of water, and if they lived in the control area but had contact with wastewater. Members of every household were assigned to the same exposure category as the members working on the land, to allow for intra-familial transmission of infection. Information was collected on the agricultural profile of every household (i.e. location of farming plot, type of irrigation water used, cultivated crops), whether and when the person had contact with wastewater, and on other risk factors that were potential confounders. Socio-economic variables collected included land tenure, maternal literacy, house roof material, number of bedrooms and number of chickens eaten per week. Hygiene- and sanitation-related characteristics included excreta disposal facility, source of drinking water, storage and boiling of drinking water, hand washing, hygienic appearance of respondent, rubbish disposal facilities, animal excreta in the yard and local source of vegetables. Exposure to wastewater was defined as having direct contact ('getting wet') with wastewater (or reservoir water) in a particular time period. Recent exposure (in the last month) was related to diarrhoeal disease and past exposure (from 1–12 months previously) was related to *Ascaris* infection. A diarrhoeal disease episode was defined as the occurrence of three or more loose stools passed in a 24-hour period and the recall period was two weeks. The prevalence of specific intestinal parasite infections was assessed by means of microscopic identification of the presence of ova or cysts in stool samples. The results for *Giardia intestinalis* and *Entamoeba histolytica* were reported separately (Cifuentes *et al.* 1993, 2000).

In the analysis, the estimates of the effect of exposure to wastewater and reservoir water were adjusted for the effects of all other variables that were potential confounders. The main results that have implications for guidelines setting are summarised in Table 7.2. Exposure to effluent from one reservoir (meeting WHO guideline level of ≤1 nematode egg per litre) was strongly associated with an increased risk of *Ascaris* infection in young children and in those over five years of age, when compared to the control group. Exposure to effluent from two reservoirs (where the quality was further improved) was not associated with an increased risk of *Ascaris* infection in young children, whereas a small risk remained for those over five years of age. Exposure to effluent from one reservoir was associated with increased diarrhoeal disease in those over five years of age (compared to the control group), whereas exposure to effluent from two reservoirs was not. The later result is not conclusive, however, since the effect of exposure to effluent from two reservoirs was only assessed in the rainy season. In the dry season the effect may be greater, as the effect of exposure to untreated wastewater was both stronger and more significant in the dry season in both age groups (compared to the control group).

Table 7.2. Effect of exposure to untreated wastewater and degree of storage of wastewater (Cifuentes 1998; Blumenthal *et al.* 2001a)

	Ascaris infection OR* (95% CI)	Diarrhoeal disease OR* (95% CI)
Effect of exposure to untreated wastewater		
0–4 years		
Dry season	18.01 (4.10–79.16)	1.75 (1.10–2.78)
Rainy season	5.71 (2.44–13.36)	1.33 (0.96–1.85)
5+ years		
Dry season	13.49 (6.35–28.63)	1.34 (1.00–1.78)
Rainy season	13.49 (7.51–23.12)	1.10 (0.88–1.38)
Effect of exposure to stored wastewater (by degree of storage)		
0–4 years		
One reservoir, dry season	21.22 (5.06–88.93)	1.13 (0.70–1.83)
Two reservoirs, rainy season	1.29 (0.49–3.39)	1.17 (0.85–1.60)
5+ years		
One reservoir, dry season	9.42 (4.45–19.94)	1.50 (1.15–1.96)
Two reservoirs, rainy season	1.94 (1.01–3.71)	1.06(0.86–1.29)

* All ORs (Odds ratios) use the control group as the reference.

The prospective cohort study of the effect of exposure to partially-treated wastewater on *Ascaris* infection was done in the same area (Peasey 2000). The study groups were the same as for the dry season study and the sample was selected from the census as outlined above. The inclusion criteria for households were: the head of the household was a farmer, male, at least 15 years old and had contact with only one quality of irrigation water, i.e. only rain-fed or only untreated wastewater. The inclusion criteria for individuals within each selected household were: at least two years old, resident in the house at least five days a week and any wastewater contact was with the same quality of wastewater as the head of the household. A baseline survey was done where the prevalence and intensity of *Ascaris* infection (as measured by the egg count) was measured on full stool samples. Subjects with *Ascaris* infection were given chemotherapy to expel the adult worms, such that the egg counts were reduced to zero. A follow-up survey was done 12 months later, and the prevalence and the intensity of reinfection after treatment measured. This design provided a more sensitive measure of prevalence of infection than the cross-sectional surveys above, as well as a measure of intensity of reinfection over a specific time period, thus reducing any misclassification of disease. Each individual was assigned a personal exposure status according to their activities involving direct contact wastewater and the frequency of that contact. This time-method further improved the classification of exposure and infection with *Ascaris* in comparison with the cross-sectional studies, and provides a more valid measure of infection related to exposure over a specific time period. Data was collected on other risk factors for *Ascaris* infection and the estimates of the effect of exposure on infection adjusted for potential confounding factors.

The main results can be summarised as follows. Contact with effluent from one reservoir was associated with an increase in prevalence of *Ascaris* infection among adults and children when compared with the control group. Multivariate analysis was done using internal comparison groups and not the external control group, since numbers of positives in the external control group were very small (due to the low prevalence of infection in the external control group and the small sample size) and a multivariate model would have been very unstable if this group had been used as a baseline. Contact with effluent from one reservoir through playing was associated with an increase in prevalence of *Ascaris* infection in children under 15 years of age, compared with those who lived in a wastewater-irrigated area but did not have contact with wastewater during play (OR = 2.61, 95% CI: 1.10–6.15). Contact with effluent from one reservoir for irrigation was not associated with a significant increase in *Ascaris* infection in children under 15 years of age when compared with children from the same area who did not irrigate. For adult men, wastewater contact during work related to chilli production was associated with an increased prevalence of *Ascaris* infection in those exposed to untreated wastewater (OR = 5.37, 95% CI: 1.79–16.10) but not in those exposed to effluent from one reservoir (OR = 1.56, 95%

CI: 0.13–18.59) when compared with adult men living in wastewater-irrigated areas who did not cultivate chilli. For adult women, contact with untreated wastewater through tending livestock in wastewater-irrigated fields was associated with increased prevalence of *Ascaris* infection (OR = 4.39, 95% CI: 1.08–17.81) but contact with effluent from one reservoir was not (OR = 0.70, 95% CI: 0.06–8.33) when compared with adult women living in wastewater-irrigated areas who did not tend livestock or who had no wastewater contact while tending livestock.

The third study was carried out mainly to assess the effect of consumption of vegetables, irrigated with partially treated wastewater, on a range of enteric infections. Infections included symptomatic diarrhoeal disease, enterotoxigenic *E. coli* infection and infection with human Norwalk-like virus (Blumenthal *et al.* 2001b). However, since a section of the study population was involved in agricultural work and were in direct contact with effluent from the second reservoir it was possible to estimate the effect of direct contact (as well as to adjust the estimate of the effect of consumption for the effect of direct contact). The effect of exposure on diarrhoeal disease was assessed through two cross-sectional surveys, in the rainy and dry seasons. The design of the surveys was similar to that used in the previous cross-sectional surveys except in two aspects where the design and analysis was improved: measures of individual exposure to effluent from the second reservoir were used (instead of the exposure of the adult male farmer) and the comparison group was individuals of the same age in the same area but who did not have contact with effluent from the reservoir (whereas earlier the comparison group was a control group from a rain-fed area). When children with contact with the effluent from the second reservoir were compared to children from the same population but with no contact with the effluent, a two-fold or greater increase in diarrhoeal disease in children aged 5–14 years was found (OR = 2.34, 95% CI: 1.20–4.57 dry season). In the first study it was found that there was no excess of diarrhoeal disease related to exposure with this water compared to the level in the control group, where rain-fed agriculture was practised (Cifuentes 1998).

Taken together, the results show that contact with wastewater retained in one reservoir and meeting WHO guidelines for restricted irrigation was associated with an increased risk of *Ascaris* infection (especially in children, in contact through play), and an increased risk of diarrhoeal disease (especially in the dry season). When the quality of the water was improved through retention in two reservoirs in series (10^3–10^4 faecal coliforms/100ml and no detectable nematode eggs), the risk of *Ascaris* infection to children was decreased, but there was still an increased risk of diarrhoeal disease to exposed children compared with those not in contact with effluent. These results indicate that the nematode egg guideline of ≤1 nematode egg per litre is adequate for the protection of farm workers but inadequate where children have contact with the wastewater (especially through play). A faecal coliform guideline for the protection of farming families is also needed. The implications of these results, and those from other studies, for modification of the 1989 WHO guidelines are discussed further elsewhere (Blumenthal *et al.* 2000a,b).

7.3.3 Drinking-water case study

In studies of drinking water, randomised control trials of interventions have been used to explore whether there is a risk of gastrointestinal (GI) disease due to consumption of drinking water meeting current microbiological standards. Payment *et al.* (1991) used a randomised controlled trial to investigate whether excess gastroenteritis was being caused by potable water supplies (outlined in greater detail in Chapter 4). The suburban area of Montreal, Canada, chosen for the study, is served by a single water treatment plant, using pre-disinfection flocculation by alum, rapid sand filtration, ozonation and final disinfection by chlorine or chlorine dioxide. The raw water was drawn from a river, which was contaminated with human sewage discharges. The study design consisted of the randomised installation of reverse-osmosis filters in study participants' households. Therefore, two groups were formed: those households with filters (control group), and those households using plain tap water. GI symptomatology was evaluated by means of a family diary of symptoms. The study lasted 15 months. The results of this study estimated the annual incidence of GI illness among tap-water drinkers to be 0.76 versus 0.50 among filtered water drinkers ($P<0.01$). In addition, the results of this study estimated that 35% of the total reported gastroenteritis among tap-water drinkers was water-related, and thus preventable. Payment *et al.* (1997) conducted a second study a few years later, altering the exposed and control groups. In this second study, two groups (tap-water group and tap-valve water group) received normal tap water through kitchen taps; the only difference between these groups was that the tap-valve water group had a valve fitted to their house to control for stagnation of water in their household plumbing. Two additional groups received bottled finished water from the plant (plant water group and purified water group) that was bottled before it entered the distribution system. The water for the purified water group was passed through a reverse-osmosis filter before it was bottled. Again, illness was assessed using a household diary. Using the purified water group as the baseline, the excess of gastrointestinal illness associated with tap water was 14% higher in the tap group and 19% higher in the tap-valve group. Children ages two to five were the most affected, with an excess of 17% in the tap-water group and 40% in the tap-valve group. Payment *et al.* concluded that their data suggest that 14–40% of the observed gastrointestinal illnesses were attributable to tap water meeting current standards, and that the water distribution system appears to be partially responsible for these illnesses. However, these studies have been criticised for failing to blind study subjects to their exposure status: those with filters knew they had filters and may have been less likely to report GI symptoms than those without filters, so biasing the results. Currently, the US Centers for Disease Control and Prevention (CDC) have started two large-scale studies of illness transmission through treated tap water to address some of the criticism of the Canadian studies.

A recent study conducted in Melbourne, Australia, is also contributing to the debate on the validity of current microbiological standards for drinking water (Hellard *et al.* 2000). The study was set up to explore whether tap water in Melbourne that was chlorinated but not filtered was associated with an increase in community gastroenteritis. Melbourne's raw water comes from large reservoirs in an unpopulated forested catchment area (markedly different from that used in the Canadian studies). A randomised double-blind controlled trial was set up. Participants in one group were given a functioning water treatment unit in the home (consisting of a filter to remove protozoa and an ultraviolet (UV) light unit to kill viruses and bacteria) while the 'tap water' group were given a mock water treatment unit, which looked identical to the functioning water treatment unit but did not alter the water. The participants were therefore 'blinded' to their exposure status. The characteristics of the two groups were the same at randomisation. Families in the study completed weekly health diaries and faecal specimens were taken when an episode of diarrhoeal disease was reported. Gastroenteritis was defined by a combination of symptoms similar to the Canadian studies, and the subject had to be symptom-free for six days before a new episode was registered. Loss to follow-up (41/600 families) was lower than in the Canadian studies. The results showed that the rate of gastroenteritis was almost the same in both groups (0.79 versus 0.82 episodes/person/year; RR = 0.99, 95% CI: 0.85–1.15). This was the case even though the tap water failed to meet the 1996 Australian Drinking Water Guidelines for water quality in terms of total coliform detection (total coliforms were present in 19% of samples, rather than <5% samples as recommended in the guidelines). The lack of an effect on community gastroenteritis of drinking this water may have been due to the cleaner catchment and better source water protection. However, it may be related to the superior epidemiological study design, using a randomised double-blinded design (with real and mock water treatment units), which may have eliminated any reporting bias present in earlier studies.

7.4 DISCUSSION

Epidemiological methods have the ability to estimate risk with a good degree of precision, but also, and perhaps just as important, have the ability to control for other risk factors and/or confounders of the outcome illness being studied. As outlined in Chapter 5, most gastrointestinal illnesses such as those related to drinking water, recreational water and wastewater reuse can be spread by more than one route. Epidemiological study is the only method that can utilise real data to separate the risk of the illness caused by the contaminated water from other risk factors for the outcome illness. Without such control, risk can be substantially overestimated.

Well designed and conducted epidemiological studies can also minimise the many biases that may occur. Experimental or intervention studies can provide the most accurate results, having minimised the potential for selection bias and confounding, but may not be suitable in some cases due to ethical or cost considerations and where subjects cannot be blinded to exposure/intervention status. Prospective cohort studies are the next best option, where the exposure precedes the disease outcome and attention is paid to selection bias and potential confounders are measured and controlled for in the analysis. Where cost, logistical or other considerations preclude the use of such studies, cross-sectional studies can provide useful results where attention is paid to measuring exposure and disease accurately and allowing for potential confounding factors (Blum and Feachem 1985). Case-control studies are not so useful in evaluating microbiological guidelines, due to recall bias in the measurement of exposure, and retrospective cohort studies are not recommended where there is bias in the measurement of exposure or disease. In the selected study types, where adequate sample sizes are used, the risk of illness related to a specific exposure can be calculated with a good degree of precision. It is clearly important that the highest quality studies are used for the setting of water-related guidelines as these can result in considerable outlay by governments and water industry.

The limitations of epidemiological studies have been thought to lie in the need for unrealistically large sample sizes to uncover very small increases in risk, and in the costs incurred and expertise needed to mount a good study. However, the case study examples show that epidemiological studies can be designed and carried out in such a way as to provide very valuable information on the validity of current guidelines and for recommending new guidelines. The sample size requirements are not unreasonable, especially if cohort studies or experimental studies are carried out. Given the cost of complying with more restrictive standards, a case can anyway be made for significant expenditure on an epidemiological study, especially if there is the chance that this will indicate that more restrictive standards are not needed.

Epidemiological studies can assess the effect of 'real' exposures and can measure the effect on more vulnerable groups (e.g. young children) as well as adults. The effect of related exposures can also be taken into account, for example children playing with wastewater as well as being exposed to it through agricultural work.

7.5 IMPLICATIONS FOR INTERNATIONAL GUIDELINES AND NATIONAL REGULATIONS

Epidemiological studies have been used in setting the guidelines for wastewater reuse (WHO 1989), and in proposing the draft guidelines for safe recreational water environments (WHO 1998) as outlined in Chapter 2. However, different approaches have been taken both in the use made of the epidemiological studies (as outlined above) and in the level of risk that was considered acceptable. In the case of wastewater reuse, evidence from a range of studies was taken into account and a guideline level proposed that was estimated to result in no measurable excess infection in the exposed population. In the case of recreational water use, an acceptable level of risk was set, and the microbiological level related to that level of risk was found, using the dose–response curve produced by the best epidemiological study available linking microbial concentrations with gastroenteritis. It seems possible, therefore, that the wastewater guidelines protect against a lower level of risk than the proposed recreational water guidelines. In contrast, the drinking water guidelines are based on '*tried and tested principles of prevention of faecal pollution and good engineering practice*' (Chapter 2). Now that more epidemiological studies of drinking water are available (see Chapter 4), it is essential that all available epidemiological evidence is taken into account in the setting of future guidelines.

7.6 REFERENCES

Baker, D., Kjellstrom, T., Calderon, R. and Pastides, H. (eds) (1999) *Environmental Epidemiology. A textbook on study methods and public health application*. World Health Organization, Geneva.

Balarajan, R., Soni Raleigh, V., Yuen, P., Wheeler, D., Machin, D. and Cartwright, R. (1991) Health risks associated with bathing in sea water. *British Medical Journal* **303**, 1444–1445.

Beaglehole, R., Bonita, R. and Kjellstrom, T. (1993) *Basic Epidemiology*, World Health Organization, Geneva.

Blum, D. and Feachem, R.G. (1985) *Health aspects of nightsoil and sludge use in agriculture and aquaculture. Part III: An epidemiological perspective*. Report No. 05/85, International Reference Centre for Waste Disposal (IRCWD), Dubendorf.

Blumenthal, U.J., Mara, D.D., Ayres, R., Cifuentes, E., Peasey, A., Stott, R. and Lee, D. (1996) Evaluation of the WHO nematode egg guidelines for restricted and unrestricted irrigation. *Water Science and Technology*, **33**(10–11), 277–283.

Blumenthal, U.J., Peasey, A., Ruiz-Palacios, G. and Mara, D.D. (2000a) Guidelines for wastewater reuse in agriculture and aquaculture: recommended revisions based on new research evidence. WELL Resource Centre, London School of Hygiene and Tropical Medicine and WEDC, Loughborough University, UK (WELL Study No. 68 Part I).

Blumenthal, U.J., Mara, D.D., Peasey, A., Ruiz-Palacios, G. and Stott, R. (2000b) Approaches to establishing microbiological quality guidelines for treated wastewater use in agriculture: recommendations for revision of the current WHO guidelines. *Bulletin of the World Health Organization* **78**(9), 1104–1116.

Blumenthal, U.J., Cifuentes, E., Bennett, S., Quigley, M. and Ruiz-Palacios, G. (2001a) The risk of enteric infections associated with wastewater reuse: the effect of season and degree of storage of wastewater. *Transactions of the Royal Society of Tropical Medicine and Hygiene* (in press).

Blumenthal, U.J., Peasey, A., Quigley, M. and Ruiz-Palacios, G. (2001b) Risk of enteric infections through consumption of vegetables irrigated with contaminated river water. *American Journal of Tropical Medicine and Hygiene* (submitted).

Cabelli, V.J., Dufour, A.P., McCabe, L.J. and Levin, M.A. (1983) A marine recreational water quality criterion consistent with indicator concepts and risk analysis. *Journal of the Water Pollution Control Federation* **55**(10), 1306–1314.

Cifuentes, E. (1995) Impact of wastewater irrigation on intestinal infections in a farming population in Mexico: the Mezquital valley. PhD thesis, University of London.

Cifuentes, E. (1998) The epidemiology of enteric infections in agricultural communities exposed to wastewater irrigation: perspectives for risk control. *International Journal of Environmental Health Research* **8**, 203–213.

Cifuentes, E., Blumenthal U., Ruiz-Palacios, G., Bennett, S., Quigley, M., Peasey, A. and Romero-Alvarez, H. (1993) Problemas de salud asociados al riego agricola con agua residual en Mexico. *Salud Publica de Mexico* **35**, 614–619. (In Spanish.)

Cifuentes, E., Gomez, M., Blumenthal U.J., Tellez-Rojo, M.M., Ruiz-Palacios, G. and Ruiz-Velazco, S. (2000) The risk of *Giardia intestinalis* infection in agricultural villages practising wastewater irrigation in Mexico. *American Journal of Tropical Medicine and Hygiene* (in press).

Esrey, S.A., Feachem, R.G. and Hughes, J.M. (1985) Interventions for the control of diarrhoeal disease among young children: improving water supplies and excreta disposal facilities. Bulletin of the World Health Organization 63(4), 757–772.

Esrey, S.A., Potash, J.B., Roberts, L. and Shiff, C. (1991) Effects of improved water supply and sanitation on ascariasis, diarrhoea, dracunculiasis, hookworm infection, schistosomiasis and trachoma. Bulletin of the World Health Organization 69(5), 609–621.

Fattal, B., Wax, Y., Davies, M. and Shuval, H.I. (1986) Health risk associated with wastewater irrigation: an epidemiological study. American Journal of Public Health 76, 977–980.

Fleisher, J.M. and McFadden, R.T. (1980) Obtaining precise estimates in coliform enumeration. *Water Research* **14**, 477–483.

Fleisher, J.M., Kay, D., Salmon, R.L., Jones, F., Wyer, M.D. and Godfree, A.F. (1996) Marine waters contaminated with domestic sewage: non-enteric illness associated with bather exposure in the United Kingdom. *American Journal of Public Health* **86**, 1228–1234.

Friis, R.H. and Sellers, T.A. (1996) *Epidemiology for Public Health Practice*, Aspen Publishers, Gaithersberg, MD.

Gorter, A.C., Sandiford, P., Smith, G.D. and Pauw, J.P. (1991) Water supply, sanitation and diarrhoeal disease in Nicaragua: results from a case-control study. *International Journal of Epidemiology* **20**(2), 527–533.

Hellard, M.E., Sinclair, M.I., Forbes, A.B., and Fairley, C.K. (2000) A randomized controlled trial investigating the gastrointestinal health effects of drinking water. 1st World Water Congress of the International Water Association, Paris, June, Session 11: Health-related water microbiology.

Hennekens, C.H. and Buring J.E. (1987) *Epidemiology in Medicine*. Little, Brown, Boston, MA.

Kay, D., Fleisher, J.M., Salmon, R.L., Jones, F., Wyer, M.D., Godfree, A., Zelanauch-Jaquotte, Z. and Shore, R. (1994) Predicting likelihood of gastroenteritis from sea bathing: results from randomised exposure. *Lancet* **344**, 905–909.

Krishnamoorthi, K.P., Abdulappa, M.K. and Aniwikar, A.K. (1973) Intestinal parasitic infections associated with sewage in farm workers, with special reference to helminths and protozoa. In *Proceedings of Symposium on Environmental Pollution*, Central Public Health Engineering Research Institute, Nagpur, India.

Payment, P., Richardson, L., Siemiatycki, J., Dewar, R., Edwardes, M. and Franco, E. (1991) A randomised trial to evaluate the risk of gastrointestinal disease due to the consumption of drinking water meeting currently accepted microbiological standards. *American Journal of Public Health* **81**, 703–708.

Payment, P., Siemiatycki, J., Richardson, L., Renaud, G., Franco, E. and Prévost, M. (1997) A prospective epidemiological study of gastrointestinal health effects due to the consumption of drinking water. *International Journal of Environmental Health Research* **7**, 5–31.

Peasey, A.E. (2000) Human exposure to Ascaris infection through wastewater reuse in irrigation and its public health significance. PhD thesis, University of London.

Pinfold, J.V., Horan, N.J. and Mara, D.D. (1995) Seasonal effects on the reported incidence of acute diarrhoeal disease in north-east Thailand. *International Journal of Epidemiology* **20** (3), 777–786.

Rothman KJ. and Greenland, S. (1998) *Modern Epidemiology*, 2nd edn, Lippincott-Raven, Philadelphia.

Shuval, H.I., Yekutiel, P. and Fattal B. (1985) Epidemiological evidence for helminth and cholera transmission by vegetables irrigated with wastewater: Jerusalem – a case study. *Water, Science and Technology* **17**, 433–442.

Shuval, H.I., Adin, A., Fattal, B., Rawitz, F. and Yekutiel, P. (1986) Wastewater irrigation in developing countries; health effects and technological solutions. Technical Paper 51, The World Bank, Washington DC.

Shuval, H.I., Wax, Y., Yekutiel, P. and Fattal, B. (1989) Transmission of enteric disease associated with wastewater irrigation: a prospective epidemiological study. *American Journal of Public Health* **79**(7), 850–852.

Snow, J. (1855) *On the Mode of Communication of Cholera*, Hafner, New York (reprinted 1965).

WHO (1998) Guidelines for safe recreational-water environments: Coastal and fresh-waters. Draft for consultation. World Health Organization, Geneva.

WHO (1989) Health guidelines for the safe use of wastewater in agriculture and aquaculture, Technical Report Series 778, World Health Organization, Geneva.

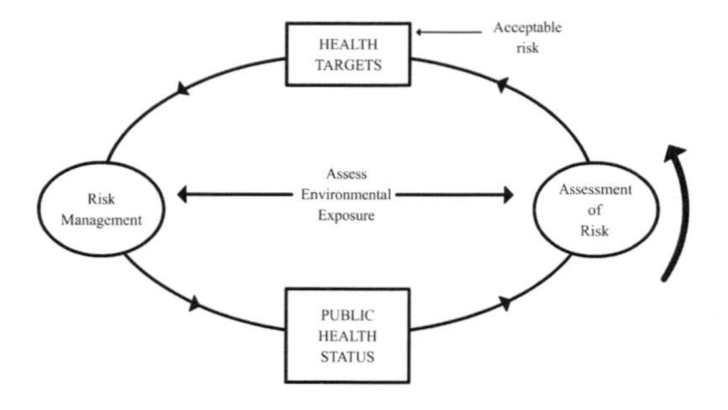

8

Risk assessment

Chuck Haas and Joseph N.S. Eisenberg

This chapter introduces the technique of microbial risk assessment and outlines its development from a simple approach based upon a chemical risk model to an epidemiologically-based model that accounts for, among other things, secondary transmission and protective immunity. Two case studies are presented to highlight the different approaches.

8.1 BACKGROUND

Quantifiable risk assessment was initially developed, largely, to assess human health risks associated with exposure to chemicals (NAS 1983) and, in its simplest form, consists of four steps, namely:

- hazard assessment
- exposure assessment
- dose–response analysis
- risk characterisation.

The output from these steps feeds into a risk management process. As will be seen in later sections this basic model (often referred to as the chemical risk paradigm) has been extended to account for the dynamic and epidemiologic characteristics of infectious disease processes. The following sub-sections elaborate on the chemical risk paradigm as outlined above.

8.2 CHEMICAL RISK PARADIGM

8.2.1 Hazard assessment

For micro-organisms, hazard assessment (i.e. the identification of a pathogen as an agent of potential significance) is generally a straightforward task. The major tasks of Quantitative Microbiological Risk Assessment (QMRA) are, therefore, focused on exposure assessment, dose–response analysis and risk characterisation. The task of risk management is one of deciding the necessity of any action based upon the risk characterisation outputs, and incorporates significant policy and trans-scientific concerns.

One outcome of the hazard analysis is a decision as to the principal consequence(s) to be quantified in the formal risk assessment. With micro-organisms, consequences may include infection (without apparent illness), morbidity or mortality; furthermore, these events may occur in the general population, or at higher frequency in susceptible sub-populations. Although mortality from infectious agents, even in the general population, cannot be regarded as negligible (Haas *et al.* 1993), the general tendency (in water microbiology) has been to regard infection in the general population as the particular hazard for which protection is required. This has been justified based on a balance between the degree of conservatism inherent in using infection as an endpoint and the (current) inability to quantify the risks to more susceptible sub-populations (Macler and Regli 1993).

8.2.2 Exposure assessment

The purpose of an exposure assessment is to determine the microbial doses typically consumed by the direct user of a water (or food). In the case of water microbiology, this may necessitate the estimation of raw water micro-organism levels followed by estimation of the likely changes in microbial concentrations with treatment, storage and distribution to the end-user (Regli *et al.* 1991; Rose *et al.* 1991). A second issue arising in exposure assessment is the amount of ingested material per 'exposure'. As a default number, two litres/person-day is used to estimate drinking water exposure (Macler and Regli 1993), although this may be conservative (Roseberry and Burmaster 1992). For contact recreational exposure, 100 ml/day has often been assumed as an exposure measure (Haas 1983a), but actual data to validate this number are lacking.

8.2.3 Dose–response analysis

It is generally necessary to fit a parametric dose–response relationship to experimental data since the desired risk (and dose) which will serve to protect public health is often far lower than can be directly measured in experimental subjects (at practical numbers of subjects). Hence it is necessary to extrapolate a fitted dose–response curve into the low-dose region.

In QMRA, for many micro-organisms, human dose–response studies are available which can be used to estimate the effects of low level exposure to micro-organisms. In prior work, it has been found that these studies may be adequately described by one of two semi-mechanistic models of the infection process. In the exponential model, which may be derived from the assumption of random occurrence of micro-organisms along with a constant probability of initiation of infection by a single organism (r), the probability of infection (P_I) is given as a function of the ingested dose (d) by:

$$P_I = 1 = \exp(= rd) \tag{8.1}$$

For many micro-organisms, the dose–response relationship is shallower than reflected by Equation 8.1, suggesting some degree of heterogeneity in the micro-organism-host interaction. This can be successfully described by the beta-Poisson model, which can be developed from Equation 8.1 if the infection probability is itself distributed according to a beta distribution (Furumoto and Mickey 1967a,b; Haas 1983b). This model is described by two parameters, a median infectious dose (N_{50}) and a slope parameter (α):

$$P_I = 1 - \left[1 + \frac{d}{N_{50}} (2^{1/\alpha} = 1) \right]^{-\alpha} \tag{8.2}$$

Figure 8.1 depicts the effect of the slope parameter on the dose–response relationship; in the limit of $\alpha \infty \infty$, Equation 8.2 approaches Equation 8.1.

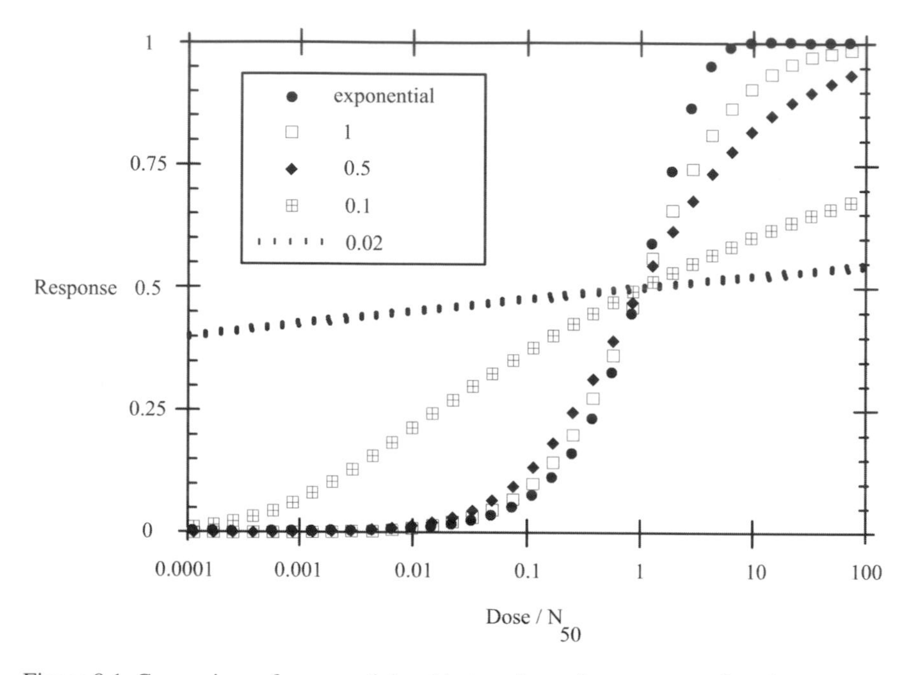

Figure 8.1. Comparison of exponential and beta-poisson dose–response functions.

The exponential and beta-Poisson models are two dose–response relationships that can be developed from biologically plausible assumptions about the infection process (Table 8.1 outlines the best-fit dose–response parameters for these models for a number of human pathogens). A general framework for plausible models can also be derived.

In addition to such quasi-mechanistic models, a variety of empirical models are possible, three models which have been used (primarily in chemical risk assessment), are the log-logistic, the Weibull, and the log-probit.

Generally, several models may fit available data in a statistically acceptable sense, and yet provide very different estimates for the risk at an extrapolated low dose. This situation is one that has frequently been encountered in chemical risk assessment (Brown and Koziol 1983). In QMRA, it may be possible to test the potential appropriateness of different dose–response functions by validating with outbreak data.

Given a set of dose–response data, i.e. exposure of populations to various doses of micro-organisms and measurement of response (such as infection), the best fitting parameters of a dose–response relationship may be computed via standard maximum likelihood techniques. The method has been illustrated for human rotavirus (Haas *et al.* 1993; Regli *et al.* 1991) and protozoa (Rose *et al.*

1991). Confidence limits to the parameters can then be found, and used as a basis for low-dose extrapolation. It should be noted, however, that in general dose–response studies have been conducted on healthy adults and may not, therefore, reflect the response of the general population.

Table 8.1. Table of best-fit dose–response parameters (human)

Organism	Exponential	Beta Poisson		Reference
	k	N_{50}	α	
Poliovirus I (Minor)	109.87			Minor *et al*, 1981
Rotavirus		6.17	0.2531	Haas *et al.* 1993; Ward *et al.* 1986
Hepatitis A virus[(a)]	1.8229			Ward *et al.* 1958
Adenovirus 4	2.397			Couch *et al.* 1966
Echovirus 12	78.3			Akin 1981
Coxsackie[(b)]	69.1			Couch *et al.* 1965; Suptel, 1963
Salmonella[(c)]		23,600	0.3126	Haas *et al.* 1999
Salmonella typhosa		3.60×10^6	0.1086	Hornick *et al.* 1966
Shigella[(d)]		1120	0.2100	Haas *et al.* 1999
Escherichia coli[(e)]		8.60×10^7	0.1778	Haas *et al.* 1999
Campylobacter jejuni		896	0.145	Medema *et al.* 1996
Vibrio cholera		243	0.25	Haas *et al.* 1999
Entamoeba coli		341	0.1008	Rendtorff 1954
Cryptosporidium parvum	238			Haas *et al.* 1996; Dupont *et al.* 1995
Giardia lamblia	50.23			Rose *et al.* 1991

[(a)] dose in grams of faeces (of excreting infected individuals)
[(b)] B4 and A21 strains pooled
[(c)] multiple (non-typhoid) pathogenic strains (*S. pullorum* excluded)
[(d)] *flexnerii* and *dysenteriae* pooled
[(e)] Nonenterohaemorrhagic strains (except O111)

8.2.4 Risk characterisation

The process of risk characterisation combines the information on exposure and dose–response into an overall estimation of likelihood of an adverse consequence. This may be done in two basic ways. First, a single point estimate of exposure (i.e. number of organisms ingested) can be combined with a single point estimate of the dose–response parameters to compute a point estimate of risk. This may be done using a 'best' estimate, designed to obtain a measure of central tendency, or using an extreme estimate, designed to obtain a measure of consequence in some more adversely affected circumstance. An alternative approach, which is currently receiving increasing favour, is to characterise the full distribution of exposure and dose–response relationships, and to combine these using various tools (for example, Monte Carlo analysis) into a distribution of risk. This approach conveys important information on the relative imprecision of the risk estimate, as well as measures of central tendency and extreme values (Burmaster and Anderson 1994; Finkel 1990).

One important outcome of the risk characterisation process using a Monte Carlo approach is the assessment of the relative contribution of uncertainty and variability to a risk estimate. Variability may be defined as the intrinsic heterogeneity that leads to differential risk among sectors of the exposed group, perhaps resulting from differential sensitivities or differential exposures. Uncertainty may be defined as the factors of imprecision and inaccuracy that limit the ability to exactly quantify risk. Uncertainty may be reduced by additional resources, for example devoted to characterisation of the dose–response relationship. Variability represents a lower limit to the overall risk distribution.

Two aspects of risk characterisation deserve further comment. In general, all available dose–response information for micro-organisms (human or animal) pertains to response to single (bolus) doses. In actual environmental (or food) exposures, doses may occur over time (or may even be relatively continuous). In the absence of specific data on the impact of prior exposure on risk, the assumption used in projecting risk to a series of doses has been that the risks are independent (Haas 1996).

8.2.5 Risk management

The results of a risk characterisation are used in risk management. The understanding of appropriate action levels for decision-making with respect to micro-organisms is still at an early stage (see Chapter 10). However, in the case of waterborne protozoa, it has been suggested (in the US) that an annual risk of infection of 0.0001 (i.e. 1 in 10,000) is appropriate for drinking water (Macler and Regli 1993).

8.3 *CRYPTOSPORIDIUM* CASE STUDY

This case study follows through the process described in the previous section and details a microbiological risk assessment focusing on *Cryptosporidium* in a US city. New York City has a central water supply reservoir that receives the flow from two watersheds (*Watershed C and Watershed D*). Oocyst levels have been determined for both watersheds since 1992. *Cryptosporidium* was chosen as the organism of interest since it is currently the pathogen most resistant to disinfection (with minimal inactivation by free chlorine alone: Finch *et al.* 1998;

Korich *et al.* 1990; Ransome *et al.* 1993). Hence, for *Cryptosporidium*, the effluent from the final water supply reservoir provides a reasonable starting point for estimating oocysts in the water as consumed.

To estimate the potential level of infection from *Cryptosporidium* present in the watershed supplies, the following inputs are needed:

- water ingestion per day (V)
- oocyst concentration at point of ingestion (C)
- dose–response relationship for *Cryptosporidium* f(V × C)

In this instance, in accordance with a number of prior risk assessments, each day of exposure (consumption of water) is considered to result in a statistically independent risk of infection (Haas *et al.* 1993; Regli *et al.* 1991).

8.3.1 Input exposure variables

Tap-water ingestion was modelled using the log-normal distribution for total tap-water consumption developed by Roseberry (Roseberry and Burmaster 1992). The natural logarithm of total tap-water consumption in ml/day is normally distributed with a mean of 7.492 and standard deviation of 0.407 (corresponding to an arithmetic mean of 1.95 l/day).

Initial examination of the time series of oocyst levels monitored to date from the two watersheds indicates a number of interesting features (Figure 8.2), namely:

- The levels of oocysts are quite variable, as is common for many microbial data sets.
- The densities appear to be higher during the earlier portion of the data record than in the more recent part of the data record (for reasons that are unclear).
- There are a substantial number of samples where no oocysts were detected. The mean detection limit for these non-detects was 0.721 oocysts/100 l.

The overall mean oocyst concentration (treating the 'non-detects' as zero's) was 0.26 and 0.31 oocysts/100 l for the watershed C and watershed D locations, respectively. Of the 292 samples taken at each location, only 45 samples at watershed C and 48 samples at watershed D were above individual daily detection limits. Of these samples, only 18 and 21, respectively, were above 0.721 oocysts/100 l (the average detection level for the non-detects). This pattern is not unusual in protozoan monitoring data, and it presents a level of complexity in assessing the risk posed by exposure to these organisms.

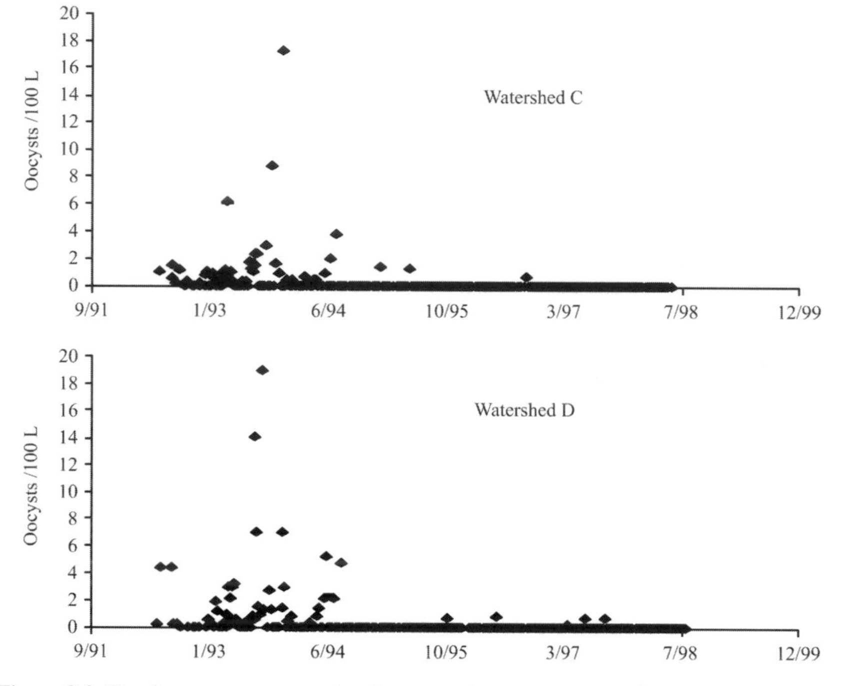

Figure 8.2. Total oocyst concentration in reservoir raw water samples.

The significant number of samples with concentrations close to or below the average detection limit must be taken into account when estimating mean oocyst densities and distribution. There are several methods that may be used when dealing with below-detection-limit (BDL) data (Haas and Scheff 1990). Two basic approaches are employed here.

- Observations that are below the detection limit are treated as if they had values equal to the detection limit, half the detection limit, or zero. The arithmetic mean of the revised data is then computed by simple averaging. These alternatives are called 'fill in' alternatives.
- The method of maximum likelihood is used. In this approach, the data are presumed to come from a particular distribution (e.g. log-normal), and standard methods for analysing data with a single censoring point are used. A likelihood function is

formulated with a contribution equal to the probability density function for all quantified values, and equal to the cumulative distribution function (up to the detection limit) for all BDL values. The values of the distribution parameters that maximise the resulting likelihood are accepted as the best estimators.

To develop the distribution for oocyst concentrations at the point of ingestion, all data from the two watersheds were examined. Using maximum likelihood, and treating all observations less than or equal to 0.721 oocysts/100 l as being censored (for all censored observations, 0.721/100 l was regarded as being the detection limit), the parameters of log-normal distributions were determined.

Table 8.2 shows the parameters of the best fitting log-normal distributions to the entire data record at each station. There is some underprediction at the extreme tails of the distribution; however, in general the fit is adequate. Investigation of alternative distributions (gamma, Weibull, and inverse Gaussian) did not yield fits superior to the log-normal distribution. The goodness of fit to the log-normal was acceptable as judged by a chi-squared test.

Table 8.2. Mean and standard deviation of best-fitting normal distribution for natural logarithm of oocyst levels (/100 l) in reservoir samples (January 1992 to June 1998)

	Watershed C	Watershed D
Mean natural logarithm	–2.752	–3.210
Std. deviation of natural log	1.828	2.177

Table 8.3 summarises the arithmetic average from both watersheds, using maximum likelihood and the various fill-in procedures (for 1992 and 1998, these averages are for portions of the year). The ‘imputed arithmetic mean’ is computed from the maximum likelihood estimates (MLEs). In more recent years, it was not possible to estimate the maximum likelihood mean densities at both locations and all times, since too few (<2) observations above the detection limit were available.

Table 8.3. Summary of mean oocyst levels (/100 l) estimated by different methods

	All years	1992	1993	1994	1995	1996	1997	1998*
Watershed C								
Imp. arith mean	0.33	0.62	1.36	0.26	0.16			
Detection limit	0.85	0.72	1.46	0.78	0.73	0.70	0.72	0.72
Half det limit	0.55	0.59	1.30	0.48	0.39	0.36	0.36	0.36
Zero det limit	0.25	0.46	1.13	0.18	0.05	0.01	0	0
Watershed D								
Imp. arith mean	0.43	1.80	1.35	0.47				
Detection limit	0.89	1.14	1.55	0.91	0.70	0.70	0.69	0.72
Half detection	0.60	0.96	1.41	0.62	0.36	0.36	0.36	0.36
Zero det limit	0.30	0.78	1.26	0.33	0.01	0.01	0.02	0

* Jan – June   Imp. – Imputed   det. - detection

The bias due to 'fill-in' methods using the detection limit and half the detection limit is quite evident in the more recent years, where the oocyst levels were generally below detection. Both of these 'fill-in' methods may overestimate total oocyst concentration in the source water. Regardless of the methods used, it is apparent that 1992 and 1993 had higher average oocyst levels than in more recent years.

In order to assess exposure, the concentrations of oocysts from each watershed were flow-weighted (to allow for relative contributions) and then combined.

The dose–response relationship for infection of human volunteers with C. parvum oocysts has been found to be exponential with a best-fit dose–response parameter (k) equal to 238 (Table 8.1). The confidence distribution to the dose– response parameter k can be determined by likelihood theory (Morgan 1992). The confidence distribution to the natural logarithm of k is then found to be closely approximated by a normal distribution with mean of 5.48 and standard deviation of 0.32.

8.3.2 Results

Given a single value of water consumption (V), oocyst concentration (C), and the dose–response parameter (k), the risk of infection to an individual can be calculated. To consider the distribution of risk, which incorporates uncertainty and variability in each of the input parameters, this calculation needs to be performed a large number of times (Monte Carlo analysis). In this technique a new set of random samples (for water consumption, oocyst concentration at each location, and the dose–response parameter) is obtained, and then individual calculations using these sets of random samples are combined to reveal an estimated distribution of risk.

Two types of results are presented below. First, the daily risk estimate is calculated for each individual year (to observe trends in risk over time), given a single water dose, dose–response parameter, and average oocyst concentration. Four oocyst concentrations are used, representing the different methods for considering data points below the detection limit. The purpose of this exercise is to observe trends in the risk estimate over time. The second set of results shows the range of estimated risk, taking into account uncertainty in all of the input parameters. This range is generated using the combined data from 1992–8.

8.3.2.1 *Point estimates*

Point estimates for the daily risk of infection from *Cryptosporidium* are presented in Table 8.4. The four columns represent different methods used to determine the average oocyst concentration, i.e. maximum likelihood and by the three 'fill in' methods. A figure of 1.95 l/day was used for the amount of water consumed and k was set equal to 238. The calculation was done using both the total (1992–8) data set and for each year individually.

Table 8.4. Computed point estimates for the daily risk of infection from *Cryptosporidium* ($\times 10^{-5}$)

	Imputed arith. mean	Fill in methods		
		Detection limit	Half detection limit	Zero detection limit
All	3.2	7.1	4.7	2.3
Years				
1992	10.7	7.8	6.5	5.3
1993	10.8	12.2	10.9	9.7
1994	3.1	6.9	4.6	2.2
1995	–	5.7	3.0	0.2
1996	–	5.7	2.9	0.1
1997	–	5.6	2.9	0.1
1998*	–	5.6	2.9	0

* (Jan – June)
(–) could not be estimated since fewer than two quantified observations are available

8.3.2.2 *Monte Carlo simulation*

While useful, point estimates of risk do not reveal the degree of uncertainty in the risk estimate. To do this, Monte Carlo simulations are necessary. Summary statistics on 10,000 iterations of the Monte Carlo model are shown in Table 8.5.

For this computation, the entire (1992–8) oocyst monitoring database was used as the water density distribution. The mean individual daily risk is estimated as 3.42×10^{-5}.

It should be noted that the results of the Monte Carlo analysis bracket the range of point estimates observed by considering each year's data set separately, whether maximum likelihood or 'fill-in' methods are used.

Table 8.5. Summary of Monte Carlo trials. Daily risk of *Cryptosporidium* infection ($\times 10^{-5}$)

Statistic	Individual daily risk
Mean	3.4
Median	0.7
Standard deviation	19.8
Lower 95% confidence limit	0.034
Upper 95% confidence limit	21.9

As part of this computation, a sensitivity analysis was conducted. The rank correlation of the individual daily risk with the various input parameters was computed. The densities of pathogens in the two effluent flows from the reservoir were found to have the greatest correlation with the estimates daily risk. The other inputs (water consumption and dose–response parameter) contributed only a minor amount to the uncertainty and variability in the estimated risk. Attention, therefore, should be paid primarily to obtaining better (more precise) estimates of the effluent oocyst concentrations.

8.3.3 Caveats

The above risk assessment has a number of caveats that should be taken into account when developing a decision based on these results.

- use of healthy volunteer data (based upon a single strain of *Cryptosporidium*)
- no account of secondary infection
- no data on oocyst viability or infectivity
- poor oocyst recovery rates
- choice of endpoint (illness may be a more important endpoint than infection).

8.3.4 Case study conclusions

An annual risk of infection of 1 in 10,000 (which has been suggested by the EPA as an acceptable level for drinking water exposure to an infectious agent) corresponds to a daily risk of 2.7×10^{-7} per person. This is below the lower 95% confidence limit to the estimated daily risk for New York based upon the calculations above. It is also below the point estimates for risk when individual years of data are treated separately. Hence, based on the assumptions used, the current risk of cryptosporidiosis infection would appear to be in excess of the frequently propounded acceptable risk level.

Microbial risk assessments should be coupled with investigation of potential future treatment decisions and watershed management strategies. For example, if information on the performance of such strategies with respect to reduction of oocyst levels is available, then the potential impact on microbial infections can be assessed. Given standard treatment efficiencies, the addition of a properly functioning water filtration plant would reduce the estimated daily and annual risk of *Cryptosporidium* infection by a factor of 100.

8.4 A DYNAMIC EPIDEMIOLOGICALLY-BASED MODEL

As outlined in the previous sections, attempts to provide a quantitative assessment of human health risks associated with the ingestion of waterborne pathogens have generally focused on static models that calculate the probability of individual infection or disease as a result of a single exposure event (Fuhs 1975; Haas 1983b; Regli *et al.* 1991). The most commonly used framework is based upon a chemical model and, as such, does not address a number of properties which are unique to infectious disease transmission, including:

- secondary (person-to-person) disease transmission
- long- and short-term immunity
- the environmental population dynamics of pathogens.

The limitations of treating infectious disease transmission as a static disease process, with no interaction between those infected or diseased and those at risk, has been illustrated in studies of *Giardia* (Eisenberg *et al.* 1996), dengue (Koopman *et al.* 1991b), and sexually transmitted diseases (Koopman *et al.* 1991a). The transmission pathways for environmentally mediated pathogens are complex. These disease processes include person-to-person, person-to-fomite to-person, person-to-water-to-person as well as food routes for those pathogens that only have human hosts, and they include animal–animal or animal–human pathways for those that have animal reservoirs. To understand the role that water plays in the transmission of enteric pathogens and to estimate the risk of disease due to drinking water within a defined population, it is important to study the complete disease transmission system.

As mentioned previously, models using the chemical risk paradigm are static and assess risks at the individual level; i.e. the risk calculation is the probability that a person exposed to a given concentration of pathogens will have an adverse health effect. The underlying assumption in this calculation is that disease occurrences are independent; that is, the chance of person A becoming infected is independent of the prevalence of disease within the population. Although this assumption is valid for disease associated with chemical exposure, in general, it is not universally appropriate for infectious disease processes. The risk of person A becoming infected is not only dependent on his direct exposure to environmental pathogens but also on exposure to other currently infected individuals (group B). Some of the group B individuals may have been infected from a previous exposure to an environmental pathogen. Therefore, in addition to direct risks of exposure, person A is indirectly at risk due to any previous exposures from group B. One implication of this secondary infection process is that risk is, by definition, manifested at a population level. Specifically, an individual is not only at risk from direct exposure to a contaminated environmental media, but also from interactions within the population that can result in exposures to infected individuals. Another implication of this secondary infection process is that risk calculations are dynamic in nature; i.e. the overall risk calculation is based not only on current exposures to a contaminated media, but also on all subsequent secondary infections.

The existence of other epidemiological states of the disease process may also affect risk estimates; e.g. post-infection status that accounts for previous exposure to the pathogen, and a carrier status that accounts for those who are asymptomatic but infectious. Post-infection status may take on different forms from long-term and complete protection to short-term and partial protection. Therefore, at any given time there may exist a portion of the population that is not susceptible to disease. Moreover, the protected portion of the population will vary in time depending on the prior prevalence levels. Asymptomatic carriers provide another potential source of infection through contact with the susceptible portion of the population. This portion of the population also varies in time.

8.5 CASE STUDY: ROTAVIRUS DISEASE PROCESS

Given the discussion above, we can conceptualise an epidemiologically-based characterisation of risk by dividing the population into distinct states with respect to disease status. States may include susceptible, diseased (infectious and symptomatic), immune (either partial or complete), and/or carrier (infectious but asymptomatic). Further, it can be understood that members of the population may move between states. Factors affecting the rate at which members move between states include:

- the level of exposure to an environmental pathogen;
- the intensity of exposure to individuals in the infectious or carrier state;
- the temporal processes of the disease (e.g. incubation period, duration of disease, and duration of protective immunity, etc.).

This conceptual model is inherently dynamic and population-based; i.e. the risk of infection is manifested at the population level. Thus, consistent with the above concepts, the initial steps prescribed by the infectious disease framework are to identify the important states for a given pathogen or class of pathogens and then develop a diagram of causal relationships among these states. From an epidemiological point of view, the population is divided into distinct states with respect to disease. Historically, when developing these types of compartmental models, members of a population have been classified as susceptible, infected, or recovered. For a pathogen such as rotavirus, however, a simple 'susceptible–infected–recovered' type model may not be sufficient to characterise the movement of the population between states. A more detailed model structure is motivated by the following properties:

- Some protection can be attained after exposure to rotavirus; however, this protective state appears to be neither absolute nor long-term; and
- It is well documented that it is possible (and in fact is common) to be infected with rotavirus without demonstrating the symptoms of the disease.

From these properties, one possible categorisation of the population with respect to the rotavirus disease process is:

- a susceptible state (S), defined by individuals susceptible to infection

- a carrier state (C), defined by individuals who are infectious but not symptomatic
- a diseased state (D), defined by individuals who are symptomatic and infectious
- a post-infection state (P), defined by individuals who are not infectious and not susceptible due to (limited and short-term) immunity.

Members of a given state may move to another state based on the causal relationships of the disease process. For example, members of the population who are in the susceptible state may move to the diseased state after exposure to a pathogenic agent. This is shown in Figure 8.3.

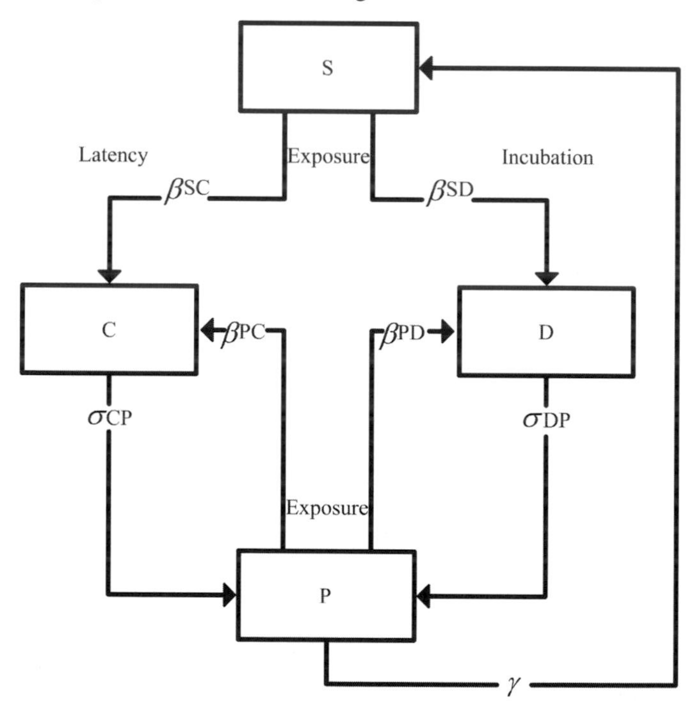

Figure 8.3. Conceptual model for rotavirus. (State variables: **S** = Susceptible = Not infectious, not symptomatic; **C** = Carrier = Infectious, not symptomatic; **D** = Diseased = Infectious, symptomatic; **P** = Post Infection = Not infectious, not symptomatic, with short-term or partial immunity.)

To describe the epidemiology of rotavirus, the conceptual model includes both state variables and rate parameters. State variables (S, C, D, and P) track the number of people that are in each of the states at any given point in time, and are defined such that S + C + D + P = N (the sum of the state variables equals the total population). The rate parameters determine the movement of the population from one state to another. In general, the rate parameters are denoted as β, γ, and γ with appropriate subscripts, where:

- $\gamma\gamma$ β is the rate of transmission from a non-infectious state, S or P, to an infectious state, C or D. These transmission rate parameters describe the movement between states due to both primary (drinking water, for example) and secondary (all other) exposure to rotavirus;
- $\gamma\gamma$ γ is the rate of recovery from an infectious state, C or D, to the post-infection state, P; and
- $\gamma\gamma$ γ is the rate of movement from the post-infection state (partial immunity), P, to the susceptible state, S.

The rate parameters may be determined directly through literature review, may be functions of other variables that are determined through literature review, or may be determined through site-specific data where possible and appropriate. One technical aspect of the approach described is that the distribution of time that members of the population spend in each of the states is assumed to be exponential (this may not always be the case and can easily be addressed; see for example Eisenberg *et al.* 1998).

The model describes movements of the population between states. Consider the susceptible portion of the population during a particular point in time. As shown in Figure 8.3, upon exposure to rotavirus three processes affect the number of susceptible individuals within the population:

- $\gamma\gamma$ some members of the population will move from the susceptible state S to the carrier state C (at rate β_{SC})
- $\gamma\gamma$ some members will move from the susceptible state S to the diseased state D (at rate β_{SD})
- $\gamma\gamma$ other members of the population will move from the post-infection state P back to the susceptible state S (at rate γ).

Analogous processes account for movement of the population between all of the states shown in Figure 8.3. Mathematical details of this model are described in detail elsewhere (Eisenberg *et al.* 1996, 1998; Soller *et al.* 1999).

8.5.1 Implementation

Using a modified version of the ILSI (International Life Sciences Institute) microbial risk framework, the implementation of a conceptual model, such as the rotavirus model, to assess the associated human health risks follows a three step process; problem formulation, analysis, and risk characterisation (ILSI 1996). This process is summarised graphically in Figure 8.4.

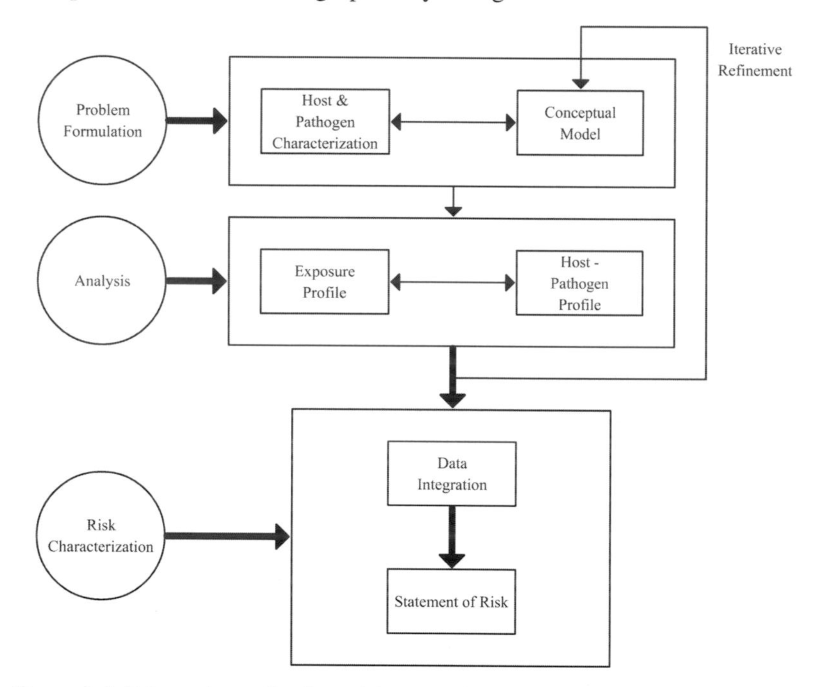

Figure 8.4. Schematic application of the ILSI framework.

8.5.1.1 Problem formulation and analysis

In addition to the development of a conceptual model in the problem formulation phase, a literature review is generally conducted to obtain relevant data. Initial host and pathogen characterisations are also developed.

The goal of the analysis phase is to link the conceptual model with the risk characterisation. This process is carried out by summarising and organising the data obtained from the problem formulation, resulting in an exposure- and host-pathogen profile that succinctly summarises data relevant to the specific problem.

8.5.1.2 Risk characterisation

In the risk characterisation phase, the exposure and host pathogen profiles are integrated to quantify the likelihood of adverse health effects due to the exposure of microbial contaminants, within the context of the uncertainties in the data and the assumptions used in the quantification process. The risk characterisation also features a data integration step. As described previously, the conceptual model is composed of both state variables and rate parameters. Data integration is the process by which the rate parameters are quantified in terms of probability distributions using available data as a foundation. Oncc the data integration step is complete, a series of simulations is conducted. A Monte Carlo simulation technique is incorporated to account for the uncertainty and variability inherent in this environmental system. The result of the simulations is a statement of risk or relative risk associated with the specific problem being addressed. Figure 8.5 illustrates how the results of these simulations can be represented.

Box plots were used to summarise each of the four scenarios shown in the graph. The first two scenarios represent the average daily prevalence of a hypothetical baseline condition for children and adults respectively. The third scenario represents children exposed to an increased contamination in drinking water compared with the baseline, and the fourth scenario represents children exposed to a decreased contamination. It is important to keep in mind that this graph is for illustrative purposes only and does not represent an actual risk assessment. With this in mind, the following information can be obtained from this plot:

- the degree of uncertainty associated with each scenario is quite large
- children experience a greater disease burden than adults
- even for very low levels of water contamination an endemic condition exists.

A detailed description of the data integration and risk characterisation processes is summarised in Eisenberg *et al.* (1996).

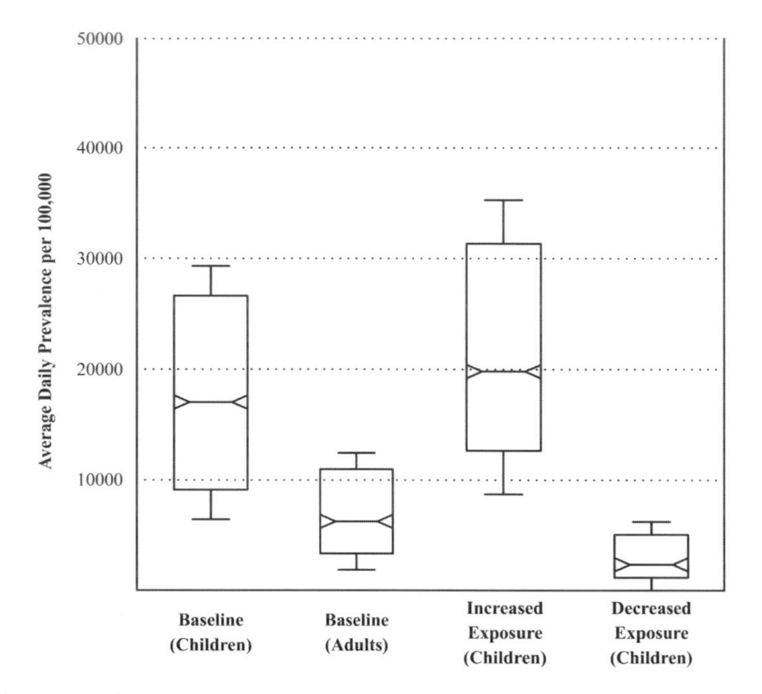

Figure 8.5. Comparison of average daily prevalence for children (1), adults (2), under baseline conditions, and for children exposed to both higher levels (3) and lower levels (4) of contamination in drinking water.

8.6 DISCUSSION

A comprehensive risk assessment methodology should account for all the important processes that affect the resultant risk estimate. One important property of an infectious disease process is the ability of an infected person to infect a susceptible person through direct or indirect contact. To rigorously incorporate this aspect of the disease transmission process, the risk calculation must account for these indirect exposures through contacts with infected individuals. The post-infection process is another property that can affect the risk estimate, since at any given time there is a group of individuals that are not susceptible to reinfection (due to previous exposures to the pathogen).

While the infectious disease process is inherently population-based and dynamic, there may be times when simplifying assumptions may be made, and the chemical risk paradigm may be appropriate. One valuable feature of the methodology presented in the rotavirus case study is that the structure can collapse into a framework analogous to the chemical risk framework (seen in previous sections) when the secondary infection rate is negligible, protection from future infection due to pathogen exposures is unimportant, and the infection process is static.

Dynamic disease process models have been used in a variety of applications. For example, Eisenberg *et al.* (1998) used this methodology to study the disease dynamics of a *Cryptosporidium* outbreak. In that study, the outcome was known and was used to determine the conditions that may have accounted for the specific outbreak. In another investigation, the same methodology was used to explore the uncertainties in assessing the risk of giardiasis when swimming in a recreational impoundment using reclaimed water (Eisenberg *et al.* 1996). In both of these studies the dynamic, population-based modelling framework was a valuable tool for providing information about the disease process in the context of uncertainty and variability.

8.7 IMPLICATIONS FOR INTERNATIONAL GUIDELINES AND NATIONAL REGULATIONS

In conjunction with epidemiology and other data sources, risk assessment can be a very powerful tool. As well as being used in partnership with epidemiology it can also provide useful insights into areas such as rare events and severe disease outcomes where epidemiology is not appropriate. The ease with which parameters can be changed within a risk assessment makes it ideal to inform both international guidelines and standards derived from specific national circumstances. It can also be used to test 'what if' scenarios, which may help target management interventions. However, the technique does have limitations and it is vital that assumptions are calibrated against real data and it is not seen simply as a substitute for other techniques. As with any model the outputs are, at best, only as good as the inputs.

8.8 REFERENCES

Akin, E.W. (1981) Paper presented at the US EPA symposium on microbial health considerations of soil disposal of domestic wastewaters.

Brown, C.C. and Koziol, J.A. (1983) Statistical aspects of the estimation of human risk from suspected environmental carcinogens. *SIAM Review* **25**(2), 151–181.

Burmaster, D.E. and Anderson, P.D. (1994) Principles of good practice for use of Monte Carlo techniques in human health and ecological risk assessment. *Risk Analysis* **14**(4), 477–481.

Couch, R.B., Cate, T., Gerone, P., Fleet, W., Lang, D., Griffith, W. and Knight, V. (1965) Production of illness with a small-particle aerosol of Coxsackie A21. *Journal of Clinical Investigation* **44**(4), 535–542.

Couch, R.B., Cate, T.R., Gerone, P.J., Fleet, W.F., Lang, D.J., Griffith, W.R. and Knight, V. (1966) Production of illness with a small-particle aerosol of Adenovirus type 4. *Bacteriology Reviews* **30**, 517–528.

Dupont, H., Chappell, C., Sterling, C., Okhuysen, P., Rose, J. and Jakubowski, W. (1995) Infectivity of *Cryptosporidium parvum* in healthy volunteers. *New England Journal of Medicine* **332**(13), 855–859.

Eisenberg, J.N., Seto, E.Y.W., Olivieri, A.W. and Spear, R.C. (1996) Quantifying water pathogen risk in an epidemiological framework. *Risk Analysis* **16**, 549–563.

Eisenberg, J.N.S., Seto, E.Y.W., Colford, J., Olivieri, A.W. and Spear, R.C. (1998) An analysis of the Milwaukee *Cryptosporidium* outbreak based on a dynamic model of disease transmission. *Epidemiology* **9**, 255–263.

Finch, G.R., Gyurek, L.L., Liyanage, L.R.J. and Belosevic, M. (1998) Effects of various disinfection methods on the inactivation of *Cryptosporidium*. AWWA Research Foundation, Denver, CO.

Finkel, A.M. (1990) Confronting uncertainty in risk management. Resources for the Future, Centre for Risk Management, Washington, DC.

Fuhs, G.W. (1975) A probabilistic model of bathing beach safety. *Science of the Total Environment* **4**, 165–175.

Furumoto, W.A. and Mickey, R. (1967a) A mathematical model for the infectivity-dilution curve of tobacco mosaic virus: Experimental tests. *Virology* **32**, 224.

Furumoto, W.A. and Mickey, R. (1967b) A mathematical model for the infectivity-dilution curve of tobacco mosaic virus: Theoretical considerations. *Virology* **32**, 216.

Haas, C.N. (1983a) Effect of effluent disinfection on risk of viral disease transmission via recreational exposure. *Journal of the Pollution Control Federation* **55**, 1111–1116.

Haas, C.N. (1983b) Estimation of risk due to low doses of micro-organisms: A comparison of alternative methodologies. *American Journal of Epidemiology* **118**(4), 573–582.

Haas, C.N. (1996) How to average microbial densities to characterise risk. *Water Research* **30**(4), 1036–1038.

Haas, C.N. and Scheff, P.A. (1990) Estimation of averages in truncated samples. *Environmental Science and Technology* **24**, 912–919.

Haas, C.N., Rose, J.B., Gerba, C. and Regli, S. (1993) Risk assessment of virus in drinking water. *Risk Analysis* **13**(5), 545–552.

Haas, C.N., Crockett, C., Rose, J.B., Gerba, C. and Fazil, A. (1996) Infectivity of *Cryptosporidium parvum* oocysts. *Journal of the American Water Works Association* **88**(9), 131–136.

Haas, C.N., Rose, J.B. and Gerba, C.P. (1999) *Quantitative Microbial Risk Assessment,* Wiley, New York.

Hornick, R.B., Woodward, T.E., McCrumb, F.R., Dawkin, A.T., Snyder, M.J., Bulkeley, J.T., Macorra, F.D.L. and Corozza, F.A. (1966) Study of induced typhoid fever in man. Evaluation of vaccine effectiveness. *Transactions of the Association of American Physicians* **79**, 361–367.

International Life Sciences Institute (ILSI) (1996) A conceptual framework to assess the risks of human disease following exposure to pathogens. *Risk Analysis* **16**, 841–848.

Koopman, J.S., Longini, I.M., Jacquez, J.A., Simon, C.P., Ostrow, D.G., Martin, W.R. and Woodcock, D.M. (1991a) Assessing risk factors for transmission of infection. *American Journal of Epidemiology* **133**, 1168–1178.

Koopman, J.S., Prevots, D.R., Marin, M.A.V., Dantes, H.G., Aquino, M.L.Z., Longini, I.M. and Amor, J.S. (1991b) Determinants and predictors of dengue fever infection in Mexico. *American Journal of Epidemiology* **133**, 1168–1178.

Korich, D.G., Mead, J.R., Madore, M.S., Sinclair, N.A. and Sterling, C.R. (1990) Effects of ozone, chlorine dioxide, chlorine and monochloramine on *Cryptosporidium parvum* oocyst viability. *Applied and Environmental Microbiology* **56**(5), 1423–1428.

Macler, B.A. and Regli, S. (1993) Use of microbial risk assessment in setting United States drinking water standards. *International Journal of Food Microbiology* **18**(4), 245–256.

Medema, G.J., Teunis, P.F.M., Havelaar, A.H. and Haas, C.N. (1996) Assessment of the dose–response relationship of *Campylobacter jejuni*. *International Journal of Food Microbiology* **39**, 101–112.

Minor, T.E., Allen, C.I., Tsiatis, A.A. Nelson, D.D. and D'Alessio, D.J. (1981) Human infective dose determination for oral Poliovirus type 1 vaccine in infants. *Journal of Clinical Microbiology* **13**, 388.

Morgan, B.J.T. (1992) *Analysis of Quantal Response Data,* Chapman & Hall, London.

National Academy of Science (NAS) (1983) *Risk Assessment in Federal Government: Managing the Process,* National Academy Press, Washington, DC.

Ransome, M.E., Whitmore, T.N. and Carrington, E.G. (1993) Effects of disinfectants on the viability of *Cryptosporidium parvum* oocysts. *Water Supply* **11**, 75–89.

Regli, S., Rose, J.B., Haas, C.N. and Gerba, C.P. (1991) Modelling risk for pathogens in drinking water. *Journal of the American Water Works Association* **83**(11), 76–84.

Rendtorff, R.C. (1954) The experimental transmission of human intestinal protozoan parasites. I. *Endamoeba coli* cysts given in capsules. *American Journal of Hygiene* **59**, 196–208.

Rose, J.B., Haas, C.N. and Regli, S. (1991) Risk assessment and the control of waterborne giardiasis. *American Journal of Public Health* **81**, 709–713.

Roseberry, A.M. and Burmaster, D.E. (1992) Log-normal distributions for water intake by children and adults. *Risk Analysis* **12**(1), 99–104.

Soller, J.A., Eisenberg, J.N. and Olivieri, A.W. (1999) *Evaluation of Pathogen Risk Assessment Framework,* ILSI, Risk Science Institute, Washington, DC.

Suptel, E.A. (1963) Pathogenesis of experimental Coxsackie virus infection. *Archives of Virology* **7**, 61–66.

Ward, R., Krugman, S., Giles, J., Jacobs, M. and Bodansky, O. (1958) Infectious hepatitis studies of its natural history and prevention. *New England Journal of Medicine* **258**(9), 402–416.

Ward, R.L., Bernstein, D.L., Young, E.C., Sherwood, J.R., Knowlton, D.R. and Schiff, G.M. (1986) human rotavirus studies in volunteers: Determination of infectious dose and serological responses to infection. *Journal of Infectious Diseases* **154**(5), 871.

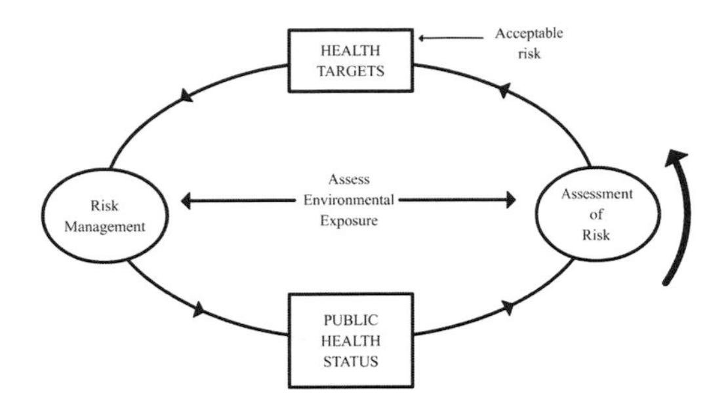

9

Quality audit and the assessment of waterborne risk

Sally Macgill, Lorna Fewtrell, James Chudley and David Kay

In order to avoid the 'garbage in, gospel out' scenario described by Burmaster and Anderson (1994) it is becoming increasingly clear that there is a need for some sort of standardised quality assessment to examine the strength of the inputs to the assessment of risk area. This chapter proposes one possible approach and notes the need for further development in this area. While the examples draw heavily on the risk assessment area, the same approach can be used for any of the tools driving the assessment of risk.

9.1 INTRODUCTION

How strong is the science for assessing waterborne health risks? Unless the answer to this question is known, then how can risk assessment or

 Water Quality: Guidelines, Standards and Health. Edited by Lorna Fewtrell and Jamie Bartram. Published by IWA Publishing, London, UK. ISBN: 1 900222 28 0

epidemiological study results be sensibly interpreted and acted upon? How can it be known with what degree of confidence, or of caution, to proceed?

These questions arise from the acknowledged limitations of science to provide definitive inputs to the assessment of waterborne risk. There are gaps and limitations in the current state of scientific knowledge. These are not identified here as limitations in competence or motivation of scientific experts. They are instead identified as intrinsic structural limitations in the fields of research which are being drawn upon.

Weinberg (1972) introduced the concept of trans-science to refer to problems which can be formulated within traditional scientific paradigms (for example as testable hypotheses) but which are beyond the capability of science definitively to resolve. Categories of problem falling within this realm include those entailing experimental set-ups that would be logistically too complex to co-ordinate in practice (owing to the sheer size and complexity of the technology or the sheer number – possibly millions – of experimental species required), problems raising ethical issues (notably the wrongs of experimentally exposing people to harmful substances), and problems where surrogate indicator species have to be studied in the absence of accessibility to true species or pathogens.

Other structural limitations, of a conceptually simpler nature, arise from the brute force of economics. Science is expensive, and it is simply not possible to fund all that would be desirable. For example, of the universe of toxic and carcinogenic chemicals that are as yet untested there is a fundamental issue in setting research priorities of whether it is better to test all of them less intensively, or intensively study a small proportion (Cranor 1995). At the same time, there are some problems that might be solved, if research priorities were such that the right team could be resourced to address itself to them. The interdisciplinary nature of some problems can of itself make them intrinsically less attractive for individual research funders to champion.

It is therefore possible to visualise a spectrum of risk assessment issues based on the strength of the available science in each case (Figure 9.1). Trans-scientific problems, by nature, lie towards the right of this range, classic laboratory science towards the left.

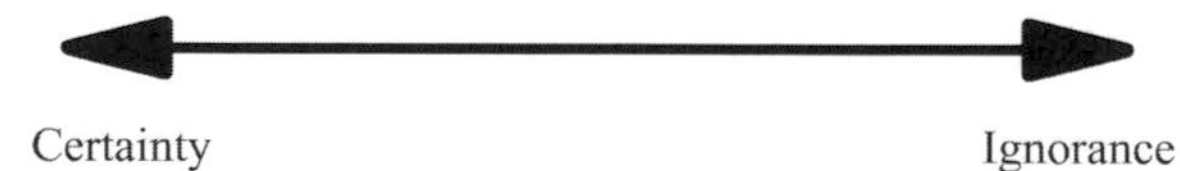

Figure 9.1. Spectrum of uncertainty.

For sensible interpretation of results, users of risk assessments and the studies that may feed into such risk assessments need to know where, along this spectrum, the science relevant to any particular issue lies. Put more strongly, as consumers of the products of science, they need a 'charter' of the quality of what they are being given.

9.2 UNCERTAINTY IN ASSESSMENT OF WATERBORNE RISKS

'One problem with quoting quantitative predicted risks is that the degree of uncertainty is quickly forgotten.' (Gale 1998, p. 1)

Uncertainties in the assessment of waterborne risks will be identified here with reference to the general paradigm for risk assessment provided by the USA National Academy of Sciences. This presents risk characterisation (the core scientific process of estimating risk) as the integration of three distinct stages (NAS 1983).

(1) Hazard assessment looks at the nature and strength of evidence that an environmental agent can potentially cause harm. The evidence may come from tests on animals, coupled with inferences about possible human effects; or from case studies of people known to have been exposed to the agent of interest; or from human volunteer experiments. There are widely recognised limitations in extrapolating animal findings to human populations. There are difficulties in being absolutely sure that the observed responses are indeed caused by the suspected substance, and not by some other cause. There are doubts about how representative an experimental group is of a population more generally, or of sub-groups that may be particularly susceptible. There are differences in treatment efficiencies.

(2) Dose–response assessment aims to specify the relationship between the dose of a substance and the extent of any resulting health effects. Calibration of dose–response models may lead to the identification of critical threshold levels below which there are no observed adverse effects, or alternatively to representation of the classic U shape of the dose–response relationship for chemical essential elements (moderate doses beneficial to health; low and high doses both harmful to health). The conclusions from dose–response assessments are often

controversial, as there can be large measurement errors, misinterpretation of symptoms and often conclusions rely on statistical analysis which is vulnerable to misuse. It is particularly difficult, perhaps impossible, to specify a dose–response model for low levels of concentration. The translation of findings from one species to another as well as from one population to another is problematic.

(3) Exposure assessment seeks to establish the intensity, duration and frequency of the exposure experienced by a human population. There is a great deal of uncertainty here, owing to difficulties in measuring dilute concentrations of substances far from their originating source, limits of detection of some substances, and lack of specific knowledge about species recovery and viability. There are also problems in predicting population distribution patterns relative to those concentrations, in knowing water consumption rates, and in lack of awareness of specific local conditions (such as plumbing or hygiene conditions).

The overall risk characterisation, as the integration of these three stages, produces an estimate of the severity and likelihood of a defined impact resulting from exposure to a specified hazard. It is sometimes expressed as a number or a range. In more sophisticated studies, Monte Carlo analysis might be included as part of the approach (e.g. Medema et al. 1995), in order to account for the full distribution of exposure and dose–response relationships in a distribution of risk. This conveys information on the relative imprecision of the risk estimate, as well as measures of central tendency and extreme values (Burmaster and Anderson 1994). There is, however, no generally accepted way of conveying the overall strength of results (for example of the confidence that one can place in estimated probability distributions, which in turn depends on the state of the science and quality of the data utilised).

Taking Cryptosporidium in tap water as an example, authors have reported a variety of risk assessment results, summarised in Table 9.1. Haas and Rose (1994) have also calculated that during the Milwaukee cryptosporidiosis outbreak people would have been exposed to 1.2 (0.42–4.5) oocysts/litre to account for the level of illness seen.

Table 9.1. Risk assessment results – *Cryptosporidium* in tap water

Risk (95% CI)	Comments	Reference
9.3×10^{-4} ($3.9 \times 10^{-4} - 19 \times 10^{-4}$)	Daily risk of infection with drinking water containing 1 oocyst/10 litres	Rose *et al.* 1995
3.6×10^{-5} ($3.5 \times 10^{-7} - 1.8 \times 10^{-3}$)	Daily risk of infection associated with drinking water supplied from a conventional surface water treatment plant in the Netherlands	Medema *et al.* 1995
0.0009 (0.0003–0.0028)	Median annual risk of infection from exposure to 1 oocyst per 1000 litres of water in non-AIDS adults	Perz *et al.* 1998
0.0019 (0.0003–0.0130)	Median annual risk of infection from exposure to 1 oocyst per 1000 litres of water in AIDS adults	Perz *et al.* 1998
3.4×10^{-5} ($0.035 \times 10^{-5} - 21.9 \times 10^{-5}$)	Daily risk of infection from exposure to New York drinking water	Haas and Eisenberg 2001
0.0001	Annual acceptable risk of infection from drinking water	Macler and Regli 1993

9.3 THE CASE FOR QUALITY AUDIT (QA) OF SCIENCE IN RISK ESTIMATES

'Quantitative risk analyses produce numbers that, out of context, take on lives of their own, free of qualifiers, caveats and assumptions that created them.' (Whittemore1983, p. 31)

Limitations in science generate uncertainty in estimates of waterborne risk. As things currently stand, this uncertainty is of unknown (or unreported) extent and degree. It is without a generally accepted published measure. This is considered to be an unsatisfactory state of affairs that could in principle be addressed if some kind of quality audit was systematically practised. In order for this to be possible, appropriate audit tools need to be developed and tested.

Quality audit has become an increasingly familiar practice in many areas. The higher education sector in the UK, for example, is now familiar with the systematic quality auditing of research and of teaching activity in all university departments, and of the different methodologies that are used in each case. Other examples include the use of certification schemes in product labelling to reflect high quality standards, and more generally the various International Organisation for Standardisation (ISO) quality initiatives.

In the context of waterborne risk management problems, the corresponding need is to know about the quality, strength, or degree of certainty of the science underpinning the risk estimates. However, whereas in the case of teaching quality assessment a low score typically indicates remediable weaknesses ('could do better'), in the case of a quality audit of science, the weaknesses identified are not necessarily remediable.

If it is recognised that uncertainty is an intrinsic quality of many of the fields of science relevant to waterborne risk assessment, then objectively it should be a matter neither of shame nor of concealment to acknowledge this position. On the contrary, it should become a matter of standard practice faithfully to reflect significant uncertainties as part of the 'findings' about how big the risks really are. At the same time, given that it is as yet not standard practice, then research is needed to investigate the best way of doing this, ultimately to develop and refine an appropriate formal protocol for representing and communicating related aspects. Tentative examples of such protocols have begun to emerge in the literature. Further development and testing is needed as a foundation for wider promotion and acceptance of their principles.

If appropriate quality audit tools could be developed and applied, then this should benefit scientific communities by meeting the need for faithful representation of the strength of the knowledge base, thereby, for example, protecting academic disciplines against over-confidence in their outputs and pre-empting accusations of overselling. It should also provide intelligence for the management of research priorities according to areas of uncertainty which are most critical for contemporary policy issues.

Correspondingly it should benefit policy communities and other users of scientific outputs, by providing a diagnostic basis from which to facilitate interpretation of scientific inputs to environmental policy. This will guard against the conferment of undue authority on findings from inherently immature fields (for example, in the setting of regulatory standards). At the same time it should guard against unfounded criticism or rejection of more definitive results. It should also reduce conflict and promote more efficient decision-making, by proactively targeting particularly critical areas:

> To improve the validity of risk estimates, quality assurance principles should be rigidly implemented, and tools for this purpose should be developed. A particular point of attention is the development of structured, transparent methods to precipitate expert opinions in the risk assessment process. (Havelaar 1998)

9.4 A PROPOSED QUALITY AUDIT FRAMEWORK

In the absence of a generally accepted quality audit (QA) procedure for risk assessment science already in the literature, a pragmatic approach, starting from first principles, is presented below. This is based on a checklist of criteria against which the strength of scientific inputs to risk characterisation can be systematically evaluated. This will pinpoint the nature of weaknesses, and provide an overall view of the strength of risk estimates.

The composition of the checklist takes its inspiration from the work of Funtowicz and Ravetz (1990) who pioneered a new numerical symbolism (notation) for representing uncertain scientific inputs to policy decisions. In demonstrating a preference for a checklist as distinct from a notational system, however, what is given below deliberately departs from the Funtowicz and Ravetz formulation. The checklist approach is preferred because it is conceptually simpler while at the same time being systematic and offering flexibility.

The starting point is a simple conceptual representation of the process of producing scientific inputs for waterborne risk assessment (Figure 9.2).

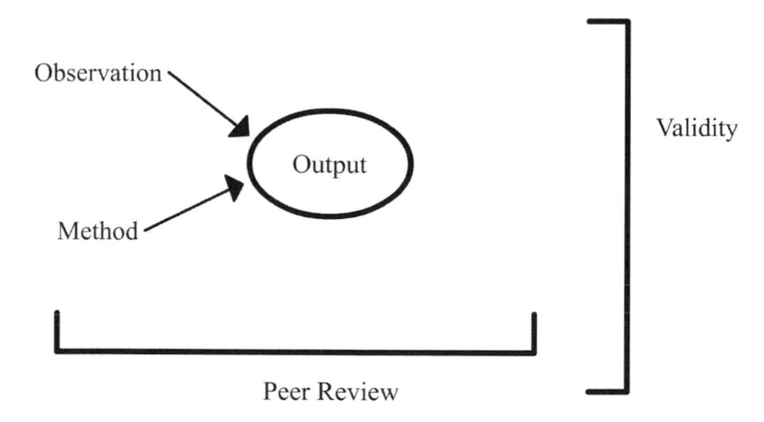

Figure 9.2. Conceptual representation of the quality audit framework components (reprinted from Macgill *et al.* 2000, with permission from Elsevier Science).

As with all scientific endeavour, this process has an empirical or observational aspect (data), and a theoretically informed methodological aspect. These two 'input' aspects combine to produce (as an 'output') an estimate of risk probability, risk magnitude or dose–response effects (or whatever) according to context. Given that the authority or standing of any such 'outputs' can only ultimately be assessed following discourse and review among a peer community, each quantification process should, in principle, be subject to peer review. Consensus, on the basis of peer review, must be a necessary condition for producing definitive quantification. Finally, the relevance (or validity) of the quantified outputs to a particular context of interest must be accounted for.

Having established a conceptual model, each of its aspects provides grounds for interrogating the strength of the scientific inputs to waterborne risk assessment.

For the observational aspect we may ask:

- How close a match is there between the phenomenon being observed to provide data input, and the measure adopted to observe it?
- How reliable is the data or empirical content ?
- How critical is the data to the stability of the result?

For the theoretical/methodological aspect we may ask:

- How strong is the theoretical base?
- How resilient is the result to changes in theoretical specification?

For the result itself we may ask:

- Has a true representation of the real world been achieved?
- Is the degree of precision appropriate?

For the process as a whole we may ask:

- How widely reviewed has it been and what is the reviewers' verdict?
- What is the degree of consensus about the state of the art of the field?

And for the appropriateness to any particular applied context we may ask:

- How relevant is it to the intended application?
- To what extent can we be assured of its completeness?

These five categories of question provide the basis of a checklist for examining the quality of scientific inputs to assessments of risk (Table 9.2). The nature of the questions that have been identified within each category will be considered more fully below, together with further background to the suggested scales for recording an evaluative response to each question.

Table 9.2. Outline quality audit framework

Dimension	Criterion	Question	Level	Score
Observation	Measure	How close a match is there between what is being observed and the measure adopted to observe it?	Primary	4
			Standard	3
			Convenience	2
			Symbolic	1
			Inertia	0
	Data	How strong is the empirical content?	Bespoke/Ideal	4
			Direct/good	3
			Calculated/limited	2
			Educated guess	1
			Uneducated guess	0
	Sensitivity	How critical is the measure to the stability of the result?	Strong	4
			Resilient	3
			Variable	2
			Weak	1
			Wild	0
Method	Theory	How strong is the theoretical base?	Laws	4
			Well-tested theories	3
			Emerging theories/comp models	2
			Hypothesis/stat processing	1
			Working definitions	0
	Robustness	How robust is the result to changes in methodological specification?	Strong	4
			Resilient	3
			Variable	2
			Weak	1
			Wild	0
Output	Accuracy	Has a true representation of the real world been achieved?	Absolute	4
			High	3
			Plausible	2
			Doubtful	1
			Poor	0
	Precision	Is the degree of precision adequate and appropriate?	Excellent	4
			Good	3
			Fair	2
			Spurious	1
			False/unknowable	0

Table 9.2 (cont'd)

Dimension	Criterion	Question	Level	Score
Peer review	Extent	How widely reviewed and accepted is the process and the outcome?	Wide and accepted	4
			Moderate and accepted	3
			Limited review and/or medium acceptance	2
			Little review and/or little acceptance	1
			No review and/or not accepted	0
	State of the art	What is the degree of peer consensus about the state of the art of the field?	Gold standard	4
			Good	3
			Competing schools	2
			Embryonic field	1
			No opinion	0
Validity	Relevance	How relevant is the result to the problem in hand?	Direct	4
			Indirect	3
			Bare	2
			Opportunist	1
			Spurious	0
	Completeness	How sure are we that the analysis is complete?	Full	4
			Majority	3
			Partial	2
			Little	1
			None	0

Also 'scores' under each criterion for unknown (–) and not applicable (n/a)

9.5 THE FIVE ASPECTS OF THE QA FRAMEWORK

9.5.1 Observation

Three types of potential empirical weakness have been identified: first, weaknesses in the appropriateness of the measure used to observe a given phenomenon of interest; second, weaknesses in the extent of empirical observation (data) available; third, sensitivity of results to changes in data inputs.

Weaknesses in the appropriateness of the measure used to observe a given phenomenon potentially arise because there is often no direct (fundamental) measure of the phenomena of interest, so an indirect measure has to be used. Well-known examples include: the use of indicator species; the spiking of laboratory samples to infer 'untraceable' elements; the use of sampling to infer characteristics of a larger (unobservable) population; the use of available (rather than desirable) levels of aggregation or resolution, for example, in measures of pollutant levels; the use of laboratory animals as 'surrogates' for human subjects; the tendency for census enumerators simply to count what is obvious to their own common sense with no guarantee of consistency from one enumerator to another. In all such cases it is desirable to know how well the given indicator represents what it is being used to depict. A qualitative scale for representing this is included in Table 9.2. Corresponding scales are suggested for all other criteria below.

A good empirical base is a prerequisite for definitive science. However, in practice, and notably in the field of environmental risk assessment, the quality of data collection can be extremely variable. Considerations of cost, for example, may mean that water quality measurement is restricted to a single sample at a given site, rather than a range of samples at different depth and spatial co-ordinates across that site.

In principle it is possible to conceive a quality range running from reliable primary data of controlled laboratory standard, or as compiled by a first-rate task force, to secondary data of lesser quality – including proxy measures and sheer guesswork (educated or otherwise). While inexpert guesses will typically be given little if any standing, educated guesses should also be interpreted with caution, because of the potential of systematic biases.

The criterion of sensitivity asks whether results are resilient to changes in inputs (data, parameter values, etc.). Formal sensitivity analysis can test this to some extent, examining the existence and impact of critical values, and framing answers in explicit probability terms. Where sensitivity analysis has not been undertaken, one may wish to judge estimates rather differently from where it has.

9.5.2 Method

There would be little more than a 'chaos of fact' if there were no coherent recognition of why certain sorts of measurement were wanted, and not others, if no general patterns could be discerned among the different elements of empirical evidence available, if there were no awareness of what constituted critical measurement, or if there were no intelligent base to the way in which empirical inputs were to be processed or combined in a model. The theoretical aspect comes into play here.

Depending on the degree of understanding of the real world, this may range (at best) from laws to (less than desirably) working definitions. The hypothesis is the elementary testable theoretical statement for the study, which may be either refuted or accepted. Even the 'emerging theory' place on the scale has only a score of '2', because of susceptibility to hypothesis errors.

Robustness calls for an examination of the resilience of the output (or estimate) to a change in theoretical specification. In some cases, change in theoretical specification may have little effect, while in others, change in model specification may be critical.

9.5.3 Output

This aspect explores possible deficiencies arising from the formal operation of theoretical approaches on empirical inputs. They include: constant and systematic errors of technical measurement instruments (lens distortion in aerial cameras, atmospheric dust distortion, optical and electromagnetic measurements, temperature change altering the length of a physical measure); random and systematic (e.g. spatial autocorrelation) errors in statistical analysis; deficiencies in specification or calibration of mathematical models (in terms of overall fit, and in terms of specific refinements). In recognition of such factors, criteria of precision and accuracy are now routinely scrutinised in a number of fields. Their inclusion in the current framework is a means of scrutinising the correctness (appropriateness) of the precision represented.

Accuracy seeks to gauge whether the science has achieved a 'true' representation of the real-world phenomena under consideration. In some cases conventional goodness of fit statistics are (or can be) built into quantification processes. A 99% confidence limit would be 'good'; 95% might be 'fair', and so on, according to context. In other cases, however, the question of accuracy cannot be answered conclusively, or even directly, either because of inability to 'observe' the reality directly (for example, in forecasting contexts), or because of lack of agreement about suitable terms in which comparisons with reality should be made. Such difficulties are better acknowledged than ignored. It is also worth noting the trade-off: a quantitative estimate given originally as a range may warrant a higher 'accuracy' rating than one given as a point estimate, or a narrower range, for the former has more scope for spanning the 'true' value.

The finer the scale of measurement, the greater degree of precision being represented (parts per billion compared to parts per million; seven versus two significant digits). From a quality assurance point of view, it is necessary to know that the scale of measurement is appropriate for the phenomenon being represented. Rogue examples include the publication of indicators to five or six significant digits when many of the source statistics were more coarsely specified, or reporting of chemical pollutants to a scale that is beyond their limits of detection. It is also necessary to know that rounding errors are valid and whether point estimates have been given when ranges or intervals would have been more appropriate.

Errors within the margins of distortion already allowed for in the degree of precision adopted for representing the result need no further consideration. For those that are not, it should be a matter of normal practice to incorporate appropriate correction factors, or specify error bars, confidence margins or other conventions in order to make due acknowledgment of them (these are automatically given in many statistical techniques, though not always rigorously implemented). Where this is done, a high precision score will be achieved. Where it is not, the score will be correspondingly low. Where precision is inherently problematic, qualitative representation of scientific outputs may be better than quantitative (numerical) expression of findings.

9.5.4 Peer review

This aspect captures one of the basic elements of the development of scientific knowledge – that of peer acceptance of the result. It is not sufficient for an individual or private agency simply to perform scientific investigations within their own terms and without a broader view. To claim a contribution to scientific knowledge, the result must be accepted across a peer community of appropriate independence and standing. The truth claim of any knowledge can only ultimately be assessed via discourse, and ultimately through consensus. Peer review is a fundamental element of the development of scientific knowledge.

In practice, review may be limited to self-appraisal, or a private group (as with consultancies and industrially funded and commercially confidential work), or it may extend quite widely to independent verification within a full, international peer community. It is also necessary to know about the outcome. The result may achieve widespread acclaim and endorsement. On the other hand, it may be severely criticised and even ridiculed.

The second of the theoretical aspects (state of the art) operates at a deeper level than the first (theoretical base) and provides a contextual backcloth for the latter. It sets out what can be expected in the light of the state-of-the art of a given field. One cannot expect to find well-tested theories in an embryonic field, and may need some convincing argument to tolerate mere speculation from an advanced field. The range is given from mature to ad hoc.

9.5.5 Validity

This invites assessment of the appropriateness of an estimate to the 'real world' problem to which it ostensibly relates, i.e. policy relevance. As is widely appreciated, model resolutions can be frustratingly deficient; models valid only for short-term projections are called on to produce long-term scenarios; highly aggregated generalised models are used for specific inferences; serious mismatches can arise between the questions that risk managers need to address and issues that science can articulate. In some cases there may be ambiguity and a lack of consensus over the appropriate measure or indicator for a given problem. Owing to an absence of definitive context-specific knowledge about particular instances of environmental risk, it is often necessary to draw on knowledge from contexts believed to be similar in deriving risk estimates.

Experiments on animals under laboratory conditions may be the best available source of knowledge about the effects of certain radioactive isotopes on human beings (to conduct corresponding experiments on humans would be forbidden on ethical grounds). However, what remains unknown is the degree of transferability of that knowledge to humans under non-laboratory conditions (or even to the same species and type of animal under non-laboratory conditions). To take a different kind of example, historical data may be the best available source of information about certain sorts of failure rates of buildings, but again the degree of transferability of that knowledge to present-day conditions is unknown. And by way of further example, simulation models are by definition an artificial representation of a phenomenon or system of interest. A trade-off here with other aspects is very evident. The requirement for policy relevance can place unachievable demands on data quality.

If the logic tree used to represent the possible pathways of risk is incomplete (i.e. possible cause-effect links are missing) then this will critically undermine the assessment of risk. Many hazard incidents have occurred because of such omissions i.e. unforeseen possibilities. For example, the Exxon Valdez oil tanker crossed over a buffer lane, a lane reserved for incoming tankers, and an additional stretch of open water, before coming to grief. These had not previously been identified as credible events. At the Three Mile Island nuclear power plant, the valve failed to close (though an instrument panel showed that it had closed); again this had not been previously identified as credible. The Cleveland industrial fire in 1944 caused 128 deaths because the consequence of a spill with no containment had not been foreseen and therefore had not been built into the risk estimates.

Circumstances that render risk assessment particularly vulnerable to 'completeness' pitfalls (Freudenberg 1992) are:

- When the system is complex
- When there are gaps in knowledge about low probability events
- When there are substantial human factors
- When the system is untestable – an inherent characteristic of the real world settings of many waterborne risk contexts.

9.5.6 Summary

The 5 aspects have generated a total of 11 criteria against which the quality of risk assessment science can be examined. Risk estimates can be evaluated with respect to each of these criteria, generating a string of 11 scores. High scores will be a cause for comfort as they indicate a strong mature science, of direct policy relevance. Low scores will be a cause for caution, as they indicate science that has acknowledged weaknesses. Although a cause for concern and caution, they should not be a cause for shame or concealment – they are simply a measure of where we are – it is not necessarily possible to do any better.

9.6 REPRESENTING THE OUTPUTS

The simplest form of representation of the outcome of applying the above framework is as a string of scores. These, in turn, might be depicted graphically by way of a more immediately accessible visual representation. Figure 9.3, for example, is a graphical representation of the results from applying the current quality audit framework to two different sets of values for drinking-water consumption. Roseberry and Burmaster (1992) report a well founded sampling method, present upper and lower bounds for monitored consumption levels, and their results are now widely quoted and accepted. The US Environmental Protection Agency (EPA) figure of two litres has filtered into relatively widespread use, although its provenance is not a matter of verifiable record. Note that Figure 9.3 (along with Figure 9.4) has a total of 12 criteria, because it was based upon an earlier version of the framework, before 'extent' and 'acceptance' were combined to form a single category.

If a single, aggregate, indicator is required, the scores from each criterion can be added together, converted into a percentage rating, and evaluated against some standard set of benchmarks, to represent the degree of comfort that can sensibly be placed in the result, for example:

0–20%	Poor
20–40%	Weak
40–60%	Moderate
60–80%	Good
80–100%	Excellent

An aggregate score of 28 out of a possible 44 would translate to 63.5%, and its strength could accordingly be reported as being 'good'. In the case of the results given in Figure 9.3, the aggregate score for the Roseberry and Burmaster study is 37.5 out of a possible 48 (based on 12 criteria) yielding a rating of 78% (good); for the US EPA data, on the other hand, the aggregate score is 10.5, yielding a rating of 22% (weak).

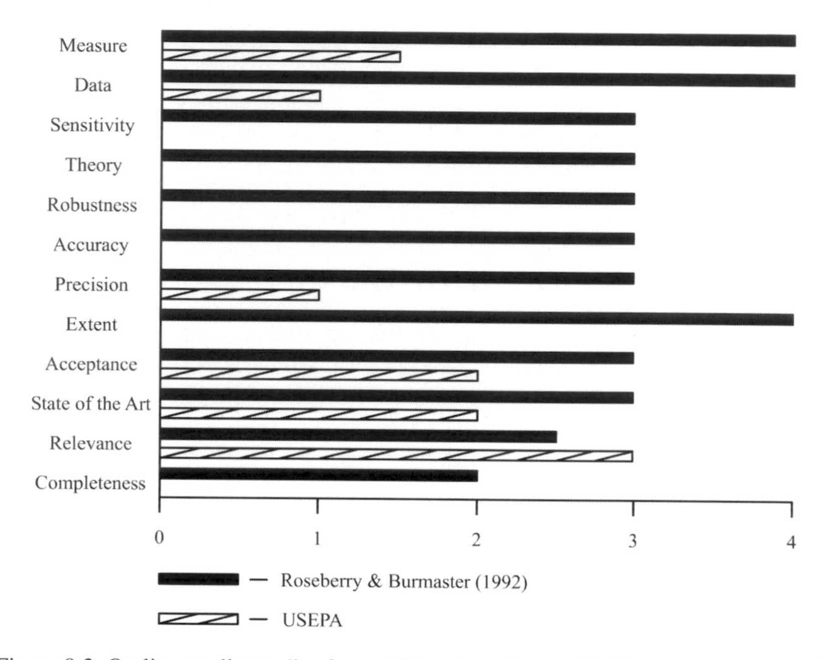

Figure 9.3. Outline quality audit of two different studies on drinking water consumption.

Not all criteria will necessarily be applicable in every context. Moreover, if they are all applicable, it may be appropriate to give a different relative weighting to each in the aggregation (as with weighted average multi-criteria methods more generally).

Where different types of scientific input are used in combination in a risk assessment, the issue arises of whether the quality audit should be applied to the composite result for the system as a whole, or whether distinct quality audits should be undertaken for individual components of the system in turn. In the former case (a composite audit) the audit process itself may be kept to manageable proportion overall, but the nature of the constituents may be so mixed as to make it difficult to apply the criteria in a meaningful way. In the latter case (a series of audits on individual components) the audit process will need to be repeated several times, but each application should have a coherent focus. The question of how best to combine the outputs of multiple audits raises further issues. For now, it is suggested that a 'weak link' rule is appropriate, in other words, the lowest score is taken for each category and the final assessment is based on a table composed of such scores (see Macgill *et al.* 2000 for an example).

9.7 APPLICATIONS

The authors have applied the framework (or variants thereof) to a range of different examples of the assessment of waterborne risks. A high degree of convergence between different experts as to the criterion scores for specific cases has been found.

Its application to the determination of *Cryptosporidium* risks in drinking water demonstrated stark differences in the strength of available knowledge at three different points along the pathways through which human health risks may be generated (Fewtrell *et al.* 2001). Notably, it is considerably weaker at the consumer's tap (the point of exposure to risk) than at the treatment works, or in terms of environmental monitoring of raw water sources. These findings are summarised in Figure 9.4 (note 12 criteria rather than 11).

9.7.1 Quality audit case study

To illustrate a quality audit in a full format, rather than the summarised results, an example is taken from the wastewater reuse field. A summary of the study is presented followed by an outline audit, showing the reasons behind individual scores.

A study of the health effects of different irrigation types (raw wastewater, reservoir-stored water and rainwater) in agricultural workers and their families was undertaken in Mexico (Cifuentes 1998). The health outcomes examined were diarrhoea and infection with Ascaris. The case control study examined a total of 9435 people over a five-month period. In addition to collecting health and water quality data, information on potential confounding factors (such as socio-economic status, water supply, sanitation provision and so on) was also collected. The raw wastewater and rainfall irrigation areas were well matched in terms of housing conditions, mother's education, water storage and toilet facilities. The principal differences between these groups were the greater proportion of landless labourers in the raw wastewater group and the greater proportion of cereals grown in the rainfall area.

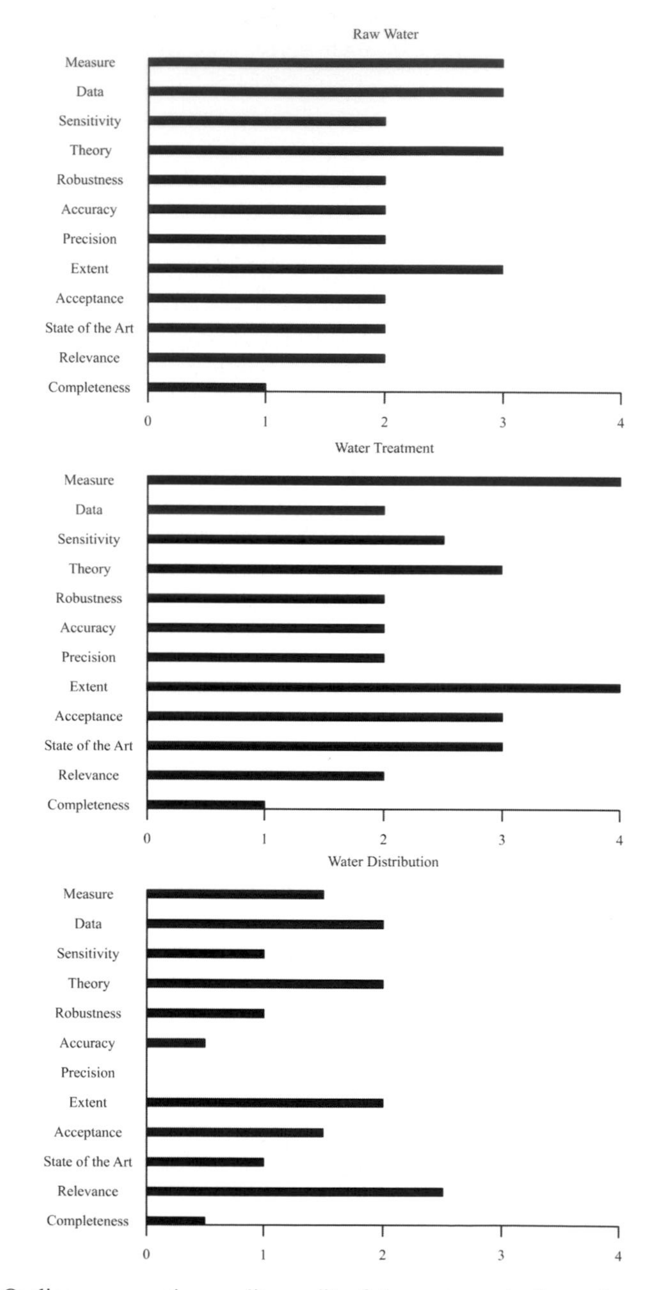

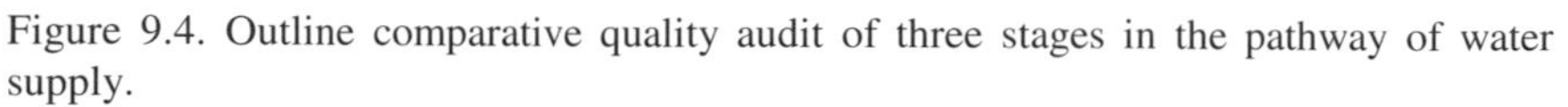
Figure 9.4. Outline comparative quality audit of three stages in the pathway of water supply.

Table 9.3 shows the outline quality audit for this study, based on the ascariasis outcome and the use of the study in terms of feeding into the guidelines process (in terms of 'Validity').

Table 9.3. Outline quality audit of wastewater reuse and levels of ascariasis

	Comments/level	Score
OBSERVATION		
Measure	Cases of ascariasis are being determined through faecal sample examination, and compared according to irrigation type.	
	Primary	4
Data	The empirical content is high, with power calculations conducted prior to the study to establish a suitable sample size.	
	Direct/Good	3
Sensitivity	Taking more than one sample per person may have increased the chances of finding positive cases. Other confounders, not accounted for, may be important.	
	Variable	2
METHOD		
Theory	The idea that pathogens can be isolated from faecal samples is well established, as is the idea that such pathogens may be transmitted via water.	
	Well-tested theory	3
Robustness	This is likely to be reasonable.	
	Variable – Resilient	2–3
OUTPUT		
Accuracy	This is certainly plausible if not better, with account taken for a number of known confounding factors.	
	Plausible	2
Precision	This is appropriate.	
	Fair	2
PEER REVIEW		
Extent	This type of cross-sectional study has been reviewed and, with appropriate note of confounding factors made, is fairly well accepted.	3
State of the Art	**Good**	3
VALIDITY		
Relevance	In terms of guidelines this study is directly relevant.	
	Direct	4
Completeness	The study examined a complete population, accounting for a number of confounding factors. It does, however, only relate to a small geographical area.	
	Partial – Majority	2–3
TOTAL		**31**

The quality audit result of 31 out of a possible 44 (i.e. 70%) demonstrates that it is considered that the study is well conducted, appropriate and can be used with a high degree of confidence. The reasoning behind each individual score is clearly laid out and can be used to stimulate discussion.

9.8 CONCLUSIONS

The case for quality audit of science for environmental policy is increasingly strong. It is not sufficient for experts intuitively to appreciate various areas of uncertainty in terms of which their findings should be qualified. Accountability calls for the evidence to be formally represented, so that all stakeholders can formulate a responsible view. Robust tools are needed for the job. In developing and testing such tools, there will inevitably be a need for compromise over the ideals of simplicity and transparency, on the one hand, and that of achieving a faithful representation of the complexities and subtleties of scientific endeavour, on the other. The framework presented here is offered as a practicable solution that can be the basis for further development and refinement in the future. Such development may include its formulation within interactive communication and information technology systems, in order to facilitate access and deliberative participation on the part of a wider group of experts in arriving at appropriate criterion scores for particular cases.

In summary, the framework outlined here allows outcomes of the risk assessment procedure to be a more transparent process open to scrutiny. Individual quality audit tables also highlight areas that could be improved and provide a platform for debate. Following the QA framework procedure through the risk assessment process should also allow decisions to be updated more easily, since only areas where there have been significant changes need to be re-examined and the results combined with the original assessment.

Widespread adoption of the QA process should prevent numbers from developing a life of their own. It is the antithesis of science to hide data imperfections and doubtful assumptions; on the contrary, there should be openness. There should be no shame in saying 'it's the best there is at the moment' (if of course it really is the best and not just something being used for convenience). If nothing else, then the foundation for the eternal plea for 'more research' will have been clearly established.

9.9 IMPLICATIONS FOR INTERNATIONAL GUIDELINES AND NATIONAL REGULATIONS

International guidelines provide a common (worldwide) scientific underpinning; as such, it is increasingly necessary to have a rigorous quality control procedure. At present, reliance is placed on the quality implied through the peer review process. The idea of a predefined and systematic quality review such as the one defined in this chapter essentially levels the playing field and allows judgements to be made from a common starting point. Such a systematic framework is also valuable at national levels as it provides a means by which unpublished data can be evaluated. Development (by the WHO) of a complementary framework or scoring system outlining the overall strength of evidence and coherence of inputs to international guidelines is underway. Together these will provide valuable input to guidelines and standards development and will also aid in the risk communication process.

9.10 REFERENCES

Burmaster, D.E. and Anderson, P.D. (1994) Principles of good practice for the use of Monte Carlo techniques in human health and ecological risk assessments. *Risk Analysis* **14**(4), 477–481.

Cifuentes, E. (1998) The epidemiology of enteric infections in agricultural communities exposed to wastewater irrigation: perspectives for risk control. *International Journal of Environmental Health Research* **8**, 203-213.

Cranor, C.F. (1995) The social benefits of expedited risk assessment. *Journal of Risk Analysis* **15**(3) 353–358.

Fewtrell, L., Macgill, S., Kay, D. and Casemore, D. (2001) Uncertainties in risk assessment for the determination of drinking water pollutant concentrations: *Cryptosporidium* case study. *Water Research* **35**(2), 441–447.

Freudenberg, W.R. (1992) Heuristics, biases and the not so general publics. In *Social Theories of Risk* (eds S. Krimsky and D. Golding), pp. 229–249, Praeger, Westport, CT.

Funtowicz, S.O. and Ravetz, J.R. (1990) *Uncertainty and Quality in Science for Policy*, Kluwer Academic Publishers, Dordrecht, the Netherlands.

Gale, P. (1998) Development of a risk assessment model for *Cryptosporidium* in drinking water. In *Drinking Water Research 2000,* Drinking Water Inspectorate, London.

Haas, C. and Eisenberg, J. (2001) Risk assessment. In *Water Quality: Guidelines, Standards and Health. Assessment of risk and risk management for water-related infectious disease* (eds L. Fewtrell and J. Bartram), IWA Publishing, London.

Haas, C.N. and Rose, J.B. (1994) Reconciliation of microbial risk models and outbreak epidemiology: The case of the Milwaukee outbreak. *Proceedings of the American Water Works Association Annual Conference, New York,* pp. 517–523.

Havelaar, A.H. (1998) Emerging microbiological concerns in drinking water. In *Drinking Water Research 2000,* Drinking Water Inspectorate, London.

Macgill, S.M., Fewtrell, L. and Kay, D. (2000) Towards quality assurance of assessed waterborne risks. *Water Research* **34**(3), 1050–1056.

Macler, B.A. and Regli, S. (1993) Use of microbial risk assessment in setting US drinking water standards. *International Journal of Food Microbiology* **18**, 245–256.

Medema, G.J., Teunis, P.F.M., Gornik, V., Havelaar, A.H. and Exner, M. (1995) Estimation of the *Cryptosporidium* infection risk via drinking water. In *Protozoan*

Parasites and Water (eds W.B. Betts, D. Casemore, C. Fricker, H. Smith and J. Watkins), pp.53–56, Royal Society of Chemistry, Cambridge.

NAS (1983) *Risk Assessment in the Federal Government: Managing the Process*, National Academy Press, Washington DC.

Perz, J.F., Ennever, F.K. and le Blancq, S.M. (1998) *Cryptosporidium* in tap water. Comparison of predicted risks with observed levels of disease. *American Journal of Epidemiology* **147**(3), 289–301.

Rose, J.B., Lisle, J.T. and Haas, C.N. (1995) Risk assessment methods for *Cryptosporidium* and *Giardia* in contaminated water. In *Protozoan Parasites and Water* (eds W.B. Betts, D. Casemore, C. Fricker, H. Smith and J. Watkins), pp. 238–242, Royal Society of Chemistry, Cambridge.

Roseberry, A.M. and Burmaster, D.E. (1992) Log-normal distributions for water intake by children and adults. *Risk Analysis* **12**(1), 99–104.

Weinberg, A. (1972) Science and trans-science. *Minerva* **10**, 209–222.

Whittemore, A.S. (1983) Facts and values in risk analysis for environmental toxicants. *Risk Analysis* **3**(1), 23–33.

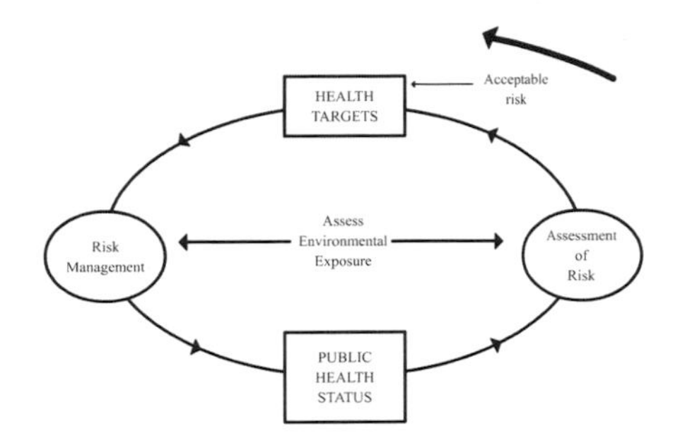

10

Acceptable risk

Paul R. Hunter and Lorna Fewtrell

The notion that there is some level of risk that everyone will find acceptable is a difficult idea to reconcile and yet, without such a baseline, how can it ever be possible to set guideline values and standards, given that life can never be risk-free? Since zero risk is completely unachievable, this chapter outlines some of the problems of achieving a measure of 'acceptable' risk by examining a number of standpoints from which the problem can be approached.

10.1 INTRODUCTION

A number of chapters within this book examine the question of what is risk and how we define it. Risk is generally taken to be the probability of injury, disease, or death under specific circumstances. However, this 'objective' measure of risk does not tell the whole story and, in determining acceptability of any particular risk, perceived risk is likely to play a large role.

The following is a list of standpoints that could be used as a basis for determining when a risk is acceptable or, perhaps, tolerable. These will be explored under broad headings.

A risk is acceptable when:

- it falls below an arbitrary defined probability
- it falls below some level that is already tolerated
- it falls below an arbitrary defined attributable fraction of total disease burden in the community
- the cost of reducing the risk would exceed the costs saved
- the cost of reducing the risk would exceed the costs saved when the 'costs of suffering' are also factored in
- the opportunity costs would be better spent on other, more pressing, public health problems
- public health professionals say it is acceptable
- the general public say it is acceptable (or more likely, do not say it is not)
- politicians say it is acceptable.

10.2 A PREDEFINED PROBABILITY APPROACH

One definition of acceptable risk that has been widely accepted in environmental regulation, although is not relevant to microbiological parameters, is if lifetime exposure to a substance increases a person's chance of developing cancer by one chance in a million or less. This level, which has come to be taken as 'essentially zero', was apparently derived in the US in the 1960s during the development of guidelines for safety testing in animal studies. A figure, for the purposes of discussion, of 1 chance in 100 million of developing cancer was put forward as safe. This figure was adopted by the Food and Drug Administration in 1973, but amended to one in a million in 1977 (Kelly and Cardon 1994). This level of 10^{-6} has been seen as something of a gold standard ever since. The US Environmental Protection Agency (EPA) typically uses a target reference risk range of 10^{-4} to 10^{-6} for carcinogens in drinking water (Cotruvo 1988), which is in line with World Health organization (WHO) guidelines for drinking water quality which, where practical, base guideline values for genotoxic carcinogens on the upper bound estimate of an excess lifetime cancer risk of 10^{-5} (WHO 1993).

Similar approaches have been adopted elsewhere and for other risks. In the UK, for example, the Health and Safety Executive (HSE) adopted the following levels of risk, in terms of the probability of an individual dying in any one year:

- 1 in 1000 as the 'just about tolerable risk' for any substantial category of workers for any large part of a working life.
- 1 in 10,000 as the 'maximum tolerable risk' for members of the public from any single non-nuclear plant.
- 1 in 100,000 as the 'maximum tolerable risk' for members of the public from any new nuclear power station.
- 1 in 1,000,000 as the level of 'acceptable risk' at which no further improvements in safety need to be made.

The HSE set these guidelines after considering risks in other contexts, with a risk of 1 in 1,000,000 being roughly the same as the risk of being electrocuted at home and a hundredth that of dying in a road traffic accident (RCEP 1998). Interestingly, although the final figure of one in a million appears to be the same as that followed in the US, the figure in the UK is an annual rather than lifetime risk.

With regards to microbiological risks from drinking water, the US EPA, using *Giardia* as a reference organism, required that the microbial risk is less than 1 infection per 10,000 people per year (Macler and Regli 1993). The logic behind the choice of *Giardia* was that it was known to be more resistant to disinfection than most other microbial pathogens (although *Cryptosporidium* sp. has since challenged this 'status'). Therefore, protection against *Giardia* infection should provide protection against other organisms, with the intention of minimising all microbial illness.

It is interesting to compare the levels of protection between microbiological and carcinogen risk. If it is assumed that there is a 50–67% frequency of clinical illness following infection with *Giardia* (Gerba *et al.* 1996) then, using the lower bound of 50%, this translates to an annual risk of illness of 1 in 20,000. Gerba and colleagues do not cite a case-fatality rate for *Giardia*, but 0.1% in the general population seems to be a reasonable level based on other pathogens causing gastrointestinal symptoms (Gerba *et al.* 1996; Macler and Regli 1993). This results in an annual risk of death of 1 in 20,000,000. Converting this to a 70-year lifetime risk to be comparable with rates cited for chemical contaminants results in a risk of 1 in 2×10^{-5}, a figure that is similar to that considered acceptable by the WHO for carcinogenic risks.

The outcome of infection, however, will vary according to a number of factors and many groups within society, such as the young, elderly, malnourished and so on are more susceptible to developing illness following infection than the general population. This is a theme that we will return to in a later section but, clearly, the level of protection will not be the same for all people.

Examination of what is currently being achieved versus what is claimed to be an acceptable risk makes for interesting and sobering reading. Haas and Eisenberg, in Chapter 8 of this book, outline a risk assessment of drinking water supplies in New York City and the risk of infection with *Cryptosporidium*. They estimate that the risk is some two orders of magnitude greater than the acceptable level. Such results back up the work of Payment and Hunter (see Chapter 4) who claimed that a very high proportion of gastrointestinal illness could be attributed to tap water, even if it met current water quality guidelines.

10.3 A 'CURRENTLY TOLERATED' APPROACH

The basic argument here is that any risk that is currently tolerated is considered to be acceptable. This approach was used by the US EPA in setting the allowable bacterial indicator densities for bathing waters (US EPA 1986). The work of Cabelli and Dufour (Cabelli *et al.* 1979, 1982, 1983; Dufour 1984) allowed health effects, in terms of swimming-associated gastroenteritis rates, to be estimated. It was established that previous standards had resulted in a gastrointestinal illness rate of 8/1000 bathers at freshwater sites and 19/1000 bathers at marine sites. These levels were considered to be tolerated (as people still used the bathing areas) and were therefore assumed to be acceptable. The new standards were based around this acceptable level.

A similar approach has been suggested by Wyer *et al.* (1999) in their experimental health-related classification for marine waters, using other risk factors as measures of acceptability. This work was based on extensive epidemiological studies conducted around the UK coastline that resulted in a dose–response relationship between the bacterial indicator faecal streptococci and gastroenteritis experienced by bathers. The dose–response relationship was found to be independent of, and not confounded by, other predictors of gastroenteritis, including the transmission of gastroenteritis from household members (termed person-to-person transmission) and a composite factor termed non-water-related risk. Each of these factors had an associated probability against which the dose–response to sea bathing could be compared. The combination of the exposure distribution (based on five years of water quality data), dose–response relationship and independent risk factors provide a standard system which is health-related. Such detailed data, however, do not exist for most countries; predictors of gastroenteritis are likely to vary markedly between different locations, and their 'acceptability' may also be culturally specific.

If an informed choice element is factored into such an approach (which is the case in the examples outlined above) such an approach may provide a promising way forward. The use of accepted risk as synonymous with acceptable risk should, however, be treated with great caution. A number of authors have noted that there is a difference between the two (Jones and Akehurst 1980; O'Riordan 1977). Using smoking as an example, until recently this has been widely accepted but is regarded today by many as unacceptably risky (Royal Society 1983). Such usage also ignores aversion behaviour on the part of the public and the fact that any risk (such as bathing in coastal waters) may only be accepted by a sub-section of the population.

10.4 A DISEASE BURDEN APPROACH

In everyday life individual risks are rarely considered in isolation. Similarly, it could be argued that a sensible approach would be to consider health risks in terms of the total disease burden of a community and to define acceptability in terms of it falling below an arbitrary defined level. For example, it may be thought that drinking water supplies should not be responsible for more than 5% and food no more than, say, 15% of cases of gastroenteritis. Such an approach is clearly useful in terms of setting priorities. In reality, attributing cases of illness to a specific cause when there is more than one route of transmission is fraught with difficulties (see Chapters 4 and 5). This, coupled with known under-reporting of gastroenteritis in countries with surveillance systems (Chapter 6) and the difficulties in extrapolating illness data to countries with limited surveillance systems experiencing very different sanitation conditions, may reduce the value of this approach.

A further problem with the disease burden approach is that the current burden of disease attributable to a single factor, such as drinking water, may not be a good indicator of the potential reductions available from improving water quality. For diseases where infection is almost universal, such as viral gastroenteritis, reducing the disease burden by one route may have little impact on the overall burden of disease. Those people who have not acquired their infection (and hence degree of immunity) from drinking water may well acquire their infection from another source (see Chapter 5).

10.5 AN ECONOMIC APPROACH

In the strict economic sense a risk is acceptable if the economic savings arising out of action to reduce a risk outweigh the cost of such action. This approach is, in effect, a simple cost-benefit analysis (Sloman 1994). For example, consider the situation that may arise over improving the quality of sea bathing waters. Following investigations it may be estimated that the cost of installing new sewage treatment facilities are some £10,000,000. The risk to bathers may be acceptable if the cost of illness from swimming in the sea over the lifetime of the new treatment works is only some £1,000,000 after taking account of inflation. The risk would be unacceptable if the cost of illness would be £20,000,000.

There are, however, many difficulties with this apparently simple approach. These include the fact that the exact amount of illness may not be known with any certainty, especially if much illness is related to specific outbreaks. Even if the amount of illness is known, the costs of that illness may be difficult to identify. Even if the costs are identifiable, the costs of illness are borne by different groups in society to those that bear the cost of the new sewage plant. Furthermore, in a humane society, we would argue that identifiable financial costs are not the only and probably not the main reason for change. These difficulties with the simple cost-benefit model will be discussed below and possible solutions identified.

Perhaps the most obvious problem is the issue of costing risk when this involves an element of probability. We may know that the probability (risk) of a major untoward event, such as an outbreak, in any given year is 0.02, say, but how does this help in deciding what to do when financial planning cycles last say five years? The most likely outcome (p = 0.904) is that no outbreak will occur in the five years and so any money spent on reducing this risk will be wasted nine times out of ten. This problem can be dealt with by simply multiplying the cost of the event saved by the probability of its occurrence (Sloman 1994). For common events, such as the risk of gastrointestinal disease in people taking part in sea bathing the annual number of cases of illness are likely to be more consistent from one year to the next and it may then be possible to do a more straightforward cost-benefit analysis. Unfortunately, other problems are less easily resolved with this simplistic economic approach.

The next problem for many societies is that the costs of risk reduction are incurred by different groups to those that benefit from the reduction in risk. Let us return to the bathing beach study. For a privatised sewage utility, the costs of the additional sewage treatment works would be incurred by the shareholders if the costs could not be passed on in higher bills and by the customers if these costs could be passed on. Identifying the groups that would benefit is more difficult. Those people who go swimming in the sea would benefit from a lower risk of illness. If such illness led to absence from work, employers would benefit. If illness led to use of health-care systems then the health service may benefit. To add further complexity to the issue, it may be the case that improvements in bathing beach quality would lead to increased tourism with further financial benefit to the local society and industry. It may, however, be possible to calculate the costs and benefits of the new treatment works to society as a whole. However, it is likely that different stakeholders will not be able to agree on the methods used to calculate these different costs and benefits. Clearly the resolution of these issues is political rather than economic in nature.

So far, in our discussion of costs we have assumed that all costs can be derived in monetary terms. Consequently we have been able to include in the costs of illness such things as loss of income due to absence from work and cost to health services from patients seeking treatment. But the major impact of illness associated with polluted beaches may not be measurable in such terms. For example, the upset from a ruined holiday or the pain and distress associated with illness cannot be directly measured in monetary terms. In any caring society, these factors must also be taken into consideration when assessing whether any risk is acceptable. For dealing with these types of issues, economists have developed a variety of cost-utility measures (McCrone 1998; Sloman 1994).

In general terms, utility can be defined as the satisfaction or pleasure that an individual derives from the consumption of a good or service (Pass and Lowes 1993). Cost-utility analysis attempts to place a value on the 'satisfaction' gained from an intervention and relate this to the cost of the intervention. In health economics one technique has tended to become a standard, that of Quality Adjusted Life Years, widely known as QALYs (McCrone 1998; Weinstein and Stason 1977). QALYs are designed to combine two independent concepts of utility, length of life and its quality. This assumes that such concepts can themselves be measured. The QALY can then be used to derive a monetary value using a marginal cost per QALY gained (National Association of Health Authorities and Trusts 1992). This financial estimate can then be inserted into the cost-benefit models described above. The problem is that QALYs have been subject to a significant amount of criticism which has led to various alternative measures being suggested (Nord 1992; McCrone 1998). An additional problem is that the allocation of a marginal cost per QALY is also highly subjective and would vary from one community to another.

A further economic insight into the issue of defining acceptable risk comes from the concept of opportunity cost. Opportunity cost can be defined as the measure of the economic cost of using scare resources to produce one particular good or service in terms of the alternatives thereby foregone (Pass and Lowes 1993). In our seawater and sewage example, if the water utility had available only £10,000,000 to spend on capital works, would it be better to spend it on improving the treatment of sewage or on another project to improve drinking water treatment to reduce risk of cryptosporidiosis? Fairley and colleagues (1999) recently used a simple form of opportunity cost analysis to argue against the introduction of regulations requiring regular monitoring of drinking water for the presence of *Cryptosporidium* oocysts. For wider issues, how can a developing nation determine how best to spend its scarce resources, between funding improved water treatment to meet stricter microbiological standards for drinking water and spending this money on improving obstetric care?

In conclusion, the science of economics does not provide society with absolute tools for determining what risks are acceptable. Nevertheless, no assessment of acceptable risk can afford to ignore economic imperatives. Economics can and should inform this debate in a very powerful way. It seems to us that cost-benefit analysis and cost-utility analysis should be part of any review of microbial standards and acceptable risk. Perhaps the most powerful economic tool in this context, however, is the issue of opportunity costs. No society can afford to tackle all risks simultaneously and thus priorities have to be set. An economic definition of acceptable risk now becomes: any risk where the costs of reducing that risk exceed the financial and utility benefits that would arise from that reduction and where such resources required in this risk reduction would not be better spent on other public health issues.

10.6 THE PUBLIC ACCEPTANCE OF RISK

This approach to determining acceptable risk is based on what is acceptable to the general public. In other words, a risk is acceptable when it is acceptable to the general public. In democratic societies, so the theory goes, the views of the general public are pre-eminent when determining what is and what is not acceptable risk. While perhaps superficially appealing as a model for determining levels of acceptable risk, this approach immediately raises a number of theoretical and practical problems.

For a public-based approach to acceptable risk to work, all sections of the community must have full access to all information required on levels of risk and have the skills to interpret that information. There must also be an effective means of reaching consensus within the community and canvassing that consensus opinion. Unfortunately, each of these preconditions are unlikely to be met in most circumstances. Some of the difficulties concerned will be addressed in this section.

Many acceptable risk decisions have to be made on the basis of incomplete information even by professionals specialising in the issues of concern (Klapp 1992). It is not surprising, therefore, that even if a society existed with a fully open government, information would not be complete. Even for information that is readily available, individuals' knowledge will often be flawed. For example, it has been known for some time that individuals' judgements about risk levels are systematically distorted. In general, people systematically overestimate the number of deaths due to uncommon causes and underestimate the numbers of deaths due to common causes (Slovic *et al.* 1979).

People's judgements about risk are frequently subject to bias (Bennett 1999); an issue from which experts are not immune. The most common sources of bias are availability bias and confirmation bias. Availability bias increases the perception of risk of events for which an example can be easily recalled. Confirmation bias occurs when individuals have reached a view and then choose to ignore additional information that conflicts with this view. In addition, public acceptability may well depend upon what Corvello (1998) has termed 'framing effect'. An individual lifetime risk of one in a million in the US is mathematically equivalent to approximately 0.008 deaths per day, 3 deaths per year or 200 deaths over a 70-year lifetime. Corvello (1998) notes that many people will view the first two numbers as small and insignificant, whereas the latter is likely to be perceived as sufficiently large to warrant societal or regulatory attention.

A further problem is that individuals perceive the nature of risk in different ways. These differences are often based on deeper societal processes. One model for describing these differences is cultural theory (Thompson *et al.* 1990). Cultural theory divides society along two axes. The first axis is the influence of the group on patterns of social relationships; the degree to which people depend on reference to socially accepted peers for influence. The second axis concerns the degree to which people feel constrained by externally imposed rules and expectations. Using these two axes, four types have been described:

- fatalists
- hierarchists
- individualists
- egalitarians.

Each of these four types differs substantially in their approach to risk (Adams 1997; Langford *et al.* 1999). For example, hierarchists believe that managing risk and defining acceptable risk is the responsibility of those in authority supported by expert advisors. Individualists scorn authority and argue that decisions about acceptable risk should be left to the individual. Egalitarians believe that definitions of acceptable risk should be based on consensus that requires trust and openness. Fatalists see the outcome of risk as a function of chance and believe they have little control over their lives.

Nevertheless there does seem to be some consistent themes in the general public's approach to identifying acceptable risk. These themes are often referred to as 'fright factors' (Bennett 1999). Risks are deemed to be less acceptable if perceived to be:

- involuntary
- inequitably distributed in society
- inescapable, even if taking personal precautions
- unfamiliar or novel
- man-made rather than natural
- the cause of hidden and irreversible damage which may result in disease many years later
- of particular threat to future generations, for example by affecting small children or pregnant women
- the cause of a particularly dreadful illness or death
- poorly understood by science
- the cause of damage to identifiable, rather than anonymous, individuals
- subject to contradictory statements from responsible sources.

While these fright factors result in different priorities amongst the general public than may be generated by professionals relying on statistical estimates of risk, they should not be dismissed as unreasonable (see also Chapter 14). The authors of this chapter would certainly agree with concerns about risk affecting future generations and causing particularly dreadful illness or death. Issues concerning the inequality of risk will be discussed below. Nevertheless, the influence of fright factors makes it very difficult to define acceptable risk based on the public's perception. Using approaches to defining acceptable risk on economic or epidemiological criteria may not be acceptable to society if fright factors are not taken into consideration.

Even if the difficulties so far described in this section can be overcome, there remains the problem of adequately canvassing the consensus of the general population. Even in democratic societies it is frequently difficult to directly gauge public opinion. In such a situation, surrogates for public opinion are usually sought. Perhaps the most powerful surrogate for public opinion is the media. However, the media is far from a perfect indicator of public opinion. Indeed, the factors that influence media interest are quite distinct from the fright factors listed above. Factors that increase media interest in an issue (media triggers) include (Bennett 1999):

- blame
- alleged secrets or cover-ups
- the presence of 'human interest' through heroes or villains
- links with other high-profile issues or people
- conflict
- whether the story is an indication of further things to come (signal value)
- many people exposed
- if there is a strong visual impact
- sex and/or crime.

The other main source of presumed public viewpoints in determining acceptable risk is the various activist or pressure groups (Grant 2000; Pattakos 1989). However, it is a mistake to believe that pressure groups necessarily reflect public opinion. Each group has its own objectives and will use science and risk assessments that support their viewpoints. Pressure groups are just as likely to be subject to confirmation bias as other members and groups in society. A key source of influence of pressure groups, especially those that use direct action, is the media. Using scientifically balanced risk assessments does not attract the media. Such pressure groups may overestimate risk in order to attract media attention or force change in public opinion in favour of their primary objectives.

In conclusion, it appears that the concept of public opinion as the primary determinant of acceptable risk has serious difficulties. Nevertheless, this does not mean that public opinion can or should be ignored. It has to play a central part in the decision-making process. How this is done can only be a political process; this is the subject of the next section.

10.7 POLITICAL RESOLUTION OF ACCEPTABLE RISK ISSUES

The reader sufficiently interested to have read this far may be forgiven for wondering how society can ever define the 'acceptable' in issues of acceptable risk. It is clear from the discussion to this point that there are many different ways to define acceptable risk and that each way gives different weight to the views of different stakeholders in the debate. No definition of 'acceptable' will be acceptable to all stakeholders. Resolving such issues, therefore, becomes a political (in the widest sense) rather than a strictly health process. This process becomes even more difficult when one considers that most of the evidence brought forward in acceptable risk decisions has wide confidence intervals. In other words, there is a considerable degree of scientific uncertainty about many risk decisions (Klapp 1992).

Whilst the, apparently, more objective approaches to acceptable risk would seem to offer a value-free option, there is still considerable uncertainty around the outcomes of these models. Klapp (1992) describes four types of uncertainty:

- extrapolation
- data
- model
- parameter.

Extrapolation uncertainty arises when experts disagree over whether findings in experimental studies can be extrapolated to real world situations. An example of this is the extrapolation of infectious dose studies to low levels of pathogens. Data uncertainty occurs when experts disagree over which data is relevant to include in risk models. This is especially important when there is conflicting data. Model uncertainty is when experts disagree over which model to use in their risk assessment models, and parameter uncertainty exists when experts disagree on how to estimate parameters for which little data is available. In general, experts are just as likely to fall prey to confirmation bias as are the lay public (Bennett 1999). Indeed, professional pressures for scientists and experts to support their original viewpoints can be immense. If an academic's reputation and future grant and consultancy income is based on his/her earlier work, then there are very strong pressures to disregard new work which devalues that early work. Expert scientific opinion is not, therefore, free from value.

In the absence of scientific certainty, Klapp (1992) argues that acceptable risk decisions arise from a process of bargaining. She draws on the rational choice theory of relations between legislators and the public but argues that legislators do not enact the wishes of the public. Instead she argues that legislators, and courts, make decisions that change the behaviour of bureaucrats. In this she draws on the economic game theory of sequential bargaining with incomplete information (Sutton 1986). This is in turn based on the Sobel-Takahashi multi-stage model of bargaining (Sobel and Takahashi 1983). Basically this revolves around a game involving two players, a buyer and seller, trying to agree on a price for an indivisible good. If both players had complete information on how much the other values the good, then a bargain could be struck immediately. The buyer knows how much both he and the seller value the good, but the seller does not know how much the buyer values the good. To discover this information, the seller has to continue to offer prices until the buyer accepts. The longer the process takes, the more information the seller has about the buyer's valuation of the good. Assume that in acceptable risk decisions, the buyer is the citizen and the seller the bureaucrat. The bureaucrat offers an initial level of risk that may or may not be acceptable to the citizen (or other stakeholder). The bureaucrat does not know at this stage what level of risk the citizen will accept. Clearly it is in the interest of the citizen to continue to reject these offers up to the point that the bureaucrat seeks an alternative route to resolving the problem.

Klapp (1992) then goes on to discuss the principal-agent model of Moe (1984). This model has the advantage in that it specifically focuses on the hierarchical relationship between citizen and bureaucrat, assuming conflict of interest and asymmetries in information. Here the citizen enters into a contract with the politician/bureaucrat in the expectation that the latter will act in the best interests of the citizen. The contract is necessary because the citizen may not have the technical information necessary to make certain regulatory decisions and the task of regulation may be too large and complex for him to undertake. However, for various reasons this relationship is problematic in that the citizen will find it difficult to control the bureaucrat's compliance with the contract. Scientific uncertainty is used by the bureaucrat to enhance his power over the public, who may not have access to such information. The bureaucrat may have his/her own interests which conflict with the citizen's ideas, and it is likely that there will be a gap in the desired and achieved performance of the bureaucrat, at least as far as the citizen is concerned. This model is also problematic in that the bureaucrat starts out as the agent of the citizen, but he subsequently gains control over the citizen.

In her own model, the bureaucratic bargain, Klapp (1992) also proposes that the bureaucrat is in a dominant position relative to the citizen, but still has an incentive to make concession in order to obtain co-operation from the citizen. Although, bureaucrats may have the power to impose their decision, they also want to gain benefit. In particular, they want the voluntary compliance of the citizen in order to avoid potential legal challenges. The bureaucrats also want to 'look good' in administering their regulatory decisions. Thus the bureaucrat has the incentive to negotiate with the public in order to obtain agreements that are mutually satisfactory. Indeed, the bureaucrat expects that such an agreement will be reached. In this model scientific uncertainty is a tool used by the citizen, or experts employed by pressure groups, to make the bureaucrat look incompetent and thus influence the debate and gain concessions.

The three models of bureaucratic bargaining that have been discussed illustrate very important points in the acceptable risk decision-making process. In particular, the hierarchical relationship between some of the key decision-makers and stakeholders, the bargaining nature of the decision-making process, and the use of uncertainty as a political tool by one side or another. The fact that the nature of this bargaining process is increasingly being superseded by recourse to the courts (Klapp 1992) does not substantially alter these conclusions.

Although not explicitly addressed by Klapp, much of the discussion about bargaining between bureaucrats and the public could also apply to bargaining with other stakeholder groups such as industry, health-care providers or other health-care groups.

If we accept this view of risk decisions arising from a bargaining process rather than formal expert analysis, two problems are raised. The first is the problem of satisficing and the second is the problem of stakeholder inequality.

10.7.1 Satisficing

A major weakness of decisions reached through the bargaining process is that frequently the optimal solution is not produced. In other words, instead of the best solution for society, one gets the solution that is acceptable to most/all stakeholders. This is known as satisficing. A problem with satisficing is that not all relevant stakeholders may be considered in defining the acceptable criteria. This will now be discussed in more detail.

10.7.2 Stakeholder inequality

In any national or international policy decision on risk, the list of stakeholders is large. This list will include academic and other experts, government agents, various pressure groups and representatives of business interests. Among the list of stakeholders will also be the public. Each stakeholder will have differing levels of power and interest in the bargaining process. One of the major concerns for the public health professional is that health differs between different sections of society. There has been considerable interest, particularly in the UK, in the issue of health inequality in society (Bartley *et al.* 1998; Townsend *et al.* 1992; Wilkinson 1996). Surprisingly, given the very obvious inequality in infection-related illnesses in both national and global societies, there has been little academic interest in the issue of inequality in infectious disease. The two areas that have been addressed in detail are probably HIV and tuberculosis (Farmer 1999), two diseases that are almost certainly not waterborne. Nevertheless, those working in the diagnosis, treatment and prevention of infectious disease are aware that the distribution and effects of infectious disease is clearly unevenly distributed within society. Different sections of society are more or less likely to suffer from various infectious diseases and, when they do acquire such diseases, they vary in their outcome. The causes of these health inequalities are various, and include genetic, geographical, behavioural and socio-economic factors (Table 10.1).

Table 10.1. Examples of factors that lead to inequality of health risk in relation to waterborne disease

Factor	Affects
Age	The very young and very old are more likely to acquire infections due to naive or waning immunity and, once infected, are more likely to develop more severe outcomes.
Pre-existing disease	A person with AIDS or severe combined immunodeficiency syndrome is likely to suffer far more severe symptoms with cryptosporidiosis and other infectious illnesses.
Genetic	People with certain genotypes are more likely to experience complications such as joint problems following gastrointestinal infections.
Gender/pregnancy	Certain infections are more severe in pregnancy, either increasing the risk of fatality for the woman (hepatitis E), or damage to the foetus (toxoplasmosis).
Behaviour	The amount of unboiled tap water an individual drinks will affect their risk of a waterborne infection.
	Foreign travel will expose an individual to risk of waterborne diseases that he will not have encountered at home.
	Other behaviours such as swimming will increase an individual's risk of acquiring infections by routes other than drinking water.
Socio-economic	The poorest members of society may suffer more severe disease due to malnourishment.
	The poorest members of society may suffer more serious economic consequences of illness because they are in jobs that do not pay sick leave or are not covered by health insurance.
	The poorest members of society may not have ready access to health care.
	Many waterborne diseases are more likely to spread to family members in overcrowded conditions.
Geography	Various waterborne diseases have marked geographical distributions; hepatitis E is largely restricted to tropical countries and tularaemia is more common in northern latitudes.
	The quality of water treatment and distribution systems differ markedly from one country to another and between locations in the same country.

One of the important conclusions that arises naturally out of any consideration of the factors that lead to inequality of health risk in relation to waterborne disease is that many of these same factors – age, gender, disability and poverty – are associated with the causes of social exclusion (Byrne 1999; Jordan 1996). The main danger of any bargaining process for risk is that of ignoring the concerns of the socially excluded groups within society. Powerful groups in the bargaining process will be industry, the wealthy and the educated. These groups will have greater access to information, and the resources and confidence to prepare their arguments. Those groups who are most likely to suffer the adverse risks are less likely to influence the debate. This is of particular concern when bargaining is resolved through satisficing. Who will know whether the solution proposed is acceptable to the socially excluded?

10.8 CONCLUSIONS

From this chapter we can conclude that acceptable risk decisions are rarely easy. In general terms one can broadly classify those approaches that emphasise formal analysis and expert opinion such as the probabilistic, economic or disease burden approaches and those that emphasise the political bargaining processes. This division of approaches could be taken to imply a clash between objectivity and subjectivity or between value-laden and value-free approaches.

The implication is that the approaches based on expert knowledge and methods are scientifically exact. Experts will be able to develop appropriate standards based on existing epidemiological and economic knowledge. Unfortunately, as we have already discussed there remains significant uncertainty around many of the processes and models that experts rely on to make their judgements. Furthermore, most experts typically do not directly express uncertainty about facts (Morgan *et al.* 1978). Indeed, professionals' opinions are frequently value-laden. Professionals derive their own values from a variety of sources (Fischhoff *et al.* 1981). As members of society, these individuals will clearly derive many of their own values from the wider society. However, professionals will also derive values from their profession. Some experts views will also be governed by pecuniary interests and take on the values of their employing organisation. These values will have a strong role in influencing the advice that experts give and the processes they go through to arrive at this advice. Consequently, we have to accept that experts form just one of several different stakeholder groups that does not necessarily have higher status over other stakeholders.

On the other hand, we have also considered the problems involved in taking a purely bargaining approach. Bargaining approaches often produce less than optimal solutions to problems especially when different stakeholders have different power, knowledge and resources. Even in societies that wish to include the public's view, it may be impossible to accurately determine the public viewpoint. The public's view on risk is often contradictory and at times may be considered irrational. There are dangers in relying on pressure groups or the media as proxies for the public view. More important for any bargaining approach was the problem of health inequality and social exclusion. Those groups most at risk are likely to have least influence in any debate in many societies.

Given all these problems, the reader may then be forgiven for despairing of ever finding an appropriate acceptable risk approach to setting standards. What can be done? We suggest that this is where public health professionals and public health organisations such as the World Health Organization have an important role.

The role of public health medicine in many societies has changed in recent years. Nevertheless, the broad responsibility of public health practitioners can be summarised as the prevention of disease and promotion of health (Connelly and Worth 1997). Given the major issues of health inequality discussed above, we would suggest that a major role of these public health professionals and organisations is one of advocacy for the disadvantaged in society. Perhaps the most important function of public health is to represent the interests of the socially excluded in policy decisions where these decisions are likely to directly or indirectly impact on health. Risks are unacceptable to public health professionals if the health gains *across society as a whole* achieved by a reduction in risk outweigh the adverse health impacts and resources required from *society as a whole* to reduce that risk. In order to make this judgement, public health practitioners will have to rely on all the models and approaches we have discussed in this chapter.

Given this approach, what are the processes in setting standards for acceptable risk? We would suggest the following systematic approach:

(1) Bring together the group of experts. Ideally this group of experts should represent a broad range of skills and professional backgrounds, and include individuals with skills and expertise in the primary area of interest of the group. In addition, there should also be individuals with broad experience of public health.
(2) Agree the objectives of the group and any constraints to which the group needs to work.
(3) Determine the strength of evidence in support of an association between the environmental factor or indicator under consideration and illness. Make explicit any uncertainties in the data and any assumptions made.

(4) Quantify the impact on the community's health of the postulated illnesses, again being explicit about assumptions and areas of uncertainty. Consider the issue of particularly susceptible groups.
(5) Model the impact of any proposed change in standards on the community, taking into consideration the wider health, the social and the economic impacts.
(6) Consider whether the resources required to implement changes in any standard are worth the improvement in health (cost-utility analysis) and, even if they are, whether the resources required would be more effectively directed at other health goals (opportunity-cost analysis). Again make explicit any assumptions and uncertainties and identify the impact on susceptible groups.
(7) Expose the analytical phase of the standard-setting process to wide scrutiny by stakeholders of every type including pressure groups, expert groups, and industry. In particular seek out views from the wider public health community.
(8) Modify proposals in the light of this consultation exercise.

It is clear that the proposed approach is based firmly on a multi-disciplinary group process. We consider this approach to be the only viable option for such complex issues. However, groups are not infallible in decision-making. One particular type of pathology is known as 'group-think' (Janis 1972). Janis identified six major defects in decision-making associated with this problem. These are paraphrased below for acceptable risk decisions:

- Limiting group discussions to a limited number of options.
- Failing to re-examine the options initially preferred by the majority for non-obvious drawbacks.
- Neglecting options initially evaluated as unsatisfactory for non-obvious benefits.
- Members make little or no attempt to obtain information from experts who can supply sound estimates of benefits and disadvantages to be expected from alternate options.
- Selective bias is shown in the way the group reacts to factual information and opinion from experts and others, spending much time discussing evidence that supports their preferred options but ignoring that which does not.
- The group spends little time discussing how the implementation of the chosen option may be hindered by others outside the group.

Given these potential defects in group decision-making, we would suggest that any proposals for acceptable risk decisions be refereed by independent experts or groups to consider whether the processes that were applied to any decisions were satisfactory.

Finally, we hope we have shown that, despite their difficulty, acceptable-risk decisions can be reached, provided individuals and groups are prepared to take a broad view of the issues, consider all groups in society and accept and confront the areas of uncertainty in their information and their own biases.

10.9 IMPLICATIONS FOR INTERNATIONAL GUIDELINES AND NATIONAL REGULATIONS

Although only making up a small input to the harmonised framework, the issue of acceptable risk is an important and extremely complex area. Acceptable risk is very location-specific and for this reason it does not fit within international guidelines, but should play an important role in adapting guidelines to suit national circumstances, where local stakeholder involvement is vital.

10.10 REFERENCES

Adams, J. (1997) Cars, cholera, cows, and contaminated land: virtual risk and the management of uncertainty. In *What Risk? Science, Politics and Public Health* (ed. R. Bate), pp. 285–304, Butterworth Heinemann, Oxford.

Bartley, M., Blane, D. and Smith, G.D. (1998) *The Sociology of Health Inequalities*, Blackwell, Oxford.

Bennett, P. (1999) Understanding responses to risk: some basic findings. In *Risk Communication and Public Health* (eds P. Bennett and K. Calman), pp. 3–19, Oxford University Press, Oxford.

Byrne, D. (1999) *Social Exclusion*, Open University Press, Buckingham.

Cabelli, V.J., Dufour, A.P., Levin, M.A., McCabe, L.J. and Haberman, P.W. (1979) Relationship of microbial indicators to health effects at marine bathing beaches. *American Journal of Public Health* **69**, 690–696.

Cabelli, V.J., Dufour, A.P., McCabe, L.J. and Levin, M.A. (1982) Swimming-associated gastroenteritis and water quality. *American Journal of Epidemiology* **115**, 606–616.

Cabelli, V.J., Dufour, A.P., McCabe, L.J. and Levin, M.A. (1983) Marine recreational water quality criterion consistent with indicator concepts and risk analysis. *Journal of the Water Pollution Control Federation* **55**(10), 1306–1314.

Connelly, J. and Worth, C. (1997) *Making Sense of Public Health Medicine*, Radcliffe Medical Press, Abingdon, Oxon.

Corvello, V.T. (1998) Risk communication. In *Handbook of Environmental Risk Assessment and Management* (ed. P. Callow), pp. 520–541, Blackwell Science, Oxford.

Cotruvo, J.A. (1988) Drinking water standards and risk assessment. *Regulatory Toxicology and Pharmacology* **8**, 288–299.

Dufour, A.P. (1984) Bacterial indicators of recreational water quality. *Canadian Journal of Public Health* **75**, 49–56.

Fairley. C.K., Sinclair, M.I. and Rizak, S. (1999) Monitoring not the answer to *Cryptosporidium* in water. *Lancet* **354**, 967–969.

Farmer, P. (1999) *Infections and Inequalities*, University of California Press, Berkely, CA.

Fischhoff, B., Lichtenstein, S., Slovic, P., Derby, S.L. and Keeney, R.L. (1981) *Acceptable Risk*, Cambridge University Press, Cambridge.

Gerba, C.P., Rose, J.B. and Haas, C.N. (1996) Sensitive populations: who is at greatest risk? *International Journal of Food Microbiology* **30**, 113–123.

Grant, W. (2000) *Pressure Groups and British Politics*, Macmillan, Basingstoke, Hampshire.

Janis, I.L. (1972) *Victims of Groupthink,* Houghton Mifflin, Boston, MA.

Jones, D.R. and Akehurst, R.L. (1980) Risk assessment: an outline. *Bull. Inst. Maths Appl.* **16**, 252–258.

Jordan, B. (1996) *A Theory of Poverty and Social Exclusion*, Polity Press, Cambridge.

Kelly, K.A. and Cardon, N.C. (1994) The myth of 10-6 as a definition of acceptable risk. *EPA Watch* **3**, 17.

Klapp, M.G. (1992) *Bargaining with Uncertainty*, Auburn House, New York.

Langford, I.H., Marris, C. and O'Riordan, T. (1999) Public reactions to risk: social structures, images of science, and the role of trust. In *Risk Communication and Public Health* (eds P. Bennett and K. Calman), pp. 33–50, Oxford University Press, Oxford.

Macler, B.A. and Regli, S. (1993) Use of microbial risk assessment in setting US drinking water standards. *International Journal of Food Microbiology* **18**, 245–256.

McCrone, P.R. (1998) *Understanding Health Economics*, Kogan Page, London.

Moe, T. (1984) The new economics of organizations. *American Journal of Political Science* **28**, 737–777.

Morgan, M. G., Rish, W. R., Morris, S. C. and Meier, A. K. (1978) Sulfur control in coal fired power plants: A probabilistic approach to policy analysis. *Air Pollution Control Association Journal* **28**, 993–997.

National Association of Health Authorities and Trusts (1992) Priority Setting in Purchasing, pp. 14–17, National Association of Health Authorities and Trusts.

Nord, E. (1992) An alternative to QALYs: the saved young life equivalent (SAVE). *British Medical Journal* **305**, 875–877.

O'Riordan, T. (1977) Environmental ideologies. *Environment and Planning* **9**, 3–14.

Pass, C. and Lowes, B. (1993) *Collins Dictionary of Economics*, 2nd edn, Harper Collins, Glasgow.

Pattakos, A.N. (1989) Growth in activist groups: how can business cope? *Long Range Planning* **22**, 98–104.

RCEP (1998) *Setting Environmental Standards. Royal Commission on Environmental Pollution*. 21st report, TSO, London

Royal Society (1983) Risk Assessment. A study group report.

Sloman, J. (1994) *Economics*, 2nd edn, Harvester Wheatsheaf, Hemel Hempstead, UK.

Slovic, P., Fischhoff, B. and Lichtenstein, S. (1979) Rating the risks. *Environment* **21**(3), 14–20.

Sobel, J. and Takahashi, I. (1983) A multi-stage model of bargaining. *Review of Economic Studies* **50**, 411–426.

Sutton J. (1986) Non-cooperative bargaining theory: an introduction. *Review of Economic Studies* **53**, 709–724.

Thompson, M., Ellis, R. and Wildavsky, A. (1990) *Cultural Theory*, Westview Press, Boulder, CO.

Townsend, P., Davidson, N. and Whitehead, M. (1992) *Inequalities in Health*, 2nd edn, Penguin, London.

US EPA (1986) Ambient water quality criteria for bacteria – 1986. PB86-158045, US Environmental Protection Agency, Washington DC.

Weinstein, M.C. and Stason, W.B. (1977) Foundations of cost-effectiveness analysis for health and medical practices. *New England Journal of Medicine* **296**, 716–721.

WHO (1993) Guidelines for drinking-water quality. Volume 1. Recommendations. World Health Organization, Geneva.

Wilkinson, R.G. (1996) *Unhealthy Societies*, Routledge, London.

Wyer, M.D., Kay, D., Fleisher, J.M., Salmon, R.L., Jones, F., Godfree, A.F., Jackson, G. and Rogers, A. (1999) An experimental health-related classification for marine water. *Water Research* **33**(3), 715–722.

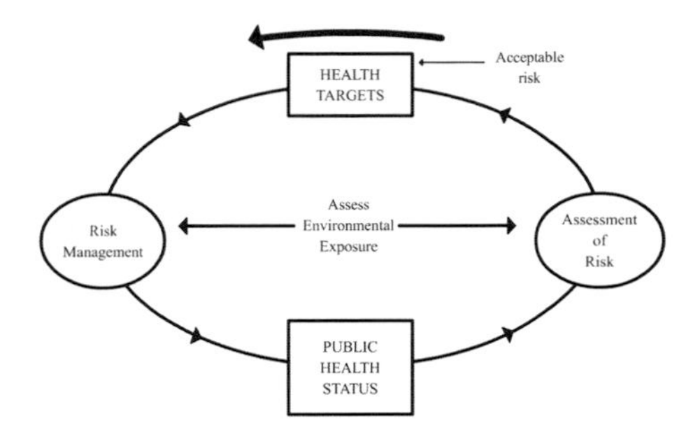

11

A public health perspective for establishing water-related guidelines and standards

Joseph N.S. Eisenberg, Jamie Bartram and Paul R. Hunter

For a number of historic reasons, the setting of water-related guidelines has become fragmented among different agencies and divorced from general public health. This runs contrary to the fundamental public health perspective that views the control of pathogens (including waterborne ones) as a more holistic activity, integrating across all exposure pathways. There are two levels at which this integration occurs. At one level, the focus is on proximal factors, such as water quality, sanitation and hygiene that have a direct causal link to disease as depicted through a 'systems' approach to transmission cycles. At another level the focus is on the distal causal factors, such as socio-economic conditions, which have an impact both on the health of a society and on individuals through their linkages to the proximal factors. The purpose of this chapter is to provide a public health perspective to motivate the need for an integrated approach to guidelines setting and, in keeping with the public health tradition, it draws together a number of 'threads' presented in earlier chapters.

This chapter, alongside Chapter 10 on acceptable risk, is especially relevant to developing understanding and approaches to the formulation of national/local objectives in terms of negotiated and agreed health targets that can be converted into implementable regulations.

11.1 INTRODUCTION

Public health has been defined as 'the science and art of preventing disease, prolonging life and promoting health through organised efforts of society'. It is concerned primarily with health and disease in populations, complementing, for example, medical and nursing concerns for the health of individual patients. Its chief responsibilities are monitoring the health of a population, the identification of its health needs, the fostering of policies that promote health, and the evaluation of health services (i.e. not only health-care services, but the totality of activities undertaken with the prime objective of protecting and improving health).

Modern public health can be traced back to the mid-nineteenth century and the work of two different men; John Snow and Edwin Chadwick. John Snow is credited as being the first person to use epidemiological methods to investigate an outbreak of cholera in the East End of London. This investigation enabled him to identify water from a single pump as the cause of the outbreak and to implement an effective control measure, namely removing the handle from the pump. Edwin Chadwick wrote *Report on an Inquiry into the Sanitary Condition of the Labouring Population of Great Britain*, one of the most important documents in the history of public health. In it, he argued that the economic cost to society of disease due to poverty, overcrowding, inadequate waste disposal and nutrition was unacceptable and greater than the cost of trying to improve these conditions. These two aspects of public health (epidemiological investigation of disease leading to effective intervention and concern with influencing social policy to improve health) emphasise that there exist distinct environmental components that impact on disease transmission. In one, the association of disease with a particular environmental source was sufficient to dictate an intervention. In the second example, Chadwick focused on the importance of social factors and therefore saw socio-political reform (such as reform of the poor law) as the major intervention in reducing disease and improving health.

As Chadwick argues, there is a wide array of social (or distal) factors in addition to biological (or proximal) factors that determine the impact of a particular pathogen on health and also the relative importance of the various transmission pathways that contribute to the disease burden. Therefore, although the proximal factors that describe the transmission cycle may be the direct cause of disease, they are often mediated by the distal factors. It is the role of public health to understand the relationship between the distal causalities (often associated with socio-economic status) and the proximal causalities (associated with biologic factors) and how these will inform intervention and control. This public health role is one that applies generally across all disease processes.

One basic feature of waterborne pathogens that makes them unique is the ability to survive in the environment outside of a host. This is a principal factor that largely dictates the possible transmission pathways that can be exploited by a waterborne pathogen in completing its lifecycle, and has implications for intervention and control. In addition to clinical controls, such as vaccination or chemotherapy, there are also a number of possible environmental controls. These include the treatment of water or other environmental media, limiting exposure to water or other environmental media, and prevention of contamination through sanitation and hygiene measures. Each of these strategies may not only reduce the disease burden associated with its pathway, it may also reduce transmission from other pathways by decreasing the amount of contamination. This interdependency of pathways suggest that to determine the most effective control requires an understanding of the complete transmission cycle. The relationship between proximal and distal factors, however, suggests that an integrated public health perspective for water-related activities should account not only for the disease transmission perspective addressing the proximal causalities, but also the distal causalities that may impact on those proximal factors.

A suitable metaphor for public health is a thermostat for the health of society. A thermostat is a negative feedback loop; one for room temperature has three components, namely:

- A sensor to measure the temperature within the room.
- A comparator to compare the room temperature with a pre-set ideal temperature.
- An actuator designed to control the flow of hot water to the radiator.

These components in public health are classed as:

- Surveillance to measure risk.
- The comparison of measured risk and predefined acceptable risk resulting in a decision on control strategies.
- Public health interventions.

'Surveillance' in this model covers the application of epidemiological tools in a descriptive manner to monitor disease incidence and in an analytic manner to assess the association of risk factors with disease incidence (see Chapters 6 and 7). Such tools may be used to investigate endemic disease in a community or outbreaks of disease as and when they occur. The comparison and subsequent decision requires a model. In public health the model is often conceptual and not necessarily explicit. Disease transmission systems, however, have been represented in the past as a mathematical model to compare data supplied by surveillance and acceptable risk values (issues on acceptable risk are covered in more detail in Chapter 10). It has also been used to provide an optimal control strategy that can then be implemented through public health intervention. Surveillance activities have little value unless they have the potential to lead to improved public health and safety by public health interventions. These interventions may be specific and small scale (removing the pump handle) or more general (development of national/international policies and strategies). Surveillance activities then come into play once more as the impact of any health intervention is evaluated.

11.2 A PUBLIC HEALTH PERSPECTIVE ON THE NATURE AND DETERMINANTS OF DISEASE

In this section we first discuss the nature of disease from a public health perspective and then go on to discuss some of the determinants of ill health.

11.2.1 Nature of disease

11.2.1.1 Health and disease

The World Health Organization (WHO) has defined health as a state of complete physical, mental, and social well-being and not merely the absence of disease or infirmity. This definition is extremely valuable. Of necessity, many of the chapters in this book concentrate on infectious disease and neglect the more holistic view of health. However, without an understanding of the impact of our efforts on health as well as on disease we may risk reducing the potential benefit of our interventions. People's quality of life is better when they have access to an adequate supply of water. If women do not have to walk many miles a day just to collect water they have more time for themselves and their families. Water also has a symbolic or spiritual meaning in many societies and the availability of water around the home adds to the sense of well-being. These and related issues are difficult to include in any formal epidemiological or risk assessment framework, but have powerful influences on health.

Turning to the nature of disease, there are several ways that disease can be categorised:

- According to the underlying aetiology (e.g. genetic, infectious, environmental, nutritional, etc.).
- According to the main disease process (e.g. inflammatory, malignant, degenerative, etc.).
- According to the main body system affected (e.g. respiratory disease, neurological disease, etc.).
- According to the course the disease follows and subsequent outcome (e.g. an acute course with recovery, acute course with death as the outcome, a chronic course etc.).

Which classification system is used depends on the purpose of the classification. Here we are primarily concerned with infectious disease spread by water. The outcomes and impacts of waterborne diseases can be acute, chronic or delayed. The distinction of these outcomes has public health importance. The effects of acute diseases occur over a short period of time whereas the effects of chronic diseases accumulate over much longer periods of time. Comparing the health of an acute versus a chronic disease can be done using Disability Adjusted Life Years (DALYs) (see Chapter 3).

As with other diseases, water-related diseases may be classified in a variety of ways, for example, according to the nature of the causative agent (protozoan, bacteria, virus etc.), or by the nature of the disease produced (diarrhoea, dysentery, typhoid, hepatitis and so on). With respect to intervention and control, however, a more appropriate classification is one based upon how changes in (largely environmental) conditions could impact on disease transmission. As such it represents a broad categorisation of principal environmental pathways.

11.2.1.2 Routes of transmission

Infectious agents have a number of options for their transmission. In general these are:

- Direct person-to-person transmission through intimate contact (such as sexually transmitted diseases).
- Direct person-to-person spread through infected body fluids (such as blood-borne viruses).
- Direct person-to-person spread through less intimate contact (such as influenza or viral gastroenteritis).

- Spread via contamination of the environment, which may include contamination of inanimate objects (fomites), water or air.
- Spread via contaminated food (such as *Salmonella*).
- Spread through an insect vector (such as malaria).
- Spread from a primary animal host to humans, either directly or indirectly via food or a contaminated environment.
- Spread to humans by environmental organisms (such as *Legionella*).

Pathogens are often able to use many of these pathways. For example, Norwalk-like viruses can be spread from person to person directly, via contaminated food, drinking water or fomites. Table 11.1 lists a classification of water-related diseases.

Table 11.1. Classification of water-related disease (after Bradley, 1974)

Category	Comments
Water-borne diseases	Caused by the ingestion of water contaminated by human or animal faeces or urine containing pathogenic bacteria or viruses; includes cholera, typhoid, amoebic and bacillary dysentery and other diarrhoeal diseases.
Water-washed diseases	Caused by poor personal hygiene; includes scabies, trachoma and flea-, lice- and tick-borne diseases in addition to the majority of waterborne diseases, which are also water-washed.
Water-based diseases	Caused by parasites found in intermediate organisms living in water; includes dracunculiasis, schistosomiasis and some other helminths.
Water-related diseases	Transmitted by insect vectors which breed in water; includes dengue, filariasis, malaria, onchocerciasis, trypanosomiasis and yellow fever.

We may wish to add to this list a fifth category, that of water-collection-related disease. This would include those diseases where spread is aided by journeying to collect water, as was found to be the case in an outbreak of meningococcal disease in a refugee camp (Santaniello-Newton and Hunter 2000). In addition, some pathogens do not infect sites within the human body but act remotely by the production of toxins that are subsequently ingested. The best water-related example of this mode of action is provided by the toxic cyanobacteria (Chorus and Bartram 1999). The role of the toxins produced by these organisms in their ecology is poorly understood. They represent a potential sixth category of water-related disease, associated with water contact.

All of the potentially waterborne pathogens share the ability for at least one of their life stages to survive, to a greater or lesser extent, outside the (human) host. The extent of that survival varies widely from presumably very short (e.g. *Helicobacter pylori*) to many years under favourable circumstances. Survival may be purely passive (many viruses), may involve robust life stages (such as the cysts and oocysts of *Giardia* and *Cryptosporidium* respectively) or may involve specific associations (such as that of *Legionella* with some free-living protozoa or *Vibrio cholerae* with certain cyanobacteria). This environmental survival distinguishes waterborne pathogens from others associated with, for example, transmission via the respiratory route (such as measles) that must infect a susceptible human host soon after leaving an infectious host.

In contrast to the situation with non-infectious disease, the risk of infection and illness is related to the level of microbial pathogens in the environment. For exclusively human pathogens, the degree of environmental contamination is related to the number of infected people. The more people with rotavirus in a community, the more likely an uninfected individual will catch it. This is because the source of all pathogens ultimately becomes the infected hosts. For many infectious diseases, the pathogen reproduces within the human host, who therefore acts as an amplifier. In order for a pathogen to persist, it must reproduce in sufficient numbers within a given host in order to allow for the infection of another host.

The specific journey a pathogen takes from host to host defines the transmission pathway and this may include non-human hosts. Diseases that are maintained within an animal population and sporadically introduced to human hosts are referred to as enzootic (c.f. endemic – upon people). For environmentally mediated pathogens, these pathways are often characterised by a significant time period outside the host. Humans can become infected through ingestion, inhalation, or dermal contact of/with pathogens.

The degree of contamination, and therefore the degree of risk, depends on the contributions of all of the different environmental transmission pathways. The transmission pathways increase in complexity when there are animal hosts that a pathogen can infect. Examples might include non-typhi *Salmonella*, *E. coli* and the bovine species of *Cryptosporidium*.

With respect to pathogen transmission, the number of cases or symptomatic individuals is not the only issue. It is also possible for an individual to be infectious but not symptomatic. These asymptomatic individuals are usually mobile due to lack of illness and have a high potential to spread a pathogen widely throughout a community.

Specific circumstances vary widely and, according to local conditions, any given pathway may dominate or make a negligible contribution to overall disease causation. Because of the importance of specific local circumstances, the relative contribution of different pathways cannot be properly/ comprehensively taken into account in the development of international norms such as WHO guidelines. The development of an understanding of local conditions and their impact on disease transmission pathways is an essential/very desirable step in adapting international guidelines to national standards (this is also an essential component of HACCP which is a generic risk-based system – see Chapter 12). A logical consequence is that national standards will progressively evolve in response to their own implementation and success. Thus, as a dominant route of exposure is partially or entirely controlled, so other routes will become of greater relative importance. If the remaining disease burden is judged to merit public health action then these routes will then become the focus of national and local regulatory activity.

11.2.1.3 Endemic disease, epidemic disease and outbreaks

Whilst the terms endemic, epidemic and outbreak may be used loosely and interchangeably in common parlance, these terms have precise meanings within the discipline of public health.

When a pathogen transmission cycle is at equilibrium within the human population the disease incidence is referred to as the endemic level, and the number of new cases remains approximately constant. An outbreak or epidemic is defined as a significant increase in the number of cases in a population over a given period of time. The term 'epidemic' is usually used for general increases in a population such as occurs with influenza (and can occur over long periods of time such as with AIDS). In contrast, the term 'outbreak' is usually used for a localised increase that occurs over a short period of time (a month or less). There are different types of outbreaks:

- point source outbreaks, in which all cases are infected at the same time;
- continual source outbreaks, in which all cases are infected over time from a source that is continually or sporadically infectious;
- propagated outbreaks, in which the disease is spread by person-to-person transmission; and
- mixed point source and propagated outbreaks, in which a point source is responsible for initial cases but then the disease is propagated to secondary cases through person-to-person spread.

Outbreaks, as compared with the endemic situation, present a large number of cases in a short period of time. Environmental health measures to control outbreaks may be very different from those intended to reduce the background (endemic) rate of disease.

As is outlined in some detail in Chapter 6, with many water-related diseases real problems are encountered in both detecting and estimating the magnitude of outbreaks and in quantifying the contribution of water to the overall disease burden.

It is generally accepted that outbreak events have special importance in public health and this should be accounted for in establishing health targets and from them, for example, water quality objectives. Thus, for example, while a public health target may be expressed in terms of a maximum tolerable disease burden, this may not be considered acceptable if it were to occur as the result of a single event. Public health target-setting may therefore make separate reference to outbreaks. Once converted to water quality objectives, this implies the need to pay special attention to extreme events (even if rare) in addition to steady-state conditions and performance.

11.2.2 Determinants of ill health

Disease is not evenly spread through society, and one of the important roles of public health is to identify the causes of this uneven distribution so that strategies can be developed to reduce risk and improve health. There are a large number of determinants of ill health. This chapter will outline four that have a significant impact on the water-related disease.

11.2.2.1 Environmental exposure

We have already discussed at length the impact of different transmission pathways (both water and non-water) on the epidemiology of waterborne disease. Clearly for any particular route of transmission to effectively transmit infection, susceptible individuals need to come into contact with the particular environmental source. The degree of such exposure is a major factor in the differential risk between individuals in a community.

For example, the amount of tap water consumed each day varies substantially from one individual to another, as does the amount of time a given individual swims. This variation in exposure has a substantial impact on the risk of infection (Hunter 2001). In a recent outbreak investigation, people who regularly went swimming were at lower risk during a drinking waterborne outbreak of cryptosporidiosis, presumably due to immunity after prior infection (Hunter and Quigley 1998). Of increasing concern, at least in developed countries, is the issue of travel-related disease (see Chapter 4), where travellers may find themselves exposed to environmental pathogens to which they have had no previous exposure.

11.2.2.2 Pre-existing health

Another important factor in the variation in ability to deal with infectious agents is an individual's existing state of health. The classic example of this is the severity of cryptosporidiosis in patients with AIDS (in whom infection may be fatal, whereas it is typically relatively mild in immunocompetent individuals). Other diseases that may affect an individual's response to a waterborne pathogen include diabetes mellitus (Trevino-Perez *et al.* 1995), malignant disease (Gentile *et al.* 1991) and organ transplantation (Campos *et al.* 2000). Perhaps the greatest impact on risk from waterborne disease worldwide is the impact on heath from malnutrition (Griffiths 1998).

11.2.2.3 Poverty

Most public health practitioners would accept that the biggest impact on human health and disease risk comes not from specific environmental factors or routes of transmission but from the social conditions in which an individual lives. Undoubtedly, poverty (both absolute and relative) is the biggest threat to health of any identifiable risk factor (Bartley *et al.* 1998; Townsend *et al.* 1992; Wilkinson 1996). People subject to poverty are more likely to suffer disease due to increased exposure to pathogens from inadequate environmental controls. Furthermore, once affected by disease, they are likely to suffer more severely because of inadequate health-care and social support systems and from poorer general health due to malnutrition and behavioural factors (such as smoking).

11.2.2.4 Acquired immunity

One of the most fundamental features that distinguishes microbiological hazards from chemical hazards in relation to human health is the phenomenon of acquired immunity i.e. the protection conferred to a host after exposure to a pathogen. For some pathogens (such as hepatitis A) once a person has been infected they will never contract the illness again (i.e. the protection is lifelong). For most waterborne pathogens the protection conferred to a host after exposure to the agent of disease is partial and temporary. For example, an individual with protective immunity due to prior exposure may require a larger dose in order for infection to occur or for symptoms to develop. Such partial protection may last for months or years. This property of infectious disease has major implications with respect to transmission both within and between populations. The greater the number of partially protected individuals, the smaller the pool of susceptible individuals that are at risk. This in turn implies that there will be a smaller pool of newly infected individuals in the future. The decreased number of infected individuals in the future means that there will be less contamination, decreasing the exposure risk.

The second aspect of complexity concerns the situation where populations from areas of low endemicity (and therefore with low immunity) travel to areas of high endemicity and therefore higher risk. Such situations occur increasingly frequently with the increasing trend in international travel. The most conspicuous example concerns the hepatitis A virus. In industrially developed nations, hepatitis A is largely controlled through water supply, sanitation, food and personal hygiene to the extent that most individuals are not exposed to the virus at all during their lifetime. In contrast, in areas where low hygiene standards prevail, hepatitis A exposure tends to occur early in life and is a relatively benign infection. However, a first exposure to hepatitis A among adults leads to a far more severe disease course. The greatest importance of hepatitis A virus is therefore to susceptible adults travelling to areas of high endemicity. For some groups of such individuals (e.g. some tourists and international aid workers) vaccination is recommended.

11.3 SKILLS AND TOOLS USED BY PUBLIC HEALTH PRACTITIONERS

Many primary scientific approaches are available to the public health practitioner in order to investigate the causes, impact and control of disease in populations. The discipline most closely associated with public health is epidemiology. Other disciplines of value include mathematical modelling, biological and physical sciences, social sciences (including economics), and demographics and vital statistics (Detels and Breslow 1997). The task of providing the best scientific information required for policy-making is difficult, due largely to the fact that environmental processes governing human health risks are complex. No single discipline can provide the information necessary to make a scientifically sound decision. Such decision making, therefore, requires careful consideration of both the information each discipline provides and their limitations. In this section we will bring together issues brought up in Chapters 6, 7 and 8 (on surveillance, epidemiology, and risk assessment modelling) from a public health perspective. More detailed descriptions of these methodologies are discussed in the respective chapters.

11.3.1 Epidemiology

Epidemiologists may utilise a number of descriptive and analytical techniques that are all based on statistical inference as a basis of proof. Epidemiological proof is built up over time as the results of various studies are added together into a body of knowledge. One of the first people to lay down principles of epidemiological proof was Bradford-Hill (1965). He suggested nine criteria from which proof of a link between human disease and exposure to a potential risk factor could be derived:

(1) Strength of association, as measured by odds ratio, relative risk or statistical significance.
(2) Consistency of finding the same association in studies conducted by many different researchers.
(3) Specificity of association such that a particular type of exposure leads to a particular disease.
(4) Temporality, in that the exposure must precede the disease.
(5) Biological gradient: people with higher exposure should get more disease.
(6) Plausibility: the proposed causative pathway must be plausible.
(7) Coherence: the hypothesis must not conflict with what else is known about the biology of the disease
(8) Experiment: can the link be supported by experiment such as intervention studies?
(9) Analogy: is there another similar disease which has a similar link?

Outcome measures from epidemiology studies are used to estimate risk. In epidemiology, risk has the connotation of probability of illness. This is, in turn, related to how common a disease is in a community. There are two measures of the commonness of disease; incidence and prevalence. The incidence of a disease is the number of new cases occurring within a certain population during a specified time period (e.g. cases per 100,000 persons per year). Prevalence is the number of cases of a disease within a specified population at a specific point in time (e.g. cases per 100,000 persons).

There are three types of epidemiological risk. Absolute risk is, in effect, the incidence of disease that tells us little about the possible causes of a disease. Attributable risk is the proportion of cases of a disease that can be linked to a risk factor, usually given as a percentage. Relative risk is the ratio between the risk of disease in one population (exposed to a particular risk factor) and a second population (not exposed).

If we know the absolute risk and either the attributable or relative risk we should have sufficient information to judge the importance of a disease and the importance of various risk factors. Unfortunately, getting accurate information on risk is not necessarily that easy given, for example, limitations in obtaining estimates of absolute risk from surveillance systems, or estimates on both attributable and relative risks, from detailed analytical epidemiological studies. Even if we were able to obtain this information on relative risk, for example, it may not tell us the impact on human health of removing a particular risk factor. For example, if the attributable risk of infection due to drinking water contaminated with Norwalk-like virus (NLV) is 20%, removing drinking water as a source of infection would not necessarily reduce disease by 20%, as people may then be infected from other sources. On the other hand, disease reduction may be greater than 20% if this reduced the risk of secondary cases in a community.

This intervention example illustrates a very important property of infectious disease transmission processes; that transmission pathways are interdependent. The traditional definition of attributable risk is based on the assumption that risk at the individual level is an independent process (i.e. the probability of an individual becoming diseased is independent of the disease status of other individuals within the community). This assumption is violated for an infectious disease process since the source of pathogens is generally other infected hosts. For example, a pathogen present in the water may infect an individual that drinks the water. This individual may then directly transmit the pathogen to others within a household, some of whom may become asymptomatic carriers who in turn transmit pathogens to a recreational water area, exposing susceptible swimmers. Other infected individuals may contaminate the wastewater that will subsequently be used in an agricultural setting, resulting in an occupational exposure. This illustration of typical causal pathways makes it clear that it is difficult to assign any of these individual cases to a specific risk factor (i.e. should these cases be considered a drinking water risk or an occupational risk) and emphasises the need for a harmonisation process, whereby water-related areas are considered together rather than in isolation.

The critical feature of the transmission process that presents us with this issue of interdependence is the fact that these pathogens cycle from host to host. This interdependence of transmission pathways is the reason that the impact of a given intervention, as mentioned above for drinking-water treatment of NLV, is not simply the attributable risk. To empirically assess the effect of treating the drinking water on disease prevalence of NLV requires an intervention study. Disease transmission models, which explicitly account for this interdependence, can provide the theoretical framework from which to address these issues and can be useful in both the design and analysis phases of intervention studies.

11.3.1.1 Surveillance

Epidemiological surveillance is the ongoing and systematic collection, analysis and interpretation of health data in the process of describing and monitoring a health event. This information is used for planning, implementing and evaluating public health interventions and programs. Surveillance data are used both to determine the need for public health action and to assess the effectiveness of programs (Klaucke *et al.* 1988). Surveillance is discussed in more detail in Chapter 6.

The discussion here will be restricted to how surveillance systems can feed into national and international public health policy and standard setting. Surveillance systems are established for a number of different reasons:

- To identify outbreaks/adverse incidents early enough to implement possible control measures.
- To identify patterns of disease in order to identify risk factors so that control measures can be implemented or standards set.
- To evaluate the impact of prevention and control programmes.
- To project future health-care needs (i.e. all activities undertaken with the prime objective of protecting and improving health).

Because in this context we are primarily concerned with setting standards for waterborne disease, the primary functions of disease surveillance are:

- To establish the incidence and severity of disease so that priorities can be set.
- To attempt to identify the association between risk of disease and exposure to environmental exposure to water.
- To assist in identifying specific contributory factors to disease transmission and thereby inform risk management (see Chapter 12).

Unfortunately, existing surveillance systems cannot necessarily provide this information. Surveillance systems capture relatively few of the cases of illness occurring in a community and hence are poor indicators of disease burden. Additionally, detection rates for enteric disease can vary dramatically from one disease to another. For example, one UK study suggested that national surveillance systems would detect only 31.8% of *Salmonella* infections, 7.9% of *Campylobacter* infections, 3.0% of rotavirus infections and 0.06% of Norwalk-like virus infections (Wheeler *et al.* 1999).

Existing surveillance systems often have very limited information about possible risk factors and the lack of data on controls makes what information that is available difficult to interpret. Furthermore, data collected by surveillance systems may not be representative of the general level of disease in the community. Differential reporting between doctors and areas may bias results.

One area in which routine surveillance can provide good information is in detecting changes over time (although whether these changes are due to changing incidence or improved diagnosis is frequently unclear). One particular aspect in this regard is the detection of outbreaks (provided that they are large enough to be obvious against the general background incidence of a disease). Outbreaks can provide very useful information about possible risk factors (such as failures in water treatment or point source pollution), hence they have often been the driving force behind changes in standards and legislation. However, care must be exercised in extrapolating from knowledge about risk factors for outbreaks to endemic disease. Outbreaks are usually responsible for a relatively small proportion of total disease burden and the risks may differ.

11.3.1.2 Descriptive and analytical epidemiology

Chapter 7 is dedicated to a discussion of epidemiological techniques and so this will not be repeated here. What we will do is remind the reader that all epidemiological methods are potentially subject to problems from bias of one type or another (Greenland 1997; Hennekens and Buring 1987) and this can potentially reduce the value of epidemiology for policy makers. Two types of bias that can adversely affect the validity of epidemiological studies are selection bias and recall bias. To a greater or lesser extent these types of bias can affect any type of study if sufficient attention is not paid to them in the design stage.

Selection bias occurs when the selected study participants differ from the population from which they are selected. This can arise in a number of ways:

- If subjects are selected in a non-random fashion, by for example using volunteers or only cases presenting to hospital.
- If hard to contact subjects, such as those without a telephone, are excluded. In many societies, the poorest sections of society are not able to afford their own telephone.
- If response rate is low because a large proportion of subjects refuses to participate.

Recall bias can also arise in different ways:

- Cases may remember exposure to a potential risk factor differently from controls. So, for example, if it is believed that a waterborne outbreak is being investigated, cases may report higher water consumption than controls even if the reality is that no such difference exists.
- Subjects are more likely to state that they have suffered from particular symptoms if they believe that they are at increased risk of such symptoms.

This latter source of recall bias has been invoked recently in a renewed debate over the size of the Milwaukee outbreak of cryptosporidiosis. This outbreak is reported as being the world's largest documented outbreak of waterborne disease, affecting some 405,000 people (MacKenzie *et al.* 1994). This estimate, however, was based on a telephone survey conducted some time after the outbreak became big news in the city. Recently, Hunter and Syed (2000) conducted a similar study during a waterborne outbreak but included control towns that were close enough to the outbreak area for people not to be sure whether or not they were part of the outbreak. Surprisingly, they found that the incidence of self-reported diarrhoea was greater in the control areas than in the outbreak areas. They suggested this was due to recall bias following the intense media coverage.

11.3.1.3 Epidemiology and policy making

By collecting population-level health risk data, epidemiology provides information crucial to policy makers who set standards and guidelines. To make best use of these data, the relevance of each study, in the context of the policy decision, must be understood. Some of the issues that should be considered are the study design, confidence intervals of risk estimates, potential biases and generalisability (see section 11.4.2). Although no single study is expected to provide perfect information, interpretation of a collection of studies may provide increased confidence in a given risk estimate. The Bradford-Hill criteria listed earlier provides a very valuable checklist in this respect. The greater the number of criteria that can be met, the more confidence there can be in the value of a change in policy/legislation.

11.3.2 Mathematical modelling (quantitative risk assessment)

Mathematical modelling of infectious disease processes has played an increasing role in the field of epidemiology. These models, which describe the disease transmission of specific pathogens, have been used to study directly transmitted diseases (such as measles), vector-borne diseases (such as malaria) and sexually transmitted disease (such as AIDS); however, they have rarely been applied to waterborne diseases. Quantitative risk assessment (QRA) has traditionally used a model structure based on a chemical risk paradigm to estimate the risk of exposure to waterborne pathogens. The limitations of using this approach to assess risk from exposure to pathogens are discussed in Chapter 8. Recently, the QRA approach has been extended using disease transmission models to account for some of those limitations (also discussed in Chapter 8).

Regardless of the model structure used, conclusions based solely on modelling studies can potentially be misleading. This is due to the fact that:

- the huge levels of uncertainty and variability inherent in these environmental systems limit the precision of model prediction
- the limited data available to assess and calibrate the model necessitates the use of a number of assumptions.

One specific concern in respect of the value of QRA relates to the fact that all models described so far concentrate on the risks associated with specific pathogens. To gain estimates of total disease burden, separate models need to be constructed for all possible pathogens. Clearly this is not a trivial task and the uncertainties associated with individual pathogen models would also be combined. Also it is more difficult within QRA to take account of health-related factors that cannot be linked to simple figures of disease numbers. Epidemiological studies can be designed to cover a range of diseases (or symptom complexes) in a single study more easily than QRA. Furthermore, epidemiological studies can be designed to investigate the impact of water on a more holistic definition of health. Thus, models are most useful when used in conjunction with epidemiology. They can provide a valuable framework from which to interpret data and elucidate processes. In this way a model can help generalise empirical findings for relevant policy making. Specifically, these models provide a theoretical framework that can:

- identify data gaps and define research goals
- aid in decision making
- define the sensitivity of these decisions.

11.3.3 Biological and physical sciences

Although rarely expert in these other sciences, public health practitioners frequently call upon the expertise of scientists and engineers from many varied backgrounds. Laboratory sciences, especially microbiology, have had a long association with public health stretching back to the time of Pasteur and Koch. Modern techniques of molecular biology have had a particularly significant impact on public health practice in recent years. With waterborne disease, such recent developments have provided techniques to improve the diagnosis of disease in humans and the detection of pathogens in environmental samples. The ability to distinguish between similar strains has also been improved by the use of molecular 'fingerprinting' methods. Such information can be vital in showing that the agent responsible for an outbreak is the same (or not) as that isolated from a drinking water supply. As is the case for cryptosporidiosis and *E. coli* infections, sub-species typing can be valuable in indicating the likely epidemiology of potentially waterborne outbreaks.

Although the management of water distribution systems is now seldom under the control of public health practitioners, a knowledge of the principles of environmental engineering is essential at times. This is particularly important during outbreaks when public health practitioners and water engineers must work closely together.

11.3.4 Social and behavioural sciences

The social and behavioural sciences have assumed increasing importance for public health practice in recent years as attention has refocused on the importance of lifestyle and social status on health. The social sciences have enabled public health professionals to describe the factors responsible for lifestyle and how these correlate with health. Sociology has also enabled a more accurate description of the factors that divide society and how these are responsible for inequalities in health (Townsend *et al.* 1992).

The behavioural sciences are also of great importance for designing public health interventions that seek to modify personal behaviour patterns through health education.

11.3.5 Demography

Demography is concerned with the structure of and changes in human populations, largely through measuring birth, death and migration. As such, demography has a significant function in public health in defining the setting in which disease occurs. Although in industrially-developed nations, deaths associated with drinking-water are relatively rare, demography is essential in identifying the size and structure of the population at risk. Without this information it would not be possible to identify the burden of disease due to water. For nations with rapid population growth, demography provides a means of predicting future demand for safe drinking water (or, more accurately, predicting the future population without access to safe drinking water).

11.4 PUBLIC HEALTH INTERVENTIONS

As discussed in the introduction to this chapter, and illustrated by the thermostat metaphor, public health practice must lead to interventions that have the potential to improve human health. There are a number of types of intervention that are available to the public health practitioner (Detels and Breslow 1997). These cover a wide variety of possible approaches that are frequently complementary and may be synergistic. Indeed, it is unlikely that any one approach will succeed when used in isolation.

11.4.1 A classification of public health interventions

11.4.1.1 Preventive medical care

One of the most important public health interventions available to society is the provision of adequate medical care. In many societies, the provision of medical care is largely controlled by public health practitioners who determine the health-care needs of their populations and then plan to provide for those needs. Medical care is essential in reducing the burden of disease by ensuring rapid diagnosis and treatment of disease so that the duration of illness, the severity of disability and the risk of death are reduced where possible. As such, the quality of medical care in society has a substantial role to play in reducing the burden of disease associated with the waterborne route. Saving young people from dehydration by provision of adequate health-care at local village level will substantially reduce the burden of disease in those societies.

Particularly for some infectious diseases, medical care can also have a larger impact on the reduction of disease incidence than simply those that benefit directly from the treatment. Rapid diagnosis and treatment of individual cases of infection should limit the time that pathogens are excreted into the environment and so reduce the total amount of infectious agent available to infect new individuals. This may be by the use of antibiotics (e.g. for Shigella dysenteriae or Salmonella typhi infections) or quarantine of infectious individuals.

Vaccination is also a form of medical intervention that has had a major impact on the burden of infectious disease worldwide (e.g. vaccination against polio). However, few vaccination campaigns have had a significant impact on the global risk of waterborne diseases. Vaccines against typhoid and hepatitis A are valuable for protecting travellers when visiting areas of increased prevalence from the risk of these infections. Unfortunately, these vaccines are currently far too costly to be used widely to reduce overall burden of disease.

11.4.1.2 Health education and behavioural modification

Health education has long been a mainstay intervention available to the public health professional. Perhaps the most obvious example of this is the anti-smoking campaigns that have been undertaken by many developed nations. Health education can be used to warn people of the dangers of one or more particularly risky behaviours or can be used to promote generally healthier lifestyles. There are numerous examples of the positive impact of health education in reducing waterborne disease. These include the promotion of breast-feeding, educating people to routinely chlorinate drinking water or to boil it during outbreaks of cholera, encouraging the use of narrow neck vessels in which to store water, and other changes in water handling. Hygiene education and household water treatment present opportunities to empower the poor and reduce their burden of water-related disease, without dependence on outside authorities, within meaningful timeframes and at low cost.

11.4.1.3 Control of the environment

Even before the advent of modern public health and the germ theory of disease, the importance of environmental control in protecting health was recognised by many different societies. Thus, the Romans built aqueducts in order to bring clean water into their cities. It could also be argued that in more recent times, the first intervention of modern public health aimed to control the environment by removing the handle from the Broad Street pump. The aim of environmental control is to protect a population from potentially infectious or noxious agents. For waterborne diseases, we are concerned with ensuring that drinking and recreational water is free from potentially infectious agents, and that human sewage and other wastes are dealt with in as safe a manner as possible. The setting, and implementation, of international guidelines has, in recent years, been a major factor in improving such quality and reducing risk.

11.4.1.4 Cultivating political will

The cost of many possible interventions for reducing waterborne disease can be enormous. For example, a large-scale water treatment works can cost several tens of millions of pounds. Getting the support to spend such large sums requires considerable political persuasion skills. Furthermore, many public health interventions require legislation and must compete with other demands for legislation time. Ever since the days of Chadwick, political skills have been central to the armamentarium of the public health professional. Without such political goodwill, few of the improvements in public health during the last century would have been possible.

11.4.2 Public health interventions and waterborne disease

In general, intervention strategies (such as improvements in sanitation) will not only reduce the disease burden associated with a targeted pathway but will also reduce the disease burden from other pathways by decreasing the amount of contamination. Similarly, many environmental health interventions and most of the water-related interventions discussed in this book act not only on a single pathogen (as would be the case with vaccination for example) but on a variety of pathogens. This is particularly clear in the case of the various pathways that contribute to faecal-oral disease transmission, a route shared by a large number of known and currently unrecognised pathogens (see Chapter 5). The fact that interventions can have effects across different pathways and multiple pathogens has certain implications.

First, dose–response relationships attempt to quantitatively describe the relationship between exposure to a given pathogen and the resulting adverse health effect (see Chapter 8). The implicit assumption when using these dose–response models in a chemical risk assessment paradigm is that by reversing the use of the curve it is possible to predict the outcome of an intervention if its impact on exposure can be estimated. This approach, however, assumes that transmission pathways are independent, which in general is not true. In particular, an intervention in one area may impact on other routes of exposure either beneficially or detrimentally. For example, increasing the volume of available water in order to facilitate hygiene behaviours may decrease infection in children and therefore may decrease recreational water transmission. On the other hand, increased water volume may also result in excess water, increasing the risk of infection from other pathogens. This limitation of the chemical risk paradigm can be addressed by incorporating transmission models as a quantitative framework for risk assessment estimate (see Chapter 8). Since transmission models represent the natural history of the disease process and have biologically-based parameters they can be effectively used alongside epidemiology.

A second implication relevant to the models discussed in Chapter 8 is that they describe pathogen-specific processes. General interventions that operate across a pathway or combination of pathways are therefore likely to have an effect underestimated by a quantitative risk assessment approach (which mostly deals with pathogens on a case by case basis) but which may be detected by epidemiological investigations if appropriate methodologies are employed. This is logical given that microbiological risk assessment focuses on causes of disease whereas epidemiological studies can aggregate causes (pathogens), both known and unknown, by looking directly at the health effects themselves (see Chapter 7).

Third, generalising epidemiological studies to different target populations under different environmental conditions should be looked at carefully. An interesting issue of study design specificity is that there is potentially a competitive effect that arises from the interdependency of transmission pathways. For example, if effective sanitation were introduced then the subsequent impact of water supply or water quality interventions may be reduced. This is supported by the available body of evidence, suggesting that one intervention may reduce the likely exposure through multiple pathways. In Esrey's review (Esrey et al. 1991), water quality was found to be a relatively inefficient intervention. However, in the studies he reviewed water quality was almost invariably an add-on or secondary intervention to, for example, water supply. In contrast, studies where water quality has been treated as a primary intervention, much higher rates of response have been detected (Quick et al. 1999).

Another example that demonstrates the effects of pathway interdependencies is the situation where exposure to a given pathogen (such as the Norwalk-like viruses) is ubiquitous. Under these conditions the impact of a single intervention on public health may be negligible or zero, since the risk has, in effect, transferred from one route of exposure to another. This is most likely to occur in relation to water-related disease where secondary transmission plays a major role and where primary introduction occurs through many different pathways.

To make the best use of available data for policy making, therefore, requires a good understanding of the specific conditions under which each study being considered was conducted. The use of mathematical models for guideline-setting has its own share of limitations, including the assumptions required to develop the model structure and define the parameters. Mathematical models, however, can be useful tools to help with the process of generalising the conclusions of epidemiology studies to other conditions. Specifically, transmission models may be useful in generalising across different environmental conditions. Integrating results from different disciplines can help to address the inevitable limitations that exist when attempting to develop scientifically sound guidelines.

11.4.2.1 Cause-effect

Considerable research efforts are expended in demonstrating cause-effect relationships, and confirmation of cause-effect is an important step in justifying (and sometimes in formulating) control measures. However, the existence of a cause-effect relationship does not mean that the cause is a dominant, or even significant, contributor to overall burden of disease. Demonstration of cause-effect may give a false sense of security regarding the ability to impose control and may discriminate against other causes of equal or potentially greater importance. Within the field of water and health management a 'rule of thumb' which is sometimes quoted is that if a cause contributes less than 5% to the burden of a disease then it should be overlooked in favour of more significant routes. While this ignores the importance of cost-effectiveness, it does illustrate an important point.

11.4.2.2 Environmental health decision-making

As is common in many other areas, the complexities of environmental and specifically water management have led to fragmentation of responsibilities and of professional areas of interest. In the field of water, and indeed within the narrower field of water and health, this means that distinct professional interest groups have developed. The lack of effective communication among these groups is remarkable and was one of the factors that became obvious at the meeting that gave rise to this book. As a result, professional communities concerned with drinking-water quality and human health may be ignorant of developing lines of thought, approaches and information in what are, in fact, closely allied areas such as recreational water use and wastewater reuse. One outcome is an inefficient multiple learning exercise, since lessons learned are not readily transmitted between the largely isolated professional groups concerned.

This trend runs contrary to much current and developing policy that is moving towards the concept and application of 'integrated' management. Failure to create linkages between key interest groups including technical/professional communities will impede the process in general and the achievement of benefits. Integration of environmental health and of water and health concerns in such management approaches has been especially poor. Chapter 15 describes some of the problems associated with cost-benefit and cost-effectiveness analysis including, for example, the difficulties in identifying the many health and non-health benefits that arise from environmental interventions and assignment of both costs and benefits to appropriate sectors. Nevertheless, an empirical basis for an integrated public health outlook on water management is readily available. Thus, for example, substantial commentary exists on the costs of sewage treatment as a public health intervention to reduce the health risks associated with recreational water use. Much of that commentary has highlighted the high costs and limited benefits. Such a viewpoint discriminates against source-related rather than use-related interventions in that it ignores benefits gained through other routes. For example, effective treatment of an upstream sewage discharge in a catchment may increase downstream drinking-water quality, downstream recreational water quality (both in the river and in the receiving coastal area) as well as the water quality in coastal areas used for shellfish farming and harvesting. This demonstrates the importance of integrating public health management across these areas and in particular the need for an integrated public health policy in order to enable rational establishment of health targets as a basis for environmental standard-setting supportive of public health.

11.5 THE PUBLIC-HEALTH-BASED CONTRIBUTION TO SETTING STANDARDS

As the reader should already have gathered, there are a few key features about the public health professional's contribution to standard setting. These are:

- The use of a broad range of skills, tools and disciplines in the standard-setting process.
- Knowledge of a wide range of disease processes and transmission routes beyond those normally considered as waterborne.
- Priority setting, by determining the importance of the adverse health effects of the issue under consideration relative to other public health needs of society.
- A commitment to, and advocacy of, the needs of the relatively disadvantaged and socially excluded sections of society.

Although not directly applicable to international standard-setting, the process of health and environmental impact assessment provides some useful insights (British Medical Association 1998). There are seven guiding principles for health impact assessments (HIA) that are worth restating in this context. Standard setting should be:

(1) Multidisciplinary, including specialists and generalists from within public health and other disciplines.
(2) Participatory: where possible key stakeholders including informed representatives of the general public should have the opportunity of expressing their views.
(3) Equity-focused, in that any changes should aim to minimise health inequalities while improving community health.

(4) Qualitative as well as quantitative, in that many important issues in public health and social well-being are not amenable to direct measurement.
(5) Multi-method, in that a variety of different models and techniques are used in the analysis so no preferred model or study dominates the debate.
(6) Explicit in both values and politics, in that the values and interests of all participants should be disclosed early in the process.
(7) Open to public scrutiny.

A public health perspective should come with the understanding that although these environmental processes are complex and although there are limitations to the tools available to the public health practitioner, decisions must be made. Some suggested activities that should be considered in the decision-making process for environmental standards are listed below:

- Determining the burden of disease (see Chapter 3), which should include the amount of illness and the severity of the impact of that illness on people's lives and the health of communities.
- Assessing the evidence of a relationship between disease and proposed environmental risk factors such as drinking water. Even if a disease has a very high impact on health its control may or may not be amenable to environmental modification by appropriate standard setting. For this we can use epidemiology methodologies (Chapter 7).
- Considering whether the risk of disease is acceptable or not, tolerable or not (see Chapter 10).
- Describing the major determinants of disease in various communities, again detailing why each determinant is important and ranking their importance.
- Considering the availability and capability of health protection and health care facilities in each community.
- Modelling the impact of proposed changes in standards on the main diseases under consideration.
- Considering whether proposed new standards are achievable in the model communities.
- Modelling the economic costs of the proposed new standards.
- Modelling the impact of proposed changes on other public health issues (including issues such as communicable disease, non-communicable disease, injury, mental health and so on).

- Considering any adverse effects on public health of changes in standards (either directly or indirectly through such things as increased unemployment).
- Considering whether there are any non-public health related (such as environmental or wildlife) benefits or drawbacks.
- Considering which other non-water-related interventions might be able to achieve the desired goal (e.g. improved housing, education, employment opportunities and health care provision) and whether available resources would be better directed at these interventions.

11.6 IMPLICATIONS FOR INTERNATIONAL GUIDELINES AND NATIONAL REGULATIONS

In this chapter we have tried to give the reader an overview of public health as it is currently practised. We have also suggested various approaches that could be used when considering the issue of international environmental guideline-setting, taking more of a public health standpoint. We had previously categorised the public health contribution as being holistic, in its sphere of interest and its use of methodologies. The public health contribution is also about priority setting and identifying those areas worthy of intervention. However, of most importance is the advocacy role of public health, particularly for the most vulnerable and socially excluded sections of the population. Today there are many powerful interest and lobby groups that seek to influence national and international governments for their own purposes. In our view, the primary contribution that public health can make to society is to provide a strong voice for those sections of society whose needs and interests may never otherwise be considered.

Despite the evident importance of the national arena, especially in standard-setting and with profound implications for more local activities (see Chapter 16), there is often very limited capacity in national public health administrations to engage adequately. This relates to the resourcing necessary to carry out basic functions (such as surveillance and outbreak investigation), the limited human resources available (both in numbers and expertise) and the fragmentation of expertise. International guidelines and their supportive background documentation provide a form of support to national public health administrations that is invaluable in this role. They provide balanced information gleaned from the overall body of evidence. Of importance may be information regarding causal relationships (between chemical contaminants of drinking water and adverse health effects, for example) and between disease and environmental risk factors. Nevertheless, many of the processes involved in standard-setting are national (even local) in character. These include, for example, determination of tolerable disease burden, and available capacities and capabilities. While local in character, there is a limited stock of high quality studies with which to inform such decision-making. The process of their collation, critical review and dissemination, which takes place during guideline derivation, is also a valuable support to national processes. Finally, many national administrations lack experience in the processes of guidelines/standards derivation and of legislative review (especially with regard to aspects of implementation). The transparent process of guidelines derivation may provide an example with which to inform national processes. Conversely, the omission of certain aspects (such as the adaptation of guidelines to prevailing social, cultural, economic and environmental circumstances, including aspects of progressive implementation) creates a vacuum, and provision of explicit guidance on these would be an asset to national public health authorities in the future.

11.7 REFERENCES

Bartley, M., Blane, D. and Smith, G.D. (1998) *The Sociology of Health Inequalities*, Blackwell, Oxford.

Bradford-Hill, A. (1965) The environment and disease: association or causation? *Proc. R. Soc. Med.* **58**, 295–300.

Bradley, D.J. (1974) Chapter in *Human Rights in Health*, Ciba Foundation Symposium **23**, 81–98.

British Medical Association (1998) *Health and Environmental Impact Assessment,* Earthscan Publications, London.

Campos, M., Jouzdani, E., Sempoux, C., Buts, J.P., Reding, R., Otte, J.B. and Sokal, E.M. (2000) Sclerosing cholangitis associated to cryptosporidiosis in liver-transplanted children. *Eur. J. Pediatr.* **159**, 113–115.

Chorus, I. and Bartram, J. (1999) *Toxic Cyanobacteria in Water: A Guide to their Public Health Consequences, Monitoring and Management,* E & FN Spon, London.

Detels, R. and Breslow, L. (1997) Current scope and concerns in public health. In *Oxford Textbook of Public Health* (eds R. Detels, W.W. Holland, J. McEwen and G.S. Omenn), pp. 3–18, Oxford University Press, Oxford.

Esrey, S.A., Potash, J.B., Roberts, L. and Shiff, C. (1991) Effects of improved water supply and sanitation on ascaris, diarrhoea, dracunculiasis, hookworm infection, schistosomiasis, and trachoma. *Bull. World Health Organ.* **69**, 609–621.

Gentile G., Venditti, M., Micozzi, A., Caprioli, A., Donelli, G., Tirindelli, C., Meloni, G., Arcese, W. and Martino, P. (1991) Cryptosporidiosis in patients with hematologic malignancies. *Rev. Infect. Dis.* **13**, 842–846.

Greenland, S. (1997) Concepts of validity in epidemiological research. In *Oxford Textbook of Public Health* (eds R. Detels, W.W. Holland, J. McEwen and G.S. Omenn), pp. 597–615, Oxford University Press, Oxford.

Griffiths JK. (1998) Human cryptosporidiosis: epidemiology, transmission, clinical disease, treatment, and diagnosis. *Advances in Parasitology* **40**, 37–85.

Hennekens, C.H. and Buring, J.E. (1987) *Epidemiology in Medicine,* Little, Brown, Boston, MA.

Hunter P.R. (2001) Modelling the impact of prior immunity, case misclassification and bias on case-control studies in the investigation of outbreaks of cryptosporidiosis. *Epidem. Infect.* (in press).

Hunter, P.R. and Quigley, C. (1998) Investigation of an outbreak of cryptosporidiosis associated with treated surface water finds limits to the value of case-control studies. *Comm. Dis. Public Health* **1**, 234–238.

Hunter, P.R. and Syed, Q. (2000) A community-based survey of self-reported gastroenteritis undertaken during an outbreak of cryptosporidiosis, strongly associated with drinking water. In *Conference Proceedings of the 10th Health-related Water Microbiology Symposium*, Paris, IWA Publishing.

Klaucke, D.N, Buehler, J.W., Thacker, S.B., Parrish, R.G., Trowbridge, F.L., Berkelman, R.L. and the Surveillance Co-ordination Group (1988) Guidelines for evaluating surveillance systems. *MMWR* **37** (S5), 1–18.

MacKenzie, W.R., Hoxie, N.J., Proctor, M.E., Gradus, M.S., Blair, K.A., Peterson, D.E., Kazmierczak, J.J., Addiss, D.G., Fox, K.R., Rose, J.B. and Davis, J.P. (1994) A massive outbreak in Milwaukee of cryptosporidium infection transmitted through the public water supply. *N. Engl. J. Med.* **331**, 161–167.

Quick, R.E., Venczel, L.V., Mintz, E.D., Soleto, L., Aparicio, J., Gironaz, M., Hutwagner, L., Greene, K., Bopp, C., Maloney, K., Chavez, D., Sobsey, M. and Tauxe, R.V. (1999) Diarrhoea prevention in Bolivia through point-of-use water treatment and safe storage: a promising new strategy. *Epidemiology and Infection* **122**, 83–90.

Santaniello-Newton, A. and Hunter, P.R. (2000) Management of an outbreak of meningococcal meningitis in a Sudanese refugee camp in Northern Uganda. *Epidemiol. Infect.* **124**, 75–81.

Townsend, P., Davidson, N. and Whitehead, M. (1992) *Inequalities in Health,* Penguin, London.

Trevino-Perez, S., Luna-Castanos, G., Matilla-Matilla, A. and Nieto-Cisneros, L. (1995) Chronic diarrhea and *Cryptosporidium* in diabetic patients with normal lymphocyte subpopulation: Two case reports. *Gaceta Medica de Mexico* **131**, 219–222.

Wheeler, J.G., Sethi, D., Cowden, J.M., Wall, P.G., Rodrigues, L.C., Tomkins, D.S., Hudson, M.J. and Roderick, P.J. (1999) Study of infectious intestinal disease in England: rates in the community, presenting to general practice, and reported in national surveillance. *Brit. Med. J.* **318**, 1046–1050.

Wilkinson, R.G. (1996) *Unhealthy Societies,* Routledge, London.

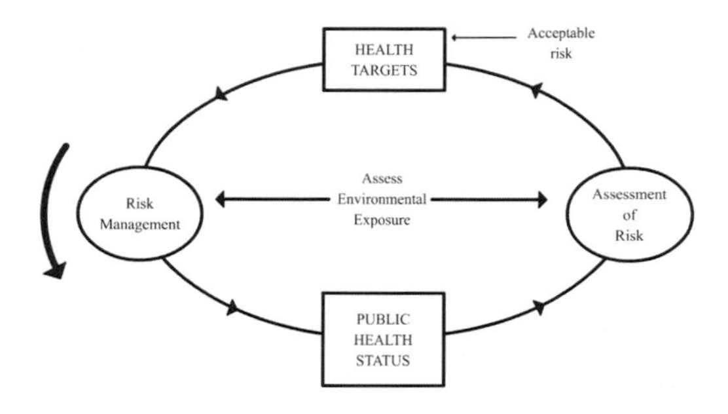

12

Management strategies

Dan Deere, Melita Stevens, Annette Davison, Greg Helm and Al Dufour

Beginning within the context of the classical risk assessment framework, this chapter discusses the origins of risks to microbiological water quality. The importance of a preventative multiple barrier approach is discussed and the advantages of controlling contamination as near to the source as possible are presented briefly. Practical, simple to use approaches are needed to identify risks and manage them at the day-to-day level. The hazard analysis and critical control points (HACCP) principles are used to illustrate such a process in relation to drinking water. Although the HACCP examples are drawn from drinking water the principles are equally applicable to the recreational water and wastewater reuse areas. A recently proposed management strategy for recreational water is also outlined.

12.1 WHAT IS RISK?

Risk is a component of everybody's lives. All activities that we are involved in carry some degree of risk. Risks can either be voluntary, such as cigarette smoking, or involuntary, such as breathing air polluted with car emissions or drinking water containing carcinogenic chemicals. There are many definitions of risk that range from broad definitions such as: 'risk is the probability of injury, disease, or death under specific circumstances' (Raman 1990) to more specific definitions such as: 'risk is the probability that an adverse outcome will occur in an individual or a group that is exposed to a particular dose or concentration of the hazardous agent' (Langley and Van Alphen 1993).

Risk Assessment is the process undertaken to evaluate whether there is a risk and, if so, how severe it is. Risk Management incorporates understanding, evaluating and prioritising risks for a given system and then implementing appropriate risk reduction strategies. In drinking-water supplies, risk assessment and risk management are essential components of ensuring the public health of consumers. Generally, risk cannot be measured accurately and is described using qualitative terms such as high, medium or low. In some instances risks can be estimated and expressed quantitatively, albeit within an uncertainty interval or probability distribution (see Chapter 8).

12.1.1 Classical risk assessment framework

Classical risk assessment involves four conceptual steps. These have already been outlined in Chapter 8, but will be revisited here taking a risk management perspective.

12.1.1.1 Hazard identification

Hazard analysis is a key component of both qualitative and quantitative risk assessment and risk management. Hazard identification is the identification of the constituents of drinking water, recreational water, wastewater reuse or whatever that may have the potential to cause harm to the user. The source of the hazard is also determined. The term hazard is usually used to refer to agents that can cause harm. An example of a microbiological hazard is the bacterium responsible for causing cholera, *Vibrio cholerae,* and the source of the hazard is faecal material from individuals infected with this agent. In terms of risk management, hazards need to be considered along with the events that result in the introduction of contamination. These event-hazards in terms of drinking water supply include storms, pipe breaks, treatment plant or disinfection plant failure.

12.1.1.2 Exposure assessment

The components of exposure assessment are:

- Identifying how and where the hazard enters the system;
- Determining who is going to be exposed to the hazard, how the hazard will reach them and what acts on the hazard within the system;
- Estimation of the concentration of the hazard that will reach the consumer; and
- The quantity and timeframe of hazard exposure.

12.1.1.3 Dose–response assessment

Dose–response assessment determines the impact that a hazard has on the population, given the concentration that the population is exposed to. Dose–response factors are calculated for many chemicals (such as lead and arsenic) and some micro-organisms based on animal and human feeding studies and studies of waterborne disease outbreaks. The results of these studies provide information on the severity of the health effects from exposure to different amounts of a given hazard.

12.1.1.4 Risk characterisation

Risk characterisation is the consolidation of information from exposure assessment and dose–response assessment. Characterising risk is determining the likelihood of an adverse effect from exposure to the specific hazard. For drinking water systems, risk characterisation has been carried out mainly for chemical contaminants. For example, for arsenic, the toxicological data is combined with the estimation of intake of water with a measured arsenic concentration to determine the risk of skin cancer and to give an acceptable 'guideline' concentration for this hazard in potable water (WHO 1993).

Risk characterisation also involves considering the uncertainty involved in each risk assessment step, for example, the extrapolation of results from animal feeding studies to humans. Other issues considered in risk characterisation include assessing the significance of the risk and whether it is acceptable, determining if action is required to reduce or eliminate the risk, and whether risk reduction can be carried out in a cost-effective manner (see Chapter 10).

A quantitative risk assessment programme is both time-consuming and subject to uncertainty. It may take years to develop a reasonable quantitative risk estimate for any particular hazard. Therefore, the management of risk should not necessarily await the outcomes of such an assessment. Instead, a more simplistic judgement-based assessment of risk would form the first action in a risk management programme, with detailed risk analyses being performed as a separate exercise (Bell 1999).

12.2 ORIGINS OF RISK

Risk management activities draw from all aspects of the classical risk assessment framework particularly the exposure assessment, which considers how it is that a person may become exposed to a contaminant. Any attempt at managing risk within a system, such as a drinking water supply system or a recreational water body, needs to start by asking what the origins of risk are within that specific system.

12.2.1 Chemical versus microbiological risk

Although this book focuses on microbiological risk it is important to note that there is a fundamental difference in the way that chemical and microbiological contamination (and therefore risks) arises, which leads to the adoption of different management strategies. Using a drinking water context the distinction is as follows:

(1) Microbiological Risk: the risk or probability of illness associated with the contamination of water supplies with bacteria, viruses, protozoa and so on. Symptoms of microbiological illness can be acute or chronic and there may also be delayed sequelae. However, in risk management terms microbiological risks are considered to have arisen from acute exposures – either an infection occurred or it did not when contaminated water was consumed.

(2) Chemical Risk: the risk of illness from chemical pollution of drinking water, or from chemicals, such as disinfection by-products that are formed within a water supply as a result of water treatment. Once again, health effects attributed to chemicals in drinking water can be acute (generally resulting from short-term exposure to high concentrations of chemical) or chronic (resulting from long-term exposure to low levels of chemical contaminant). However, due to the huge dilution factors involved, few chemicals are likely to reach concentrations in water that would result in discernible health effects due to a short period of exposure (where they do reach high concentrations, the water is generally likely to be undrinkable due to foul taste). Therefore, in risk management terms, chemical risks tend to be considered to have arisen from long-term, even lifetime, exposures.

In practice, as the above paragraphs illustrate, the picture is not black and white. However, for the purposes of this chapter we shall focus on microbiological risks due to acute exposures. As will become clear, this distinction is not academic and has significant implications in terms of risk management.

12.3 ORIGINS OF MICROBIOLOGICAL RISK

Using drinking water as an example, microbiological contamination can arise at many points in the catchment to tap supply chain. Figure 12.1 gives a generic catchment to tap flow diagram for microbiological risk that illustrates points that have been well established as sources of risk in many systems.

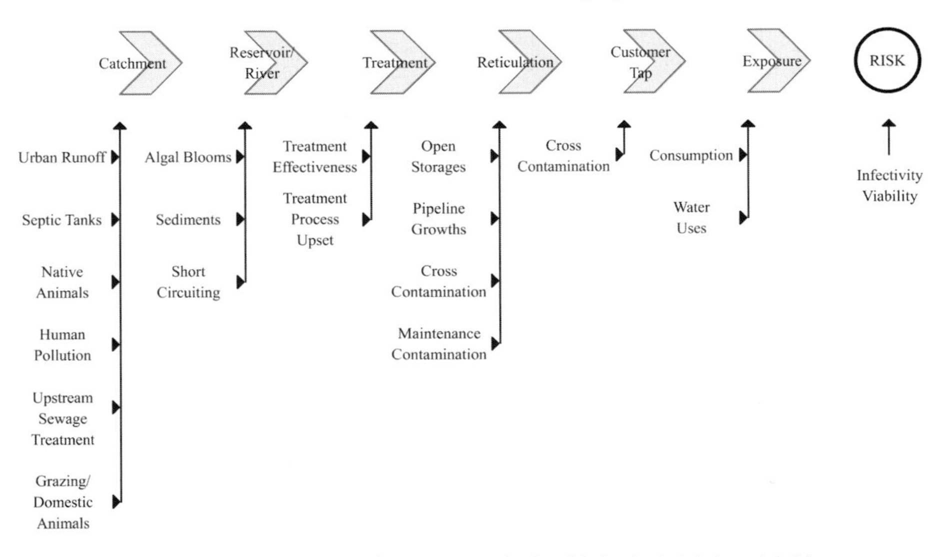

Figure 12.1. Generic flow diagram for sources of microbiological risk in a drinking water context (adapted from Stevens *et al.* 1995).

We have already discussed key terms such as hazard and exposure. Two additional terms and concepts will now be introduced. The first is *events*; this term will be used to describe an occurrence that leads to an increase in the risk of exposure. An example might be a storm in the catchment of a water supply system that leads to increased faecal material being washed into a reservoir (or equally a storm that leads to discharge of faecal material into a bathing area).

The second important concept related to events is that events need to be considered together as scenarios – *the fault-tree concept*. As can be seen by considering Figure 12.1, an event such as a storm is only likely to lead to a health risk to a community if other events occur as well. For example, there would first need to be significant levels of infectious pathogens in faeces in the catchment. Second, the storm would need to be severe enough to wash significant levels of contamination into the source water. Furthermore, water would need to be abstracted before significant levels of pathogens have lost viability and/or settled from the water body. Finally, an appropriate treatment barrier would either need to be absent or overwhelmed by the pathogen load. Such an example illustrates that in most cases events should not be considered in isolation but as part of a chain of events. A simple diagram illustrating this is given in Figure 12.2 (Stevens *et al.* 1995).

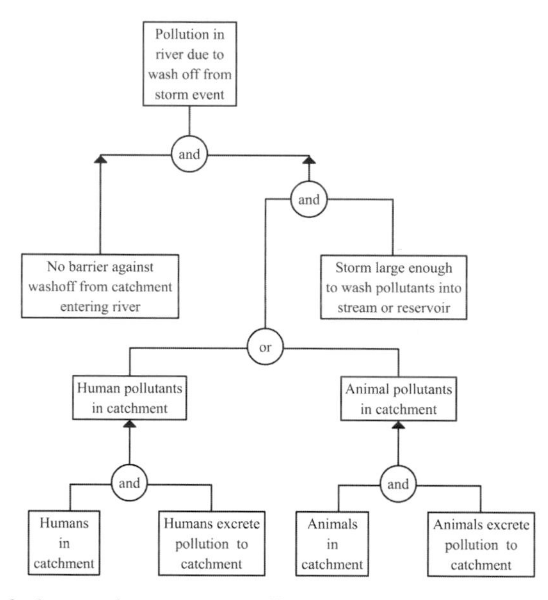

Figure 12.2. Generic fault tree for storm runoff polluting a drinking water source (adapted from Stevens *et al.* 1995).

12.3.1 Multiple barriers

Drawing from the above basic concepts it is important to move on to consider magnitudes of effect and probabilities of occurrence rather than simply presence or absence of risk or occurrence or otherwise of events. For example, animal faeces containing pathogens are not best considered as being either present or absent in significant amounts, but rather they are present at a range of contamination levels distributed in time. Equally, storms can have a range of severities and treatment barriers a range of capabilities or degrees of failure. Thus, the objective of risk management is to consider the events that contribute to risk and to focus on mitigating factors – barriers to risk. The objective is to reduce the risk to an acceptable level and/or to minimise risk by optimising the risk reduction throughout the system and by optimising the available barriers.

The use of multiple barriers works at two levels. First, in most cases, barriers act to reduce rather than completely eliminate risk. Therefore, since events are linked, the use of multiple barriers provides multiple levels of protection that act together to reduce the total risk by more than the reduction achieved by any one barrier. Second, where a barrier is reduced in its effectiveness, the presence of other barriers helps to maintain a reduced level of risk throughout the failure. This is the first of several reasons why the acute nature of the exposure timeframes relevant to microbiological risk is important. Even a short barrier failure where that barrier is a major factor in risk reduction could lead to unacceptable levels of risk exposure – maybe even a disease outbreak. However, where there are multiple barriers that are each capable of giving major risk reductions, failure of any one barrier is less significant. To give an advanced theoretical drinking water example, in a detailed assessment of microbiological risk, Teunis *et al.* (1997) considered the microbiological risk exposure for a population depending on a single high efficiency barrier for protection (filtration plant). The authors illustrated that in such scenarios, almost all the risk to the consumers during any one-year time period arises during the summation of the very brief periods, perhaps less than one day in total during that year, when the treatment barrier is operating poorly.

Another implication of the need for multiple barriers is that barriers need to be effective when they are most needed. For example, if most septic tanks in a catchment overflow during storms, and most treatment plant failures also occur during storms due to overloading, how well the treatment plant and septic tanks work most of the time becomes relatively unimportant if most of the risk exposure occurs during these occasional storms. Thus, another implication of the acute exposure of relevance to microbiological risk is the need for barriers to be effective during the short exposure to extreme event periods.

12.3.2 Outbreaks don't just happen

So far we have taken a theoretical perspective. We have considered microbiological risk from first principles by going through the thinking associated with predicting and understanding exposure pathways, sources of contaminants, events that lead to increased risk and the use of multiple barriers and the multiple benefits associated with these. It is also useful to take a practical perspective and look at the types of events and scenarios of linked events that have led to actual disease outbreaks, these being extreme examples of microbiological risk exposure. Table 12.1 (Davison *et al.* 1999) illustrates deficiencies in system operation, management or risk identification that were responsible for outbreaks of cryptosporidiosis from drinking water supplies in the US (Rose *et al.* 1997).

Table 12.1. Some shortcomings identified in some cryptosporidiosis outbreaks in the US

Deficiency	Comment
Monitoring equipment for filtration optimisation during periods of rapid change in source water.	Equipment was improperly installed, poorly maintained, turned off, ignored or temporarily inoperable.
Treatment plant personnel did not respond to faulty or inoperable monitoring equipment.	Deficiencies in the equipment were not compensated for by increasing the type and frequency of monitoring.
Filter backwash was returned to the head of the treatment process.	This process results in the possibility of concentrating oocysts, which may be put back into the system during a filtration breach.
Sources of high contamination were found near the treatment facility.	No mitigating barriers were in place to protect against introduction of oocysts into receiving waters (streams and groundwater) during periods of high runoff.
Sources of *Cryptosporidium* were unknown in the catchment prior to the outbreak event.	Knowledge of the sources of *Cryptosporidium* could have facilitated mitigation of the risk.
Natural events may have been instrumental in flushing areas of high oocyst concentrations into receiving waters.	Heavy rain can flush/carry oocysts into waters upstream of the treatment plant.
Filtration processes were inadequate or altered.	During periods of high turbidity, altered or suboptimal filtration resulted in turbidity spikes and increased turbidity levels being noted in the finished water.

Similar observations were made regarding the UK outbreaks reviewed by the UK Group of Experts (McCann 1999). Table 12.2 gives examples of disease outbreaks and their causes grouped according to cause to show the variety of scenarios that can lead to disease outbreaks.

Table 12.2. Scenarios affecting municipal drinking water implicated as causes of disease outbreaks

Causal event(s)	Aetiology	Water type	Cases	Reference
Pre abstraction and treatment				
Surface run off from contaminated catchment after heavy rain. Increased Cl demand due to turbidity.	*Campylobacter*	Chlorinated surface water	3000	Vogt *et al.* 1982
Contaminated surface run off from meltwater and heavy rain entering municipal wells.	*Campylobacter*	Untreated groundwater	241	Millson *et al.* 1991
Drought followed by heavy rain agricultural surface run off and poor coagulation and mixing.	*Cryptosporidium*	Cl'ed + package filtered river water	34	Leland *et al.* 1993
Poor mixing and flocculation with filters started up without backwashing.	*Cryptosporidium*	Surface water (CT)	13,000	Rose *et al.* 1997
Increase in turbidity, poor coagulation and backwash recycling.	*Cryptosporidium*	Surface water (CT)	403,000	Rose *et al.* 1997
Catchment contaminated by higher than realised population, Cl dosage too low.	*Giardia*	Cl'ed surface water	350	Shaw *et al.* 1977
Post abstraction and treatment				
Backflow of farm-contaminated river water due to low mains pressure.	*Campylobacter*	Sand filtered groundwater	2000	Mentzing 1981
Switch to unchlorinated stagnant reservoir subject to animal contamination.	*Campylobacter*	Untreated tank water	150	
Agricultural runoff entering unsealed supply.	*Cryptosporidium*	Surface water (CT)	27	Badenoch 1990
Deliberate contamination of water storage tank.	*Giardia*	Municipal supply	9	Ramsay and Marsh 1990
Cross-connection between pressure dropped potable and wastewater lines at pump wash.	*Giardia* & *Entamoeba*	Surface water (CT)	304	Kramer *et al.* 1996
Sewage overflow entering pipes after repairs of ice breaks made without post chlorination.	*E coli* O157	Municipal supply	243	Swerdlow *et al.* 1992
Birds entering water storage tank.	*Salmonella*	Untreated groundwater	650	Angulo *et al.* 1997

CT: Conventionally treated Cl: Chlorine Cl'ed: chlorinated

These scenarios are similar to the ones identified in Sweden (Chapter 6). They also represent the 'tip of the iceberg', as outbreak detection is notoriously difficult, as outlined in Chapters 4 and 6. Thus, given the problems of under-reporting and outbreak detection it is useful that we learn about the source of microbiological risk from detected outbreaks. This can then be extrapolated to sub-detectable-outbreak scenarios. The study of outbreaks provides useful insights into the origins of waterborne disease risks in total and illustrates that events and scenarios or chains of events involving barrier failures and/or other unusual events are the key risk sources and, therefore, targets for risk management.

12.4 MANAGING RISK

At its most simple, waterborne disease risk management involves:

- identifying potential sources of contamination; and
- managing barriers to prevent contamination reaching end-users.

In an ideal system this would be satisfactory since:

- all scenarios by which contamination could enter would be understood;
- barriers would be effective at eliminating the risk from these sources;
- any barrier failure would be detected and corrective actions taken; and
- individuals with the power to manage risk would have this as their primary interest and would behave appropriately.

In reality:

- the arrangement of waterborne contamination sources and barriers is very complex and is never perfectly understood;
- barriers are rarely absolute barriers and function primarily to reduce risk, not eliminate it;
- finite resources limit the ability of contamination sources to be controlled at source or via barriers; and
- individuals with the power to manage risk may have conflicting interests and people cannot be controlled and relied upon totally.

This detail and complexity prevents an individual from fully understanding and managing the risks to waterborne contamination, and means that simplistic approaches to risk management will be ineffective. In reality, arrangements are complicated and multiple individuals and stakeholders are required to be involved both for identifying contamination scenarios and managing barriers. This complexity necessitates the use of systems to manage risk.

12.4.1 A systems approach

Managing risk in real systems requires a systems approach. This section provides a checklist of steps used in managing risk in a water supply and in producing a risk management plan. The terminology used is kept consistent with that of Hazard Analysis and Critical Control Points (HACCP). This has been found to be an acceptable framework for guiding the process of risk management in water supplies (Barry *et al.* 1998; Deere and Davison 1998; Gray and Morain 2000; Havelaar 1994). There are a number of other frameworks that are similar and many factors (Davison *et al.* 1999), such as the practical experience with, and widespread knowledge about, HACCP that make it a potential model of choice. The principles of HACCP are shown in Figure 12.3.

HACCP has, as its basis, a focus on controlling hazards as close to their source as possible. It has been described as a 'space age' system for assuring food safety due to its development during the 1960s US Space Program for protecting astronauts from unsafe food and beverages. An effective quality assurance system that addresses these principles has become the benchmark means to assure food and beverage safety since its codification in 1993 by the United Nations Food and Agricultural Organization (FAO) and World Health Organization (WHO) Codex Alimentarius Commission.

12.4.1.1 Assemble team and resources

Complex systems cannot be understood and managed by any one person. A team of individuals needs to be identified that will have the collective responsibility for identifying risks and barriers from contamination source to the point of exposure. This team needs to be made up of individuals with the skill required to identify risks and barriers as well as the authority to ensure barrier management is developed. Experts, not normally associated with the system in question, may need to be brought into the team as the occasion arises.

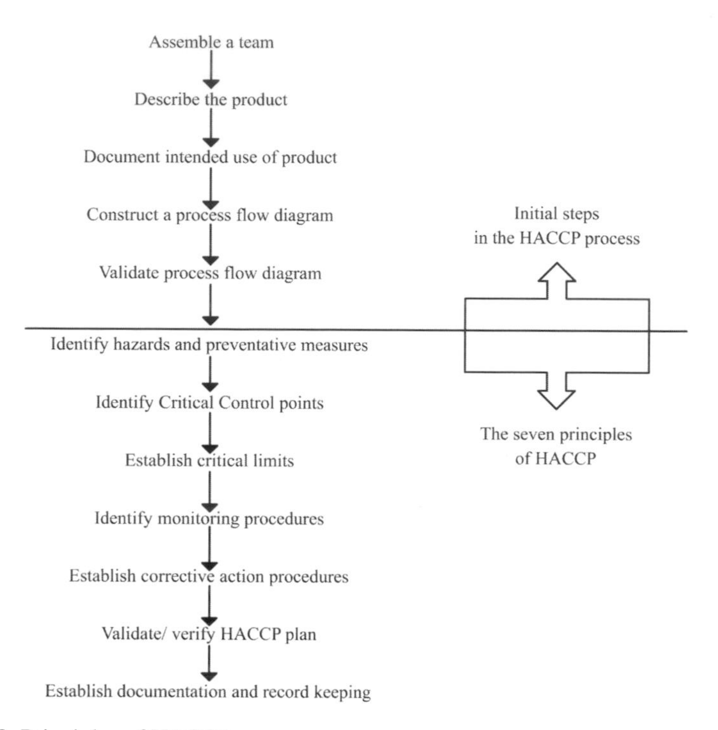

Figure 12.3. Principles of HACCP.

This could include veterinary and human infectious disease specialists, scientists, engineers or an independent team facilitator. For example, the catchment HACCP plan described by Barry *et al.* (1998) involved such a multi-disciplinary team. The team needs to have the resources, in terms of time and equipment, to perform the task.

12.4.1.2 Flow chart and flow chart verification

Complex systems cannot easily be visualised. A working representation, such as a hierarchical series of flow charts, needs to be produced describing possible sources of contamination, transfer pathways by which contamination can reach end-users, and barriers. Representations of systems can be inaccurate. Verification of the representation could involve field audits and cross-checking by others with specific system knowledge.

12.4.1.3 Describe the water and its use

Risks cannot be completely eliminated. There needs to be an understanding of the health status of the exposed population and the level of risk to which they can acceptably be exposed. This enables the most relevant contaminants to be identified and, in some cases, their allowable concentrations to be determined. This equates to a description of the nature of the water at the point of exposure. It is important to be realistic about the communities' likely uses of the water. Thus, if drinking water is intended only for consumption by healthy adults with all others required to boil their water, is this understood by these end-users? Is this in fact likely? If not, unboiled water consumption among these other groups needs to be considered as a likely end-use. Once these foundation steps have been performed, a logical process for identifying risks and barriers should be followed.

12.4.1.4 Hazard analysis

Using the systematic representation (e.g. flow chart) as a guide, hazards, their sources and scenarios by which they could contaminate the water need to be identified. Ideally, some assessment of the risk from each of these hazards and events needs to be made. This is useful because priorities can be assigned to each potential risk. Some can simply be ignored, making the overall job of risk management simpler. Others can be assigned as important in terms of aesthetics and quality but not necessarily of public health significance. Finally, those that are of public health significance can be focused on as the first priority. Two examples of approaches used for assigning risks to hazards are given in Table 12.3 and Figure 12.4.

Table 12.3. Example of a common, simple risk assessment framework as used in South East Water (Melbourne, Australia) HACCP plan (Risk Factor = Likelihood × Severity of Consequences. If a risk factor is 6 or greater, the hazard is to be considered further in the HACCP plan and monitoring and corrective actions should be devised.)

	Severity of Consequences				
Risk Factor Matrix **Likelihood**	Insignificant No impact or not detectable	Minor Significant impact	Moderate Impact of target levels	Major Impact on franchise levels	Catastrophic Public health risk
Almost certain (daily)	5	10	15	20	25
Likely (weekly)	4	8	12	16	20
Moderate (monthly)	3	6	9	12	15
Unlikely (annually)	2	4	6	8	10
Rare (every five years)	1	2	3	4	5

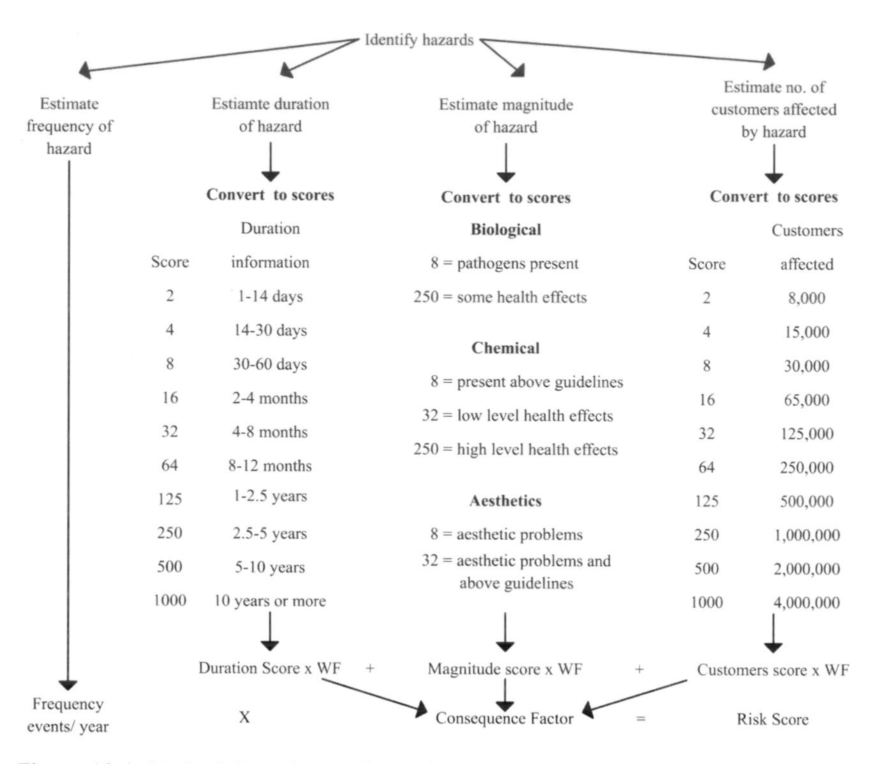

Figure 12.4. Methodology for scoring risks used in the Sydney Water hazard analysis developed by Parametrix/AWT (courtesy of Carl Stivers, Parametrix, Australia).

12.4.1.5 Critical control points

The identification of barriers to contamination and preventative measures is the first step in managing risk. Generally, microbiological risks are best controlled at or as near as possible to the source of contamination, because multiple benefits arise from control at source that do not arise from control once systems are already contaminated (end of pipe treatments). These are as follows:

- Amplification: once released into the aquatic environment, microbiological contaminants can cause infections and multiply. This can have detrimental effects by increasing the pathogen load on end of pipe barriers as well as leading to increased prevalence of pathogens in the environment generally. This contamination also reduces the value of the water upstream of the end-of-pipe treatment point.

- Multiple barrier protection: reliance on end-of-pipe treatments can lead to almost total reliance on a single barrier for protection. Unless this barrier is extremely reliable and effective, this will expose end-users to risk during barrier failure.
- Polluter pays: an emphasis on controlling pollution at source reduces the cost transfer from the polluter to the end-user of the water. Instead, the polluter is more likely to bear a greater share of the cost of preventing contamination or of cleaning it up. As well as being ethically attractive, prevention of contamination and treatment at source may in fact turn out to be the lowest community-cost solution for some contaminants.

Barriers are Control Points (CP), that is, points that control the risk by reducing or eliminating the transfer of pathogens to end-users. To ensure an appropriate prioritisation some of these points can be singled out as the most significant and can be termed Critical Control Points (CCP). These are points at which barrier effectiveness is essential for safe water use. Some barriers are involved in control of aesthetics; these can be termed Quality Control Points (QCP). These points are important or even critical for acceptable quality, but not necessarily for safety.

Critical control points identified in recently produced HACCP plans cover a variety of areas from the consumers' properties (e.g. fitting of backflow prevention devices) to disinfection and raw material control at treatment plants and control of catchment animals (Barry *et al.* 1998; Gray and Morain 2000). Note that some activities are not designated as critical control points but are instead delegated to the status of supporting programmes. An example would be the use of best practice management in catchments (Ashendorff *et al.* 1997) or the procedure used for repairing burst water mains. This is discussed further below.

12.4.1.6 Critical limits

Risk management activities should be focused on control points. Procedures and targets need to be determined such that control activities can meet an appropriate specification. This specification might refer to a measurable physical property of the water, such as turbidity, or to an observable property in a catchment area. These measurable/observable factors can have limits assigned to them such that provided the control point is operating within these prescribed limits, the hazards can be taken as being under control. This is an important concept. The hazards themselves usually are not the measurable factor. Instead, some feature of the barrier that can be observed or measured is chosen that can act as a surrogate for control of the hazard. This is important for several reasons:

- Hazards are often not practically measurable at concentrations that represent an acceptable risk. This makes the use of surrogates more protective.
- Hazards are generally not measurable in real-time or on a continuous basis. Ideally, limits such as observations or measurable properties of water will be available at the time and point of inspection. This permits rapid interrogation and response, which is more protective and preventative.
- There are numerous possible hazards that may vary rapidly in both time and space. Such hazards are likely to be present as a result of a scenario of particular events. They may not be present in a relatively steady-state situation. As a result, the absence of detectable concentrations of hazards at one point in time does not necessarily provide assurance of its absence later. The use of surrogates is, therefore, more conservative.

The limits set will usually be grouped at two levels. Operational limits may be set at a point where a response is required, for example as an early warning, but where water quality is not likely to be significantly compromised. Critical limits will be set at points where urgent action is required to ensure that water quality and safety remains acceptable.

12.4.1.7 Monitoring and corrective actions

Managing risks at control points requires the detection of control point failures. Monitoring is required to pick up operational and critical limit exceedances. The nature and frequency of this monitoring will depend on what is being monitored. Thus, when selecting and setting operational and critical limits it is important to consider the practicalities of monitoring these. If they can't be monitored with sufficient frequency and practicality to reveal barrier failures in good time, there is little point setting them. In some cases, a combination of observations and/or measurables may together constitute what is taken as a limit (e.g. a critical disinfection envelope may consist of a combination of pH, chlorine concentration, temperature and time and be determined using an appropriate algorithm; Smith *et al.* 1995). The frequency of monitoring depends on the speed at which barrier failure can occur and the rate at which contamination can build up after failure. For example, it may take only a matter of hours for source water turbidity levels to change and increase beyond acceptable levels such that this parameter may need to be checked at intervals of minutes. In contrast, the density of feral mammals present in a catchment area could take five years to change significantly and this density could, therefore, be checked at intervals of years.

When exceedances to limits are detected by monitoring activities some action needs to be taken to bring the control point back into specification. Ideally this action will be predetermined and will be tested for its effectiveness.

12.4.1.8 Record keeping, validation and verification

Documenting the key aspects of the plan is in itself a useful discipline to ensure clarity and completion. It also provides a written record of the plan for use by others and as a basis for updating. Recording monitoring activities and significant events provides a body of information for long-term trend analysis and auditing.

The complete plan (hazards, control points, limits, monitoring and corrective actions) becomes a guide on how to operate a specific water supply system for a safe, high-quality water supply. To be relied upon this plan needs to be supported by accurate technical information. Assumptions about sources and barriers need to be checked to validate this accuracy. Validation combines system-specific information with published scientific information. In many cases there will be generic or system-specific unknowns and professional judgement will need to be applied. In the longer term, research can be used to fill these data gaps. There is a need to pick up and incorporate new information as it becomes available to ensure that the plan remains valid. Collating, synthesising and disseminating such information could be a potentially important role of international organisations (such as the WHO).

Once a plan is implemented, there needs to be some verification that it is being followed in practice. Furthermore, there needs to be some verification that the water reaching consumers, or the bathing water (and so on) is in fact of an appropriate quality.

12.4.2 Managing people and processes

The development of a realistic plan describing how things should operate is only the first step in managing risk. If the plan is to work it must be followed in practice. There needs to be a supporting programme of good operating practice. Furthermore, the people and processes responsible for managing risk need to follow the plan as intended. To achieve this involves leaving the realm of hard science and HACCP theory and entering the world of quality management systems – systems for gaining control of people and processes to ensure the desired outcome. The importance of this control of day-to-day activities cannot be overstated and it is worth noting that the organisations that have implemented HACCP plans, be they food, drink or tap-water suppliers, have seen it as essential to underpin the process with a quality management approach. Key elements of a supporting quality management programme are:

- Strong commitment at all organisational levels;
- Good operational practices described in standard operating procedures for repair, maintenance and operation;
- Ongoing education and training of employees in good operational practices;
- Product and raw material traceability;
- Control and use of key documents, checklists and data records; and
- A compliance culture with strong auditing to ensure procedures are followed.

In Europe and Australia the standard approach to drinking water quality management, ISO 9000, is appropriate and is by far the most commonly used model. In the US, where ISO systems have not been widely adopted, an alternative system is under development building on a treatment plant control system termed Partnerships for Safe Water (Pizzi *et al.* 1997). Simpler HACCP/ISO 9000-based quality assurance systems have been developed for small food and beverage organisations. It is anticipated that such systems could also be applied to smaller water authorities, which might not have the resources to implement a full ISO 9000 and HACCP system but would nevertheless benefit from these systems being in place.

12.5 DRINKING WATER CASE STUDIES

In 1996, the Australian Drinking Water Guidelines (NHMRC/ARMCANZ 1996) stated the need to follow a quality system and multiple barrier approach. To enable water suppliers to adopt the system most consistent with their organisational practices, the guidelines did not single out any particular system. Australian water companies have responded to these guidelines by undergoing HACCP-type risk assessment processes and implementing quality management systems to control their processes and people. Example case studies could be drawn from most of Australia's major cities (such as Brisbane Water (Gray and Morain 2000), Sydney Water, Melbourne Water and South East Water in Melbourne) as well as a number of rural supplies (DLWC 1999).

12.5.1 Sydney Water, Sydney, New South Wales, Australia

Over a period of approximately a year, hazard assessment and management workshops were carried out by Sydney Water to evaluate the risks to each of its 14 water supply systems. Various stakeholders (including State Health Officials) and Sydney Water employees were invited to the workshops and asked to contribute their knowledge in ascertaining the hazards (from catchment to customer) that had happened to or were likely to happen to Sydney's water. The team was divided into groups that concentrated on the various aspects of the water supply system including catchment, storage, treatment and distribution. The identified hazards were then scored based on the methodology given in Figure 12.4, which was developed by a consultant team of risk assessment and water quality specialists (with inputs from Sydney Water) for Sydney Water. This methodology is very flexible as it can be adapted to specific systems (based on number of customers for instance) and provides a more sophisticated approach to hazard assessment when compared to risk assessment matrices often quoted in HACCP methodologies. An example of the generic types of hazards and the scoring results is given in Table 12.4.

Table 12.4. Example hazards and ranking scheme based on Sydney Water hazard assessment methodology. For weighting factors, see Figure 12.4

Hazard	Dur.	Mag.	Cust.	Conseq factor	Freq.	Risk score	Total risk
	Weighting factors						
	0.24	0.47	0.29				
Filter breakthrough	2	250	16	122.6	3.50	429	11.5
Pathogen ingress through mains breaks	2	8	2	4.8	60.0	289	19.2
Incorrect dam screen filters	2	250	16	122.6	2.00	245	25.7
Flushing causing resuspension of sediment in pipes	2	32	4	16.7	12.0	200	31.1
Backwash supernatant returning to treatment system	2	250	16	122.6	1.50	184	36.0
Inappropriate treatment train for high pathogen contamination	8	250	16	124.1	1.20	149	40.0

Dur. - duration; Mag. - magnitude; Cust. - customers affected; Conseq. - consequence; Freq. - frequency (events/year).

Those hazards that were identified as posing approximately 80% of the cumulative risk were chosen for hazard management. It should be noted that the hazards in Table 12.4 are example hazards only and are not exhaustive. It should also be noted that, based on the concept of multiple barriers, the absence of a barrier or management option constitutes a hazard to water quality in its own right.

The same workshop participants were then asked to revise the chosen hazards to make sure that important ones had not been missed. This process is vital as it, again, utilises the knowledge and intuition of the people who operate the water systems on a day-to-day basis rather than relying purely on a 'numbers approach'. Participants were also asked to provide information on the management options in place or those required.

12.5.2 South East Water, Melbourne, Victoria, Australia.

Over a period of around nine months, South East Water implemented an ISO 9000 quality management system across the whole business. This included all aspects of operational practice affecting water quality. After this system had become established, South East Water implemented a HACCP system over a four-month period. Both systems are externally audited and have been certified to international standards (in the case of HACCP this is the National Sanitation Foundation standard).

Representatives from internal operational areas, health authorities, other water companies and consultants were included in HACCP teams. In total, six relatively small teams developed plans relevant to their specialist areas, such as 'backflow prevention' or 'disinfection'. One of the most commonly used risk assessment matrices, given in Table 12.3, was applied. This approach is simpler than that used by Sydney Water, reflecting the relatively simpler water supply system. An example worksheet illustrating the level of detail adopted for the HACCP plan is given in Table 12.5. In essence, any operational aspect thought to have an impact on public health was considered significant and underwent the HACCP process.

A number of studies were undertaken to validate that particular operational practices were effective at maintaining safe water. Some of these were very specific. For example, South East Water repairs around 1500 burst water mains per year. A detailed follow-up of the water quality after a number of these burst main repairs was used to validate the mains burst repair procedure. In a more comprehensive validation exercise, a study similar to the epidemiological studies described by Payment (1997) was performed using epidemiologists from Monash University Medical School. Around 600 houses receiving water supplied by South East Water were selected and half were fitted with water filter/UV units and the other half with sham units.

Table 12.5. Example based upon a HACCP worksheet for a hazard-event (South East Water HACCP plan)

Step	Hazard	Preventative measures	Risk	CCP/ QCP/ CP	Target level	Action level	Monitoring procedure	Corrective action
Storages	Pathogen contamination of closed storages	Scheduled maintenance program. Cleaning of storages Bird proofing Roofing of open storages	Unlikely/ catastrophic 10	CCP	Intact bird proofing mesh	Any breach in bird proofing	Water Operations field technicians carry out inspections during site visits. These inspections occur on average fortnightly with reports being recorded in personal diaries or rung in immediately.	Any breaches are notified to Water Operations engineers who repair bird proofing and undertake any or all of the following actions: Scour contaminated water Drain and clear storage Flush affected area Increase disinfection dosing Bypass storage Alternative supply Actions undertaken are recorded

This double-blinded randomised clinical trial study design was used to follow the health status of these 600 families for 18 months (Hellard *et al.* submitted) to look for the effect of filtering/disinfecting water as a means of validating the normal water delivery process.

12.6 A NEW APPROACH FOR MONITORING AND ASSESSING RECREATIONAL WATER QUALITY

The traditional management approach to maintaining the safety of bathing beach waters has been to sample the water on a routine basis to determine if the microbiological quality meets certain predetermined limits. This approach has several shortcomings, some of which are directly related to the analytical methods used to measure water quality. Currently used methods involve the measurement of micro-organisms commonly associated with excreta from the intestinal tract of warm-blooded animals (see Chapter 13). The measurement of indicator bacteria provide a retrospective assessment of water quality. Usually 24 hours are required before the density of indicator bacteria in the water sample can be determined. Under these circumstances the hazard associated with the excreta-contaminated water may no longer be present by the time the indicator bacteria are detected. This type of temporal delay in detecting hazards in water is not effective for managing risks associated with beach waters. A second shortcoming of indicator bacteria is that some of the micro-organisms used to measure water quality may have extra-enteral sources. Industrial wastes are known to provide an environment that can produce coliforms and faecal coliforms. These shortcomings mean that the organisms do not adequately provide warning of potential risk from enteric pathogens and may serve to confuse the picture.

Although these shortcomings are not catastrophic, they do present a situation where frequent routine sampling may not be economically effective when compared to the benefits that may accrue with regard to maintaining minimum health risks for swimmers. The costs of frequent routine monitoring can be burdensome to small communities and currently there is no acceptable alternative to the traditional approach.

In 1998, a tentative alternative approach to the testing currently used was proposed at an expert consultation that was co-sponsored by the World Health Organization and the United States Environmental Protection Agency (Bartram and Rees 2000; WHO 1999). The approach (which requires filed testing and possibly further adaptation) is ideally suited to amendment to account for specific local circumstances, and leads to a classification scheme based upon health risk. It presents two significant elements, namely:

- A classification scheme based on an inspection of various sources of faecal contamination (i.e. a sanitary survey), the extent to which faeces affect beach waters and the density of faecal indicator bacteria in beach water samples.
- The possibility of reclassifying a beach to a higher class if effective management interventions are instituted to reduce exposure and thereby lower the risk of swimming-associated illness.

The elements of the proposed scheme are illustrated in Figure 12.5. The advantage of the classification scheme, as opposed to the usual pass/fail approach, lies in its flexibility. A large number of factors can influence the condition of a beach and the classification system reflects this, allowing regulators to invoke mitigating approaches for beach management.

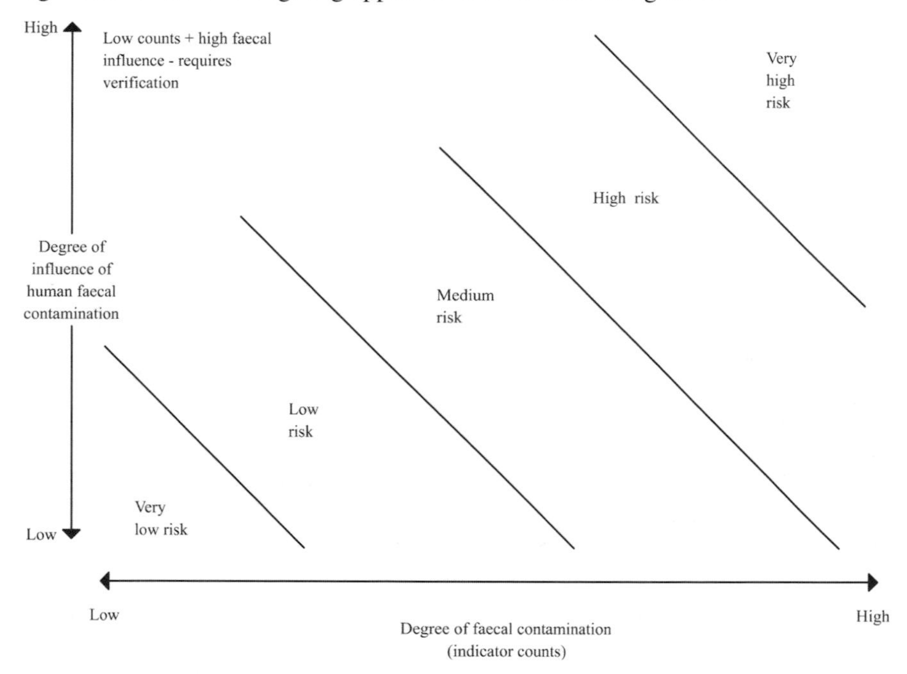

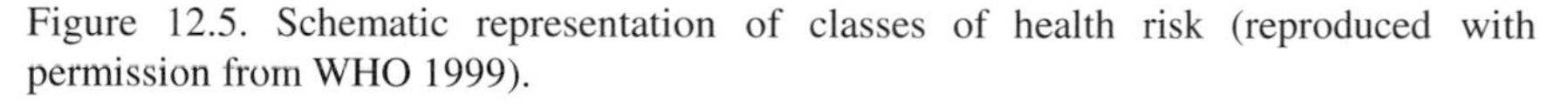

Figure 12.5. Schematic representation of classes of health risk (reproduced with permission from WHO 1999).

The horizontal axis of the figure shows the degree of faecal contamination as measured with indicator bacteria. The vertical axis shows the degree of influence of human faecal contamination. The degree of faecal contamination has a direct influence on the classification of risk and the indicator densities provide a means of monitoring beach waters to determine changes in classification.

12.6.1 Principle sources of human faecal contamination

The most important sources of human faecal contamination that affect bathing beaches are:

- discharges from sewage treatment plants including those from combined sewer overflows;
- riverine sources, where rivers receive sewage discharges, and the river is used directly for recreation or it flows to coastal or lake areas used for recreation; and
- contamination from bathers themselves.

The faecal contaminants from these sources can be graded based on either the distance from the beach of the discharge outfall, the distance of travel or travel time to the beach in river systems or the density of swimmers at beaches.

12.6.1.1 Sewage discharges

The risk potential associated with sewage discharges can be estimated if information regarding the length of the outfall and the degree of treatment of the wastewater is available. For example, raw untreated sewage discharged directly on to the beach would carry a very high risk potential. If the discharge is carried far from the beach through a long distance outfall, the risk potential would become negligible. At the opposite end of the risk gradient, a very low risk potential results if the sewage receives tertiary treatment plus disinfection even if the treated sewage is discharged directly on to the beach. The matrix of the degree sewage treatment and the outfall distance from the beach is a key element in determining the influence of faecal contamination as it relates to risk classification. The risk potential is outlined in Table 12.6.

Table 12.6. Risk potential to human health through exposure to sewage (reproduced with permission from WHO 1999)

Treatment	Discharge type		
	Directly on beach	Short outfall[1]	Effective outfall[2]
None[3]	Very high	High	Not applicable
Preliminary	Very high	High	Low
Primary (inc. septic tanks)	Very high	High	Low
Secondary	High	High	Low
Secondary + disinfection	Medium	Medium	Very low
Tertiary	Medium	Medium	Very low
Tertiary + disinfection	Very low	Very low	Very low
Lagoons	High	High	Low

[1] The relative risk is modified by population size. Relative risk is increased for discharges from large populations and decreased for discharges from small populations.
[2] This assumes that the design capacity has not been exceeded and that climatic and oceanic extreme conditions are considered in the design objective (i.e. no sewage on the beach).
[3] Includes combined sewer overflows.

12.6.1.2 Riverine discharges

Riverine and estuarine beaches and beaches near the mouth of rivers can be affected by faecal contamination from point sources, such as sewage treatment plants which discharge into the river. The risk potential associated with discharges of sewage into rivers can be estimated by determining the size of the discharging population and the flow rate of the river receiving the sewage. The flow rate influences the dilution of the sewage as it enters the river. The dilution effect is related to dry weather and wet weather flow. Dry weather flow is associated with low dilution and this usually occurs during the bathing season. Low flow rivers provide very little dilution effect and, therefore, the size of the discharging population takes on increased significance because of the high volume of waste produced. Under all conditions plug flow with no dispersion is assumed. To form a data set from which risk potential can be estimated, various combinations of population size and river flow rate are used in conjunction with the type of treatment applied to the sewage influent. These estimates can be used to classify beaches on rivers or near coastal waters affected by riverborne faecal contamination.

The risk potential classification system for riverine systems is similar to that used for ocean outfalls. The dilution effect gradient runs from a high population density with low river flow to a medium population density with a medium flow river to a low population density with a high river flow. The risk potential is greatest with high population and low river flow, and the lowest risk potential with low population and high river flow. This pattern of risk holds for effluents of all types of treatment. The type of treatment does affect the risk potential. As the treatment process becomes more complex, the risk potential for each dilution effect decreases. In practice several discharges into a single river are likely to occur and where larger discharges are treated to a higher level, then smaller sources (including septic tank discharges) and combined sewer overflows may represent the principal source of concern.

The classification system can be used directly for freshwater river beaches and for beaches in estuarine areas. The system may also be appropriate for beaches near the mouth of rivers contaminated with faecal wastes.

12.6.1.3 Bather shedding

Bathers have been shown to shed high densities of *E. coli*, enterococci and *Pseudomonas aeruginosa* in tank studies where total body immersion was examined under controlled conditions (Breittmayer and Gauthier 1978; Smith and Dufour 1993). Other studies have demonstrated the accumulation of faecal indicator bacteria over the course of a day at populated beaches (Cheng *et al.* 1991). Two elements, bather density and water bodies with very little water movement contribute to bather-to-bather transmission of illness. These two elements can be used to develop a risk potential matrix which lists low risk for high bather density and high dilution, and a very low risk in the case of low bather density and high dilution. Medium risk results from high bather density and low dilution, which becomes low risk if both the bather density and dilution are low. These risks may be higher if the beach is populated with high numbers of young children or if no sanitary facilities are available.

12.6.2 Assessing microbiological quality

Sewage and faeces contain a number of harmless bacteria, such as enterococci and some types of *E. coli*, and chemicals, such as coprostanol, which can be used to detect the presence of faecal material in water (see Chapter 13). These indicator bacteria or chemicals can be used to quantify the amount of faeces at a beach. They have been used to show the relationship between beach water quality and swimming-associated illness (WHO 1998). Primary indicators, such as *E. coli*, faecal streptococci and enterococci have been used for years as measures of faecal contamination. Other micro-organisms, such as clostridia or coliphage, are also associated with faecal contamination but have not received broad acceptance as traditional indicator bacteria. These organisms are designated as secondary indicators and they are used mainly for follow-up analyses or for their instructive value.

Faecal streptococci or enterococci are used as marine water quality indicators in temperate climates. These micro-organisms can be placed in categories that describe percentiles of water quality densities that are associated with health effects. These bacteria are considered primary indicators and they have been used on a routine basis for many years. It has been suggested that these indicators have sources other than the gut of warm-blooded animals, such as soil or plants, in tropical environments. In this situation sulphite-reducing clostridia or Clostridium perfringens have been proposed as secondary indicators.

In temperate freshwater environments, *E. coli* is an effective indicator of faecal contamination, in addition to faecal streptococci and enterococci. Secondary indicators such as clostridia are suggested for use in freshwater tropical climates. The percentile values associated with health effects in swimmers would not necessarily be the same in marine and in fresh waters.

The percentiles can be categorised based on the relationship to illness. For instance, five categories (labelled A to E) could be segregated based on upper 95 percentile values of <10, 11–50, 51–200, 201–1000 and >1000 for faecal streptococci or enterococci, where all categories above 50 are associated with swimmer illness. The use of these categories is advantageous for classifying risks and for reclassifying risks associated with faecal contamination.

12.6.3 Primary classification of beaches

Primary classification of a beach involves conducting a sanitary inspection of potential pollution sources to determine the susceptibility of the beach to faecal influence and a microbiological assessment of beach waters using primary indicators. Once the appropriate categories are determined by sanitary inspection and microbiological assessment they can be fitted into a table, such as Table 12.7, to determine the primary classification for a beach. For example, if the microbiological quality of beach water, as indexed by faecal streptococci or enterococci, is in the 11–50 indicator density range (category B) and the faecal influence category was found to be moderate then the particular beach would be classified as 'good'.

Reclassification of a beach to a higher or lower level might result from a number of events, such as a major break in an outfall pipe or a significant modification of the treatment process. Either one of these events could dramatically affect the quality of the bathing water with respect to faecal contamination of a beach. A break in an outfall pipe could deliver much higher levels of faecal contamination to a beach and thereby increase the risk of swimming-associated illness over a potentially long time period until the break is repaired. This condition would necessarily change the microbiological assessment category and the sanitary inspection category with a resultant change to a lower classification. Similarly, an improvement in the treatment process, wherein a primary treatment process was upgraded to a secondary treatment process, would improve the quality of the water reaching a beach. This would not only affect the sanitary inspection category, but also the microbiological assessment category. The long-term effect of changing these two categories would be to change the classification of a beach to a higher, more desirable level.

Table 12.7. Primary classification matrix (reproduced with permission from WHO 1999)

Sanitary Inspection Category#	Microbiological Assessment Category (indicator counts)				
	A	B	C	D	E
Very low	Excellent	excellent	good	good+	fair+
Low	Excellent	good	good	fair	fair+
Moderate	good*	good	fair	fair	poor
High	good*	fair*	fair	poor	very poor
Very High	fair*	fair*	poor*	very poor	very poor

= susceptibility to faecal influence; + implies non-sewage sources of faecal indicators (e.g. livestock) and this should be verified; * indicates unexpected results requiring verification.

Some events are much more variable than intervention measures or system breakdowns. Rain events are situations where the sanitary inspection categories and the microbiological assessment category can be significantly modified. Sewage treatment plant effluents can bypass the treatment system because wastewater sewage drains are frequently also used for stormwater. Under heavy rain conditions the combined system can overwhelm the treatment regime and effluents are discharged without treatment. Combined sewer overflows to riverine environments affect the categorisation of potential risk associated with the treatment category. The overflows will change the category from treated to untreated sewage and will lead to an increase in the risk potential. Although rainfall events are not predictable, the effect of these events on beaches are predictable. For example, it can be determined that the microbiological assessment category may increase to the 101 to 200 indicator density range from the 51 to 100 range with a half-inch of rainfall. This change, along with the faecal influence category change, might lower the beach classification from good to poor. Since this change can be predicted based on the amount of rainfall, a temporary management action would be appropriate at the beach.

12.6.4 Management actions

Routine monitoring may be a key element in maintaining the safety of bathing waters. It is a direct measure of microbial quality of beach waters and closely related to beach classification. It is also a means of measuring beach change in status over time. If the beach has been properly classified and the faecal influence is fairly constant, it should be possible to substantially reduce the monitoring requirements.

Direct action should be a principal management approach. Improvement of the treatment process or other remedial actions such as diversion of the sewage discharges away from a beach by constructing long-distance outfalls will significantly lower the potential risk associated with human excreta. The retention of wastewater combined with stormwater so that it can be treated later rather than discharged to receiving waters untreated, would also significantly lower risks at bathing beaches.

The management of intermittent events, such as stormwater runoff, can usually be addressed by informing the public, either through the media or by posting signs at the beach, that a short-term health hazard exists. Other means of dissuading the public from exposure to contaminated water may be to close nearby car parks or not to provide public transport to the bathing beach.

Under the proposed classification scheme, routine monitoring would always require that some type of annual sanitary inspection be performed. The monitoring scheme would be variable, with those beaches classified as 'excellent' or 'very poor' requiring the least amount of monitoring, whereas beaches classified as 'good', 'fair' or 'poor' might require more frequent sampling because their quality would be most likely to change with small changes in faecal influence.

Although many of these management options are practised to some extent, their use with the proposed scheme for beach classification has not been implemented. The value of the suggested scheme for classifying beach waters is that it may lead to a number of activities that will decrease the risk of exposure to faeces-contaminated bathing waters. It will allow individuals to make informed choices about their personal risks. It will encourage local risk management because the system is simple and economically feasible. It will minimise the monitoring effort and thereby minimise costs. It will encourage local decision-making with regard to public health actions. Lastly, it will encourage incremental improvement in local water quality management because the categories and the priorities for improving public health are well defined.

The validity and value of the proposed risk classification scheme should be evaluated through extensive field testing to verify the scientific soundness of the approach.

12.7 IMPLICATIONS FOR INTERNATIONAL GUIDELINES AND NATIONAL REGULATIONS

Management frameworks such as those outlined in this chapter can be generalised and, as such, are amenable to incorporation into international guidelines. It is increasingly being recognised that management through the use of end product standards only, while still important, is limited. Complementing end product standards with measures and indicators of safe process and practice, as outlined here, is a powerful tool in health protection. Many management strategies represent common sense and good housekeeping. They are relatively easy to implement, are cost-effective and can be especially valuable, in national terms, where the cost of water testing is a major impediment to the adoption of local standards.

12.8 REFERENCES

Angulo, F.J., Tippen, S., Sharp, D.J., Payne, B.J., Collier, C., Hill, J.E., Barrett, T.J., Clark, R.M., Gelreich, E.E., Donnell, H.D. and Swerlow, D.L. (1997) A community waterborne outbreak of salmonellosis and the effectiveness of a boil water order. *American Journal of Public Health* **87**(4), 580–584.

Ashendorff, A., Principe, M.A., Seeley, A., LaDuca, J., Beckhardt, L., Faber Jr, W. and Mantus, J. (1997) Watershed protection for New York City's supply. *Journal of the American Water Works Association* **89**, 75–88.

Badenoch, J. (1990) *Cryptosporidium* in water supplies. Report of the Group of Experts. Department of the Environment, Department of Health. HMSO.

Barry, S.J., Atwill, E.R., Tate, K.W., Koopman, T.S., Cullor, J. and Huff, T. (1998) Developing and Implementing a HACCP-Based Programme to Control *Cryptosporidium* and Other Waterborne Pathogens in the Alameda Creek Watershed: Case Study. American Water Works Association Annual Conference, 21–25 June 1998, Dallas, Texas.

Bartram, J. and Rees, G. (2000) *Monitoring Bathing Waters,* E & FN Spon, London.

Bell, G. (1999) Managing food safety: HACCP and risk analysis. *The New Zealand Food Journal* **29**(4), 133–136.

Breittmayer, J.P. and Gauthier, M.J. (1978) Contamination, bacterium d'une zone balneaire liee a sa frequentation. *Water Research* **12**, 193–197. (In French.)

Cheng, W.H.S., Chang, K.C.K. and Hung, R.P.S. (1991) Variations in microbial indicator densities in beach waters and health-related assessment. *Epidemiology and Infection* **106**, 329–344.

Davison A, Davis, S. and Deere, D. (1999) Quality assurance and due diligence for water – Can HACCP deliver?. Paper presented AWWA/WMAA Cleaner Production in the Food and Beverage Industries Conference, Hobart, 1–3 September.

Deere, D.A. and Davison, A.D. (1998) Safe drinking water. Are food guidelines the answer? *Water* November/December: 21–24.

DLWC (1999) The management of *Giardia* and *Cryptosporidium* in town water supplies: Protocols for local government councils. NSW Department of Land and Water Conservation, Australia.

Gray, R. and Morain, M. (2000) HACCP application to Brisbane water. *Water* **27**, January/February, 41–42.

Havelaar, A.H. (1994) Application of HACCP to drinking water supply. *Food Control* **5**, 145–152.

Hellard, E., Sinclair, M., Forbes, A. and Fairley, C. (submitted) A randomized controlled trial investigating the gastrointestinal effects of drinking water. *American Journal of Public Health.*

Kramer, M.H., Herwaldt, B.L., Craun, G.F., Calderon, R.L. and Juranek, D.D. (1996) Waterborne disease – 1993 and 1994. *Journal American Water Works Association* **88**(3), 66–80.

Langley, A. and van Alphen, M. (eds) (1993) *Proc. 2nd National Workshop on the Health Risk Assessment and Management of Contaminated Sites*, South Australian Health Commission, Adelaide.

Leland. D., Mcanulty. J., Keene. W., and Stevens. G. (1993) A cryptosporidiosis outbreak in a filtered-water supply. *Journal American Water Works Association* **85**(6), 34–42.

McCann, B. (1999) UK counts cost of Crypto protection. *Water Quality International* May/June: 4.

Mentzing, L.O. (1981) Waterborne outbreaks of *Campylobacter* enteritis in central Sweden. *Lancet* **ii**, 352–354.

Millson, M., Bokhout, M., Carlson, J., Speilberg, L., Aldis, R., Borczyk, AZ. and Lior, H. (1991) An outbreak of *Campylobacter jejuni* gastroenteritis linked to meltwater contamination of a municipal well. *Canadian Journal of Public Health* **82**, 27–31.

NHMRC/ARMCANZ (1996) Australian Drinking Water Guidelines. National Health and Medical Research Council and Agriculture and Resource Management Council of Australia and New Zealand.

Payment, P. (1997) Epidemiology of endemic gastrointestinal and respiratory diseases – incidence, fraction attributable to tap water and costs to society. *Water Science & Technology* **35**, 7–10.

Pizzi, N., Rexing, D., Visintainer, D., Paris, D. and Pickel, M. (1997) Results and Observations from the Partnership Self Assessment. Proceedings 1997 Water Quality Technology Conference, Denver, Colorado, 9–12 November.

Raman, R. (1990) Risk perceptions and problems in interpreting risk results. Paper presented at the 18th Australasian Chemical Engineering Conference CHEMECA '90. Auckland, New Zealand, August.

Ramsay, C.N. and Marsh, J. (1990) Giardiasis due to deliberate contamination of water supply. *Lancet* **336**, 880–881.

Rose, J.B., Lisle, J.T. and LeChevallier, M. (1997) Waterborne Cryptosporidiosis: Incidence, outbreaks and treatment strategies. In *Cryptosporidium* and Cryptosporidiosis (ed. R. Fayer), CRC Press, Boca Raton, FL.

Shaw, P.K., Brodsky, R.E., Lyman, D.O., Wood, B.T., Hibler, C.P., Healy, G.R., Macleod, K.I., Stahl, W., and Schultz, M.G. (1977) A community-wide outbreak of giardiasis with evidence of transmission by a municipal water supply. *Annals of Internal Medicine* **87**, 426–432.

Smith, B.G. and Dufour, A.P. (1993) *Effects of Human Shedding on the Quality of Recreational Water,* American Society of Microbiology, Atlanta, Georgia.

Smith, D. B., Clark, R. M., Pierce, B. K. and Regli, S. (1995) An empirical model for interpolating C*t* values for chlorine inactivation of *Giardia lamblia*. *Aqua* **44**, 203–211.

Stevens, M., McConnell, S., Nadebaum, P. R., Chapman, M., Ananthakumar, S. and McNeil, J. (1995) Drinking water quality and treatment requirements: A risk-based approach. *Water* **22**, November/December 12–16.

Swerdlow, D.L., Mintz, E.D., Rodriguez, M. *et al.* (1992) Severe life-threatening cholera in Peru: predisposition for persons with blood group O. Abstract 941. Program, 32nd Interscience Antimicrobial Agents Chemotherapy conference, 267.

Teunis, P.F.M., Medema, G.H. and Havelaar, A.H. (1997) Assessment of the risk of infection by *Cryptosporidium* or *Giardia* in drinking water from a surface water source. *Water Research* **31**, 1333–1346.

Vogt, R.L., Sours, H.E., Barrett, T., Feldman, R.A., Dickinson, R.J. and Witherell. L. (1982) *Campylobacter* enteritis associated with contaminated water. *Annals of Internal Medicine* **96**, 292–296.

WHO (1993) Guidelines for drinking-water quality. Volume 1. Recommendations. World Health Organization, Geneva.

WHO (1998) Guidelines for safe recreational-water environments: coastal and freshwaters. Consultation Draft, World Health Organization, Geneva.

WHO (1999) Health-based monitoring of recreational water: the feasibility of a new approach (the 'Annapolis protocol'). Protection of the Human Environment. Water, Sanitation and Health Series, World Health Organization, Geneva.

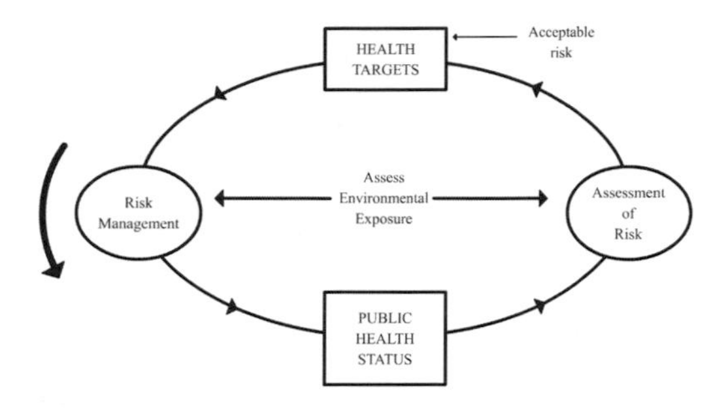

13

Indicators of microbial water quality

Nicholas J. Ashbolt, Willie O.K. Grabow and Mario Snozzi

Current guidelines in the three water-related areas (drinking water, wastewater and recreational water) assess quality, in microbiological terms, by measuring indicator organisms. This chapter looks at the history and examines some of the methods used to assess the microbiological quality of water, highlighting the current limitations and also possible future developments.

13.1 INTRODUCTION

Traditionally, indicator micro-organisms have been used to suggest the presence of pathogens (Berg 1978). Today, however, we understand a myriad of possible reasons for indicator presence and pathogen absence, or vice versa. In short, there is no direct correlation between numbers of any indicator and enteric pathogens (Grabow 1996). To eliminate the ambiguity in the term 'microbial indicator', the following three groups (outlined in Table 13.1) are now recognised:

- General (process) microbial indicators,
- Faecal indicators (such as *E. coli*)
- Index organisms and model organisms.

Table 13.1. Definitions for indicator and index micro-organisms of public health concern (see Box 13.1 for definitions of microbial groups)

Group	Definition
Process indicator	A group of organisms that demonstrates the efficacy of a process, such as total heterotrophic bacteria or total coliforms for chlorine disinfection.
Faecal indicator	A group of organisms that indicates the presence of faecal contamination, such as the bacterial groups thermotolerant coliforms or *E. coli*. Hence, they only **infer** that pathogens may be present.
Index and model organisms	A group/or species indicative of pathogen presence and behaviour respectively, such as *E. coli* as an index for *Salmonella* and F-RNA coliphages as models of human enteric viruses.

A direct epidemiological approach could be used as an alternative or adjunct to the use of index micro-organisms. Yet epidemiologic methods are generally too insensitive, miss the majority of waterborne disease transmissions (Frost *et al.* 1996) and are clearly not preventative. Nonetheless, the ideal is to validate appropriate index organisms by way of epidemiological studies. A good example is the emerging use of an enterococci guideline for recreational water quality (WHO 1998; Chapter 2 of this volume). Often epidemiologic studies fail to show any relationship to microbial indicators, due to poor design (Fleisher 1990, 1991) and/or due to the widely fluctuating ratio of pathogen(s) to faecal indicators and the varying virulence of the pathogens.

The validity of any indicator system is also affected by the relative rates of removal and destruction of the indicator versus the target hazard. So differences due to environmental resistance or even ability to multiply in the environment all influence their usefulness. Hence, viral, bacterial, parasitic protozoan and helminth pathogens are unlikely to all behave in the same way as a single indicator group, and certainly not in all situations. Furthermore, viruses and other pathogens are not part of the normal faecal microbiota, but are only excreted by infected individuals. Therefore, the higher the number of people contributing to sewage or faecal contamination, the more likely the presence of a range of pathogens. The occurrence of specific pathogens varies further according to their seasonal occurrence (Berg and Metcalf 1978).

In summary, there is no universal indicator, but a number, each with certain characteristics. Therefore, this chapter focuses on elucidating the appropriate uses for indicator micro-organisms with a view to their role in the management of waterborne microbial risks. To understand the current use of indicators, however, it is necessary to first understand their historical development.

13.2 DEVELOPMENT OF INDICATORS

13.2.1 The coliforms

The use of bacteria as indicators of the sanitary quality of water probably dates back to 1880 when Von Fritsch described *Klebsiella pneumoniae* and *K. rhinoscleromatis* as micro-organisms characteristically found in human faeces (Geldreich 1978). In 1885, Percy and Grace Frankland started the first routine bacteriological examination of water in London, using Robert Koch's solid gelatine media to count bacteria (Hutchinson and Ridgway 1977). Also in 1885, Escherich described *Bacillus coli* (Escherich 1885) (renamed *Escherichia coli* by Castellani and Chalmers (1919)) from the faeces of breast-fed infants.

In 1891, the Franklands came up with the concept that organisms characteristic of sewage must be identified to provide evidence of potentially dangerous pollution (Hutchinson and Ridgway 1977). By 1893, the 'Wurtz method' of enumerating *B. coli* by direct plating of water samples on litmus lactose agar was being used by sanitary bacteriologists, using the concept of acid from lactose as a diagnostic feature. This was followed by gas production, with the introduction of the Durham tube (Durham 1893). The concept of 'coli-form' bacteria, those bacteria resembling *B. coli*, was in use in Britain in 1901 (Horrocks 1901). The colony count for bacteria in water, however, was not formally introduced until the first Report 71 (HMSO 1934).

Therefore, the sanitary significance of finding various coliforms along with streptococci and *C. perfringens* (see Box 13.1) was recognised by bacteriologists by the start of the twentieth century (Hutchinson and Ridgway 1977). It was not until 1905, however, that MacConkey (1905) described his now famous MacConkey's broth, which was diagnostic for lactose-fermenting bacteria tolerant of bile salts. Nonetheless, *coli-forms* were still considered to be a heterogeneous group of organisms, many of which were not of faecal origin. The origins of the critical observation that *B. coli* was largely faecal in origin while other coliforms were not, could be claimed by Winslow and Walker (1907).

13.2.1.1 Coliform identification schemes

Various classification schemes for coliforms have emerged. The earliest were those of MacConkey (1909) which recognised 128 different coliform types, while Bergey and Deehan (1908) identified 256. By the early 1920s, differentiation of coliforms had come to a series of correlations that suggested indole production, gelatin liquefaction, sucrose fermentation and the Voges

Proskauer reaction were among the more important tests for determining faecal contamination (Hendricks 1978). These developments culminated in the IMViC (Indole, Methyl red, Voges–Proskauer and Citrate) tests for the differentiation of so-called faecal coliforms, soil coliforms and intermediates (Parr 1938); these tests are still in use today.

Water sanitary engineers, however, require simple and rapid methods for the detection of faecal indicator bacteria. Hence, the simpler to identify coliform group, despite being less faecal-specific and broader (for which *Escherichia, Klebsiella, Enterobacter* and *Citrobacter* were considered the most common genera) was targeted. One of the first generally accepted methods for coliforms was called the Multiple-Tube Fermentation Test.

Box 13.1. Definitions of key faecal indicator micro-organisms.

Coliforms: Gram-negative, non spore-forming, oxidase-negative, rod-shaped facultative anaerobic bacteria that ferment lactose (with β-galactosidase) to acid and gas within 24–48h at 36±2°C. **Not** specific indicators of faecal pollution.

Thermotolerant coliforms: Coliforms that produce acid and gas from lactose at 44.5± 0.2°C within 24±2h, also known as faecal coliforms due to their role as faecal indicators.

Escherichia coli (E. coli): Thermophilic coliforms that produce indole from tryptophan, but also defined now as coliforms able to produce β-glucuronidase (although taxonomically up to 10% of environmental *E. coli* may not). Most appropriate group of coliforms to indicate faecal pollution from warm-blooded animals.

Faecal streptococci (FS): Gram-positive, catalase-negative cocci from selective media (e.g. azide dextrose broth or m Enterococcus agar) that grow on bile aesculin agar and at 45°C, belonging to the genera *Enterococcus* and *Streptococcus* possessing the Lancefield group D antigen.

Enterococci: All faecal streptococci that grow at pH 9.6, 10° and 45°C and in 6.5% NaCl. Nearly all are members of the genus *Enterococcus*, and also fulfil the following criteria: resistance to 60°C for 30 min and ability to reduce 0.1% methylene blue. The enterococci are a subset of faecal streptococci that grow under the conditions outlined above. Alternatively, enterococci can be directly identified as micro-organisms capable of aerobic growth at 44±0.5°C and of hydrolysing 4-methlumbelliferyl-β-D-glucoside (MUD, detecting β-glucosidase activity by blue florescence at 366nm), in the presence of thallium acetate, nalidixic acid and 2,3,5-triphenyltetrazolium chloride (TTC, which is reduced to the red formazan) in the specified medium (ISO/FDIS 7899-1 1998).

Sulphite-reducing clostridia (SRC): Gram-positive, spore-forming, non-motile, strictly anaerobic rods that reduce sulphite to H_2S.

Clostridium perfringens: As for SRC, but also ferment lactose, sucrose and inositol with the production of gas, produce a stormy clot fermentation with milk, reduce nitrate, hydrolyse gelatin and produce lecithinase and acid phosphatase. Bonde (1963) suggested that not all SRC in receiving waters are indicators of faecal pollution, hence *C. perfringens* is the appropriate indicator.

Bifidobacteria: Obligately anaerobic, non-acid-fast, non-spore-forming, non-motile, Gram-positive bacilli which are highly pleomorphic and may exhibit branching bulbs (bifids), clubs, coccoid, coryneform, Y and V forms. They are all catalase-negative and ferment lactose (except the three insect species; *B. asteroides, B. indicum* and *B. coryneforme*) and one of the most numerous groups of bacteria in the faeces of warm-blooded animals.

Bacteriophages (phages): These are bacterial viruses and are ubiquitous in the environment. For water quality testing and to model human enteric viruses, most interest in somatic coliphages, male-specific RNA coliphages (F-RNA coliphages) and phages infecting *Bacteroides fragilis*.
Coliphages: Somatic coliphages attack *E. coli* strains via the cell wall and include spherical phages of the family *Microviridae* and various tailed phages in 3 families. The F-RNA coliphages attack *E. coli* strains via the sex pili (F factor) and are single-stranded RNA non-tailed phages in four groups (Table 13.3).
Bacteroides fragilis **bacteriophages**: These infect one of the most abundant bacteria in the gut, belong to the family *Siphoviridae* with flexible tail (dsDNA, long non-contractile tails, capsids up to 60 nm). Phages to the host strain, *B. fragilis* HSP40 are considered to be human-specific, but phages to *B. fragilis* RYC2056 are more numerous and not human-specific (Puig *et al.* 1999).

13.2.1.2 Most probable number method

In 1914, the first US Public Health Service Drinking Water Standard adopted a bacteriological standard that was applicable to any water supply provided by an interstate common carrier (Wolf 1972). It specified that not more than one out of five 10 ml portions of any sample examined should show the presence of the *B. coli* group by the specified Multiple-Tube Fermentation procedure (now referred to as the Most Probable Number or MPN procedure).

Although this test is simple to perform, it is time-consuming, requiring 48 hours for the presumptive results. There are a number of isolation media each with its bias and the bacteria enriched are not a strict taxonomic group. Hence, the total coliforms can best be described as a range of bacteria in the family *Enterobacteriaceae* varying with the changing composition of the media.

Following presumptive isolation of coliforms, further testing is required for confirmation of the coliform type. During the late 1940s there was a divergence between the UK and US approaches to identifying the thermotolerant or so-called ‘faecal’ coliforms. In the UK, Mackenzie *et al.* (1948) had shown that atypical fermentors of lactose at 44°C were indole-negative, whereas *E. coli* was indole-positive. Confirmation of *E. coli* with the indole test was undertaken in the UK, but lactose fermentation at 44°C alone was used in the US (Geldreich 1966). Thus over a period of some 50 years, water bacteriologists developed the concept of *B. coli* (later *E. coli*) as the indicator of faecal pollution, but continued to attach significance to the total lactose fermenters, known variously as ‘coli-aerogenes’ group, *Escherichia-Aerobacter* group, colon group or generally referred to as the ‘total coliforms’ group.

The range of non-faecal bacteria represented in the coliform group and the environmental growth of thermophilic (faecal) coliforms *Klebsiella* spp. and *E. coli* (Ashbolt *et al.* 1997; Camper *et al.* 1991) have concerned bacteriologists and sanitary engineers since the 1930s (Committee on Water Supply 1930). At the other extreme, recent outbreaks of cryptosporidiosis in the absence of coliforms (per 100 ml) are well known (Smith and Rose 1998), and many earlier classic failures of coliforms to identify waterborne pathogens have also been reported.

Despite the obvious failings of the total coliform group to indicate health risk from bacterial pathogens, they provide valuable information on process efficiency which is clearly important in relation to health protection.

13.2.1.3 Membrane filtration method

Until the 1950s practical water bacteriology relied almost exclusively, for indicator purposes, on the enumeration of coliforms and *E. coli* based on the production of gas from lactose in liquid media and estimation of most probable numbers using the statistical approach initially suggested by McCrady (1915). In Russia and Germany, however, workers attempted to culture bacteria on membrane filters, and by 1943 Mueller in Germany was using membrane filters in conjunction with Endo-broth for the analysis of potable waters for coliforms (Waite 1985). By the 1950s membrane filtration was a practical alternative to the MPN approach, although the inability to demonstrate gas production with membranes was considered a major drawback (Waite 1985).

The arbitrary definitions adopted for *E. coli* and the related coliforms were all based upon cultural characteristics, including the ability to produce gas from lactose fermentation (HMSO 1969). Hence, the thermotolerant coliforms include strains of the genera *Klebsiella* and *Escherichia* (Dufour 1977), as well as certain *Enterobacter* and *Citrobacter* strains able to grow under the conditions defined for thermotolerant coliforms (Figureras *et al.* 1994; Gleeson and Gray 1996). This phenotypic approach has also resulted in *E. coli* or a related coliform being ignored simply because they failed to ferment lactose, failed to produce gas from lactose or were indole-negative at 44.5°C. The approach had been repeatedly questioned (Waite 1997), and was only resolved in the UK in the 1990s (HMSO 1994).

It has long been recognised that artificial culture media lead to only a very small fraction (0.01–1%) of the viable bacteria present being detected (Watkins and Xiangrong 1997). Since MacConkey's development of selective media for *E. coli* and coliforms at the beginning of the twentieth century (MacConkey 1908), various workers have shown these selective agents inhibit environmentally or oxidatively stressed coliforms.

13.2.1.4 Defined substrate methods

Media without harsh selective agents but specific enzyme substrates allow significant improvements in recoveries and identification of target bacteria. In the case of coliforms and *E. coli*, such so-called defined substrate methods were introduced by Edberg *et al.* (1988, 1990, 1991). What has evolved into the Colilert® technique has been shown to correlate very well with the traditional membrane filter and MPN methods when used to test both fresh and marine water (Clark *et al.* 1991; Eckner 1998; Fricker *et al.* 1997; Palmer *et al.* 1993). Furthermore, these enzyme-based methods appear to pick up traditionally non-culturable coliforms (George *et al.* 2000).

These developments have led to further changes in definitions of total coliforms and *E. coli*. In the UK, for example, total coliforms are members of genera or species within the family *Enterobacteriaceae*, capable of growth at 37°C, which possess β-galactosidase (HMSO 1989, 1994). In an international calibration of methods, *E. coli* was enzymatically distinguished by the lack of urease and presence of β-glucuronidase (Gauthier *et al.* 1991). Furthermore, the International Standards Organisation has recently published miniaturised MPN-based methods for coliforms/*E. coli* and enterococci based on the defined substrate approach (ISO/FDIS 1998, 1999).

13.2.2 Faecal streptococci and enterococci

Parallel to the work on coliforms, a group of Gram-positive coccoid bacteria known as faecal streptococci (FS) were being investigated as important pollution indicator bacteria (Houston 1900; Winslow and Hunnewell 1902). Problems in differentiating faecal from non-faecal streptococci, however, initially impeded their use (Kenner 1978). Four key points in favour of the faecal streptococci were:

(1) Relatively high numbers in the excreta of humans and other warm-blooded animals.
(2) Presence in wastewaters and known polluted waters.
(3) Absence from pure waters, virgin soils and environments having no contact with human and animal life.
(4) Persistence without multiplication in the environment.

It was not until 1957, however, with the availability of the selective medium of Slanetz and Bartley (1957) that enumeration of FS became popular. Since then, several media have been proposed for FS and/or enterococci to improve on the specificity.

Taxonomically FS are represented by various *Enterococcus* spp. and *Streptococcus bovis* and *S. equinus* (WHO 1997). Of the faecal streptococci, the preferred indicators of faecal pollution are the enterococci. The predominant intestinal enterococci being *E. faecalis*, *E. faecium*, *E. durans* and *E. hirae*. In addition, other *Enterococcus* species and some species of *Streptococcus* (namely *S. bovis* and *S. equinus*) may occasionally be detected. These streptococci however, do not survive for long in water and are probably not enumerated quantitatively. Thus, for water examination purposes enterococci can be regarded as indicators of faecal pollution, although some could occasionally originate from other habitats.

13.2.2.1 Significance of the thermotolerant coliform:faecal streptococci ratio

Geldreich and Kenner (1969) proposed that a faecal coliform:faecal streptococci ratio of four or greater may indicate human pollution, whereas ratios of two or less may indicate animal pollution. There are many factors, however, that can jeopardise the usefulness of this ratio. Foremost are the quicker die-off of coliforms in the environment and different counts from various media used for bacterial isolation (Geldreich 1976). Hence, the use of this ratio is no longer recommended unless very recent faecal pollution is being monitored (Howell *et al.* 1995).

13.2.3 Sulphite-reducing clostridia and other anaerobes

Until bifidobacteria were suggested as faecal indicators (Mossel 1958), *C. perfringens* was the only obligately anaerobic, enteric micro-organism seriously considered as a possible indicator of the sanitary quality of water (Cabelli 1978). Despite the first isolation of bifidobacteria in the late 1800s Tissier 1889) and very high numbers in human faeces (11% of culturable bacteria), their oxygen sensitivity (as with most other strict anaerobes; Loesche 1969) has limited their role as useful faecal indicators in waters (Cabelli 1978; Rhodes and Kator 1999).

The anaerobic sulphite-reducing clostridia (SRC, see Box 13.1) are much less prevalent than bifidobacteria in human faeces, but their spore-forming habit gives them high environmental resistance (Cabelli 1978). *C. perfringens* is the species of clostridia most often associated with the faeces of warm-blooded animals (Rosebury 1962), but is only present in 13–35% of human faeces (Table 13.2).

Although *C. perfringens* has been considered a useful indicator species for over one hundred years (Klein and Houston 1899), its use has been largely limited to Europe, and even then as a secondary indicator mixed in with other SRC (Bonde 1963; HMSO 1969; ISO 1975). The main criticism of the use of *C. perfringens* as a faecal indicator is its long persistence in the environment, which is considered to be significantly longer than enteric pathogens (Cabelli 1978). Bonde (1963) suggested that all SRC in receiving waters are not indicators of faecal pollution, hence *C. perfringens* is the appropriate indicator.

Table 13.2. Microbial indicators (average numbers per gram wet weight) excreted in the faeces of warm-blooded animals (adapted from Geldreich 1978)

Group	Thermotolerant coliforms	Faecal streptococci	*Clostridium perfringens*	F-RNA Coliphages[b]	Excretion (g/day)

Farm animals					
Chicken	1,300,000	3,400,000	250	1867	182 (71.6)[c]
Cow	230,000	1,300,000	200	84	23,600 (83.3)
Duck	33,000,000	54,000,000	–	13.1	336 (61.0)
Horse	12,600	6,300,000	<1	950	20,000
Pig	3,300,000	84,000,000	3980	4136	2700 (66.7)
Sheep	16,000,000	38,000,000	199,000	1.5	1130 (74.4)
Turkey	290,000	2,800,000	–	–	448 (62.0)
Domestic pets					
Cat	7,900,000	27,000,000	25,100,000		
Dog	23,000,000	980,000,000	251,000,000	2.1	413
Human	13,000,000	3,000,000	1580[a]	<1.0–6.25	150 (77.0)
Ratios in raw sewage	50	5	0.3	1	–

[a] Only 13–35% of humans excrete
[b] F-RNA coliphage data from Calci *et al.* (1998). Note low numbers in human faeces, and only excreted by about 26% of humans, about 60% of domestic animals (including cattle, sheep, horses, pigs, dogs and cats), and 36% of birds (geese and seabirds) (Grabow *et al.* 1995).
[c] Moisture content

13.2.4 Bacteriophages

Viruses which infect bacteria, known as bacteriophages or simply as phages, were first described from the intestinal tract of man in the early 1900s (D'Herelle 1926; Pelczar *et al.* 1988). The use of phages as models for indicating the likely presence of pathogenic enteric bacteria first appeared in the 1930s, and direct correlations between the presence of certain bacteriophages and the intensity of faecal contamination were reported (several references cited by Scarpino 1978).

The evolving role for phages to coliforms, known as coliphages (Box 13.1; Table 13.3) however, has been to model human enteric viruses. The DNA-containing tailed coliphages (T type) and RNA-containing phages that infect via the F-pili (sex factor) (F-RNA coliphages) have been the most used.

Table 13.3. Major groups of indicator coliphages (adapted from Leclerc *et al.* 2000) (ds = double stranded; ss = single stranded)

Gp	Family	Nucleic acid	Tail type	Location of attack	Phage examples	Size (nm)
A	*Myoviridae*	ds DNA	contractile	cell wall	T2, T4, T6 (even numbers)s	95 × 65
B	*Siphoviridae*	ds DNA	Long, non-contractile	cell wall	λ, T5	54
C	*Podoviridae*	ds DNA	Short, non-contractile	cell wall	T7, T3	47

D	*Microviridae*	ss DNA	None, large capsomeres	cell wall	ϕX174, S13	30
E	*Leviviridae*	ss RNA	None, small capsomeres	F+ pili	**Group 1**: MS-2, f2, R-17, JP501 **Group 2**: GA, DS, TH1, BZ13 **Group 3**:Qβ, VK, ST, TW18 **Group 4**: SP, F1, TW19, TW28	20–30
F	*Inoviridae*	ss DNA	No head, flexible filament	F+ pili	SJ2, fd, AF-2,M13	810 × 6

13.2.4.1 Phages in water environments

Studies on the incidence of phages in water environments have been reported from most parts of the world for some decades now. Unfortunately the data are not particularly consistent and comparisons are generally not meaningful. One reason for this is that there are many variables that affect the incidence, survival and behaviour of phages in different water environments, including the densities of both host bacteria and phages, temperature, pH and so on.

Another important reason is the inconsistency in techniques used for the recovery of phages from water environments, and eventual detection and enumeration of the phages. This is not altogether surprising because virology, including phages, is a young and rapidly developing science. Phages can be recovered and detected by many techniques and approaches, and much of this work is still in a research or developmental stage. A major reason for discrepancies in results is the host bacteria used for the detection of various groups of phages. Nonetheless, international collaboration is now leading to meaningful, universally accepted guidelines for the recovery and detection of phages in water environments (such as those produced by the International Organisation for Standardisation).

13.2.5 Faecal sterol biomarkers

The presence of faecal indicator bacteria gives no indication of the source, yet it is widely accepted that human faecal matter is more likely to contain human pathogens than animal faeces. The detection of human enteric viruses is specific, however; the methods are difficult and expensive, and not readily quantifiable. Vivian (1986), in his review of sewage tracers, suggested that using more than one method of determining the degree of sewage pollution would be prudent and advantageous. The use of alternative indicators, in this case faecal sterols as biomarkers, in conjunction with existing microbiological indicators, offers a new way to distinguish sources of faecal contamination and monitor river 'health' (Leeming *et al.* 1998).

Coprostanol has been proposed as a measure of human faecal pollution by a large number of researchers since the late 1960s, however, coprostanol has never really been embraced as a sanitary indicator for sewage pollution because its presence was not considered as indicative of a health risk. However, Leeming *et al.* (1996) showed that herbivores have a different dominant form (24-ethyl coprostanol) and it was later shown that these differences could be exploited to determine the contribution of faecal matter from herbivore and human sources relative to each other (Leeming *et al.* 1998).

13.3 PATHOGEN MODELS AND INDEX MICRO-ORGANISMS

The similar morphology, structure and behaviour of F-RNA coliphages, as well as other phages to that of human enteric viruses, suggests that they should be better models for faecal pollution than the faecal indicator bacteria when human viruses are the likely aetiological agents. The same applies to properties such as removal by water treatment processes and resistance to disinfection processes.

While one would expect a poor correlation of phage numbers to the level of human enteric virus titre (phages are always in sewage, but pathogen numbers vary widely based on human infection), what is important for a model organism is that many phages are as resistant as (human) enteric viruses. Laboratory experiments with individual coliphages confirmed that their survival in natural water environments resembles that of enteric viruses and that some phages are at least as resistant as certain enteric viruses to water environments and to commonly used disinfectant such as chlorine (Grabow 1986; Kott *et al*, 1974; Simkova and Cervenka 1981; Stetler 1984; Yates *et al.* 1985).

The value of phages as models/surrogates for viruses has been applied in the routine monitoring of raw and treated drinking water supplies (Grabow *et al.* 2000), and in the assessment of the efficiency of domestic point-of-use water treatment units (Grabow *et al.* 1999b). While they are useful and meet many of the basic requirements as surrogates for enteric viruses, a number of deficiencies are noted in Box 13.2.

As a result of the deficiencies outlined in Box 13.2 phages cannot be regarded as absolute indicators, models or surrogates for enteric viruses in water environments. This is underlined by the detection of enteric viruses in treated drinking water supplies which yielded negative results in tests for phages, even in presence-absence tests on 500 ml samples of water (Grabow *et al.* 2000). Phages are probably best applied as models/surrogates in laboratory experiments where the survival or behaviour of selected phages and viruses are directly compared under controlled conditions (EPA 1986; Grabow *et al.* 1983, 1999b; Naranjo *et al.* 1997).

In addition to enteric viruses, parasitic protozoa are important disinfection-resistant pathogens. When sewage is the source of these pathogens, the anaerobic spore-forming bacterium *Clostridium perfringens* appears to be a suitable index for enteric viruses and parasitic protozoa (Payment and Franco 1993). Spores of *C. perfringens* are largely of faecal origin (Sorensen *et al.* 1989), and are always present in sewage (about 10^4–10^5 cfu $100ml^{-1}$). Their spores are highly resistant in the environment, and vegetative cells appear not to reproduce in aquatic sediments, which can be a problem with traditional indicator bacteria (Davies *et al.* 1995).

Like spores to *C. perfringens*, *Bacillus* spp. spores can also be used as models for parasitic protozoan cysts or oocysts removal by water treatment (Rice *et al.* 1996). Furthermore, since vegetative bacterial cells are inadequate models for disinfection, phages or clostridial spores may provide useful models (Tyrrell *et al.* 1995; Venczel *et al.* 1997).

Box 13.2. Limitations of phages.

Phages are excreted by a certain percentage of humans and animals all the time while viruses are excreted only by infected individuals for a short period of time. The excretion of viruses heavily depends on variables such as the epidemiology of various viruses, outbreaks of viral infections and vaccination against viral infections (Grabow *et al.* 1999a). Consequently there is no direct correlation between numbers of phages and viruses excreted by humans (Borrego *et al.* 1990; Grabow *et al.* 1993; Vaughn and Metcalf 1975).

Methods for somatic coliphages detect a wide range of phages with different properties (Gerba 1987; Yates *et al.* 1985).

At least some somatic coliphages may replicate in water environments (Borrego *et al.* 1990; Grabow *et al.* 1984; Seeley and Primrose 1982).

Enteric viruses have been detected in water environments in the absence of coliphages (Deetz *et al.* 1984; Montgomery 1982; Morinigo *et al.* 1992).

Human enteric viruses associated with waterborne diseases are excreted almost exclusively by humans (Grabow 1996). Phages used as models/surrogates in water quality assessment are excreted by humans and animals. In fact, the faeces of animals such as cows and pigs generally contains higher densities of coliphages than that of humans (Furuse *et al.* 1983; Osawa *et al.* 1981), and the percentage of many animals that excrete phages tends to be higher than for humans (Grabow *et al.* 1993, 1995).

The microbiota of the gut, diet, and physiological state of animals seems to affect the numbers of coliphages in faeces. Osawa *et al.* (1981) reported that stools from zoo animals contained a higher percentage of positive tests for phages than those from domestic farm animals.

The composition and numbers of phages excreted by humans is variable. Patients under antibiotic treatment were found to excrete lower numbers of phages than comparable patients and healthy individuals not exposed to antibiotics (Furuse *et al.* 1983).

Differences between phages and enteric viruses are also reflected by differences in the efficiency of adsorption-elution techniques for their recovery. These differences are due to differences in adsorption properties, which have major implications for behaviour in water environments and at least some treatment processes.

13.4 EMERGING MICROBIOLOGICAL METHODS

13.4.1 Fast detections using chromogenic substances

The time required to perform tests for indicator organisms has stimulated research into more reliable and faster methods. One result is the use of chromogenic compounds, which may be added to the conventional or newly devised media used for the isolation of the indicator bacteria. These chromogenic substances are modified either by enzymes (which are typical for the respective bacteria) or by specific bacterial metabolites. After modification the chromogenic substance changes its colour or its fluorescence, thus enabling easy detection of those colonies displaying the metabolic capacity. In this way these substances can be used to avoid the need for isolation of pure cultures and confirmatory tests. The time required for the determination of different indicator bacteria can be cut down to between 14 to 18 hours.

13.4.2 Application of monoclonal and polyclonal antibodies

Antibodies (glycoproteins produced by mammals as part of their defence system against foreign matter) possess highly specific binding and recognition domains that can be targeted to specific surface structures of a pathogen (antigen). Immunological methods using antibodies are widely used to detect pathogens in clinical, agricultural and environmental samples. Antisera or polyclonal antibodies, the original source of immune reagents, are obtained from the serum of immunised animals (typically rabbits or sheep). Monoclonal antibodies which are produced *in vitro* by fusing plasma cells of an immunised animal (usually a mouse or rat) with a cell line that grows continuously in culture (so that the fused cells will grow continuously and secrete only one kind of antibody molecule (Goding 1986)), can be much better standardised (Torrance 1999).

Such monoclonal antibodies have been successfully used for the detection of indicator bacteria in water samples (Hübner *et al.* 1992; Obst *et al.* 1994). In these studies the water sample was subjected to a precultivation in a selective medium. In this way the complication of detecting dead cells was avoided. Another option for the detection of 'viable' indicators is the combination of immunofluorescence with a respiratory activity compound. This approach has been described for the detection of *E. coli* O157:H7, *S. typhimurium* and *K. pneumoniae* in water (Pyle *et al.* 1995). Detection of *Legionella* from water samples has also been achieved with antibodies (Obst *et al.* 1994; Steinmetz *et al.* 1992). In general, immunological methods can easily be automated in order to handle high sample numbers.

Antibody technology is often used in medicine with enzyme amplification (ELISA – enzyme linked immunosorbent assay), to allow the development of an antigen signal readable by the naked eye. Such an approach is under development for the rapid identification of coliform microcolonies (Sartory and Watkins 1999). As always with immunological techniques, the specificity of the reagents and optimisation of their use is paramount. Although total coliforms are a broad group and likely to be unsuitable immunological targets in environmental waters, *E. coli* could be identified from other coliforms.

Until reliable index organisms are identified for the parasitic protozoa *Cryptosporidium* and *Giardia*, their detection is also relevant when describing methods for important faecal organisms. Current methods for their detection rely on antibodies to assist in the microscopic identification amongst other environmental particles (Graczyk *et al.* 1996). In addition, magnetic beads coated with antibodies are used for concentration and separation of the oocysts and cysts (Rochelle *et al.* 1999) as described below for immunomagnetic separation (IMS) methods.

13.4.3 IMS/culture and other rapid culture-based methods

Immunomagnetic separation offers an alternative approach to rapid identification of culturable and non-culturable micro-organisms (Safarik *et al.* 1995). The principles and application of the method are simple, but rely on suitable antibody specificity under the conditions of use. Purified antigens are typically biotinylated and bound to streptoavidin-coated paramagnetic particles (e.g. Dynal™ beads). The raw sample is gently mixed with the immunomagnetic beads, then a specific magnet is used to hold the target organisms against the wall of the recovery vial, and non-bound material is poured off. If required, the process can be repeated, and the beads may be removed by simple vortexing. Target organisms can then be cultured or identified by direct means.

The IMS approach may be applied to recovery of indicator bacteria from water, but is possibly more suited to replace labour-intensive methods for specific pathogens. An example is the recovery of *E. coli* O157 from water (Anon 1996a). Furthermore, *E. coli* O157 detection following IMS can be improved by electrochemiluminescence detection (Yu and Bruno 1996).

13.4.4 Gene sequence-based methods

Advances in molecular biology in the past 20 years have resulted in a number of new detection methods, which depend on the recognition of specific gene sequences. Such methods are usually rapid and can be tailored to detect specific strains of organisms on the one hand or groups of organisms on the other. The methods have a substantial potential for future application in the field of drinking water hygiene (Havelaar 1993). An international expert meeting in Interlaken concluded (OECD 1999) that the application of molecular methods has to be considered in a framework of a quality management for drinking water. The new methods will influence epidemiology and outbreak investigations more than the routine testing of finished drinking water.

13.4.4.1 PCR (polymerase chain reaction)

With the polymerase chain reaction and two suitable primer sequences (fragments of nucleic acid that specifically bind to the target organism) trace amounts of DNA can be selectively multiplied. In principle, a single copy of the respective sequence in the assay can produce over a million-fold identical copies, which then can be detected and further analysed by different methods.

One problem with PCR is that the assay volume is in the order of some micro-litres, whereas the water sample volume is in the range of 100–1000 ml. Bej *et al.* (1991) have published a filtration method to concentrate the sample, but another problem is that natural water samples often contain inhibitory substances (such as humic acids and iron) that concentrate with the nucleic acids. Hence, it is critical to have positive (and negative) controls with each environmental sample PCR to check for inhibition and specificity.

It may also be critical to find out whether the signal obtained from the PCR is due to naked nucleic acids or living or dead micro-organisms (Toze 1999). One solution has been established by using a three-hour pre-incubation period in a selective medium so that only growing organisms are detected (Frahm *et al.* 1998). Other options under development include targeting short-lived nucleic acids such as messenger or ribosomal RNA (Sheridan *et al.* 1998).

A most important advantage of PCR is that the target organism(s) do not need to be culturable. A good example is the specific detection of human *Bacteroides* spp. to differentiate human faecal pollution from that of other animals (Kreader 1995).

13.4.4.2 FISH (fluorescence in situ hybridisation)

This detection method uses gene probes with a fluorescent marker, typically targeting the 16S ribosomal RNA (16S rRNA) (Amann *et al.* 1995). Concentrated and fixed cells are permeabilised and mixed with the probe.

Incubation temperature and addition of chemicals can influence the stringency of the match between the gene probe and the target sequence. Since the signal of a single fluorescent molecule within a cell does not allow detection, target sequences with multiple copies in a cell have to be selected (e.g. there are 10^2–10^4 copies of 16S rRNA in active cells). A number of FISH methods for the detection of coliforms and enterococci have been developed (Fuchs *et al.* 1998; Meier *et al.* 1997; Patel *et al.* 1998).

Although controversial for many pathogens, low-nutrient environments may result in cells entering a non-replicative viable but non-culturable (VBNC) state (Bogosian *et al.* 1998). Such a state may not only give a false sense of security when reliant on culture-based methods, but may also give the organisms additional protection (Caro *et al.* 1999; Lisle *et al.* 1998). An indication of VBNC *Legionella pneumophila* cell formation was given by following decreasing numbers of bacteria monitored by colony-forming units, acridine orange direct count, and hybridisation with 16S rRNA-targeted oligonucleotide probes (Steinert *et al.* 1997). It was concluded that FISH detection-based methods may better report the presence of infective pathogens and viable indicator bacteria.

13.4.5 Future developments

The future holds numerous possibilities for the detection of indicator and pathogen index organisms. On the horizon are methods based on microarrays and biosensors. Biosensors in the medical area have largely been based on antibody technology, with the antigen triggering a transducer or linking to an enzyme amplification system. Biosensors based on gene recognition, however, look very promising in the microarray format for detecting micro-organisms.

Microarrays using DNA/RNA probe-based rRNA targets may be coupled to adjacent detectors (Guschin *et al.* 1997). Eggers *et al.* (1997) have demonstrated the detection of *E. coli* and *Vibrio proteolyticus* using a microarray containing hundreds of probes within a single well ($1 cm^2$) of a conventional microtiter plate (96 well). The complete assay with quantification took less than a minute.

DNA sensing protocols, based on different modes of nucleic acid interaction, possess an enormous potential for environmental monitoring. Carbon strip or paste electrode transducers, supporting the DNA recognition layer, are used with a highly sensitive chronopotentiometric transduction of the DNA analyte recognition event. Pathogens targeted to date include *Mycobacterium tuberculosis, Cryptosporidium parvum* and HIV-1 (Vahey *et al.* 1999; Wang *et al.* 1997a,b).

13.5 THE CURRENT APPLICABILITY OF FAECAL INDICATORS

Many members of the total coliform group and some so-called faecal coliforms (e.g. species of *Klebsiella* and *Enterobacter*) are not specific to faeces, and even *E. coli* has been shown to grow in some natural aquatic environments (Ashbolt *et al.* 1997; Bermudez and Hazen 1988; Hardina and Fujioka 1991; Niemi *et al.* 1997; Solo-Gabriele *et al.* 2000; Zhao *et al.* 1997). Hence, the primary targets representing faecal contamination in temperate waters are now considered to be *E. coli* and enterococci. For tropical waters/soils, where *E. coli* and enterococci may grow, alternative indicators such as *Clostridium perfringens* may be preferable.

Numerous epidemiological studies of waterborne illness in developed countries indicate that the common aetiological agents are more likely to be viruses and parasitic protozoa than bacteria (Levy *et al.* 1998). Given the often lower persistence of vegetative cells of the faecal bacteria compared to the former agents, it is not surprising that poor correlations have been reported between waterborne human viruses or protozoa and thermotolerant coliforms (Kramer *et al.* 1996). Such a situation is critical to understand, as evident from recent drinking water outbreaks where coliform standards were met (Craun *et al.* 1997; Marshall *et al.* 1997). Nonetheless, water regulatory agencies have yet to come to terms with the inherent problems resulting from reliance on faecal indicator bacteria as currently determined.

Fortunately, new index organisms for some pathogens look promising as performance organisms in the HACCP-type management approaches (see Chapter 12). Examples of such index organisms are *C. perfringens* and the phages. *C. perfringens* for parasitic protozoa, but only if derived from human faecal contamination (Ferguson *et al.* 1996; Payment and Franco 1993). Their resistance to disinfectants may also be an advantage for indexing disinfectant-resistant pathogens. In Europe, the European Union (EU) recommends the absence of *C. perfringens* in 100ml as a secondary attribute to drinking waters (EU 1998), while in Hawaii, levels are laid down for marine and fresh waters (Anon 1996b). Also F-RNA coliphages or *Bacteroides fragilis* bacteriophages are preferred to assess the removal or persistence of enteric viruses (Calci *et al.* 1998; Puig *et al.* 1999; Shin and Sobsey 1998; Sinton *et al.* 1999). As these index organisms are relatively untested worldwide, extensive trials are necessary before their general acceptance in microbial risk assessment. It should be noted that useful index organisms in one system are not necessarily of value in a different environment.

A further confusion over the use of indicator organisms arises from the fact that some indicator strains are also pathogens. This is perhaps best illustrated by the toxigenic *E. coli* strains (Ohno *et al.* 1997). *E. coli* O157:H7 has been responsible for illness to recreational swimmers (Ackman *et al.* 1997; Keene *et al.* 1994; Voelker 1996) and several deaths have been documented through food- and waterborne outbreaks (HMSO 1996; Jones and Roworth 1996). Such toxigenic *E. coli* are also problematic to detect, as they may form viable but non-culturable cells in water (Kogure and Ikemoto 1997; Pommepuy *et al.* 1996).

13.6 IMPLICATIONS FOR INTERNATIONAL GUIDELINES AND NATIONAL REGULATIONS

Indicators have traditionally played a very important role in guidelines and national standards. Increasingly, however, they are being seen as an adjunct to management controls, such as sanitary surveys, and there is a move away from a specified indicator level end product. In other words, indicators are being replaced by on-line analyses (say for chlorine residual or particle sizes) at critical control points (Chapter 12).

A single indicator or even a range of indicators is unlikely to be appropriate for every occasion and therefore it is useful to tailor indicator choice to local circumstances when translating international guidelines into national standards. Additionally, with the change in management paradigm, more indicators of process efficiency are required rather than reliance on the 'old-style' faecal indicators.

13.7 REFERENCES

Ackman, D., Marks, S., Mack, P., Caldwell, M., Root, T. and Birkhead, G. (1997) Swimming-associated haemorrhagic colitis due to *Escherichia coli* O157:H7 infection: evidence of prolonged contamination of a freshwater lake. *Epidemiol. Infect.* **119**, 1–8.

Amann, R.I., Ludwig, W. and Schleifer, K.-H. (1995) Phylogenetic identification and *in situ* detection of individual microbial cells without cultivation. *Microbiol. Rev.* **59**, 143–169.

Anon (1996a) Outbreaks of *Escherichia coli* O157:H7 infection and cryptosporidiosis associated with drinking unpasteurized apple cider – Connecticut and New York, October 1996. *Morbid. Mortal. Weekly Rep.* **46**, 4–8.

Anon (1996b) Proposed amendments to the Hawaii Administrative Rules Chapter 11-54-08, Recreational Waters. In *Water Quality Standards,* pp. 54–86, Department of Health, State of Hawaii.

Ashbolt, N.J., Dorsch, M.R., Cox, P.T. and Banens, B. (1997) Blooming *E. coli*, what do they mean? In *Coliforms and E. coli, Problem or Solution?* (eds D. Kay and C. Fricker), pp. 78–85, The Royal Society of Chemistry, Cambridge.

Bej, A.K. and McCarthy, S. (1991) Detection of coliform bacteria and *Escherichia coli* by multiplex PCR: comparison with defined substrate and plating methods for monitoring water quality. *Appl. Environ. Microbiol.* **57**, 2429–2432.

Berg, G. (1978) The indicator system. In *Indictors of Viruses in Water and Food* (ed. G. Berg), pp. 1–13, Ann Arbor Science Publishers, Ann Arbor, MI.

Berg, G. and Metcalf, T.G. (1978) Indicators of viruses in waters. In *Indicators of Viruses in Water and Food* (ed. G. Berg), pp. 267–296, Ann Arbor Science Publishers, Ann Arbor, MI.

Bergey, D.H. and Deehan, S.J. (1908) The colon-aerogenes group of bacteria. *J. Med. Res.* **19**, 175.

Bermudez, M. and Hazen, T.C. (1988) Phenotypic and genotypic comparison of *Escherichia coli* from pristine tropical waters. *Appl. Environ. Microbiol.* **54**, 979–983.

Bogosian, G., Morris, P.J.L. and O'Neil, J.P. (1998) A mixed culture recovery method indicates that enteric bacteria do not enter the viable but nonculturable state. *Appl. Environ. Microbiol.* **64**, 1736–1742.

Bonde, G.J. (1963) *Bacterial Indicators of Water Pollution,* Teknisk Forlag, Copenhagen.

Borrego, J.J., Cornax, R., Morinigo, M.A., Martinez-Manzares, E. and Romero, P. (1990) Coliphages as an indicator of faecal pollution in water. Their survival and productive infectivity in natural aquatic environment. *Wat. Res.* **24**, 111 116.

Cabelli, V.J. (1978) Obligate anaerobic bacterial indicators. In *Indicators of Viruses in Water and Food* (ed. G. Berg), pp. 171–200, Ann Arbor Science, Ann Arbor, MI.

Calci, K.R., Burkhardt, W., III, Watkins, W.D. and Rippey, S.R. (1998) Occurrence of male-specific bacteriophage in feral and domestic animal wastes, human feces, and human-associated wastewaters. *Appl. Environ. Microbiol.* **64**, 5027–5029.

Camper, A.K., McFeters, G.A., Characklis, W.G. and Jones, W.L. (1991) Growth kinetics of coliform bacteria under conditions relevant to drinking water distribution systems. *Appl. Environ. Microbiol.* **57**, 2233–2239.

Caro, A., Got, P., Lesne, J., Binard, S. and Baleux, B. (1999) Viability and virulence of experimentally stressed nonculturable *Salmonella typhimurium. Appl. Environ. Microbiol.* **65**, 3229–3232.

Castellani, A. and Chalmers, A.J. (1919) *Manual Tropical Medicine*, 3rd edn, Bailliere, Tyndall and Cox, London.

Clark, D.L., Milner, B.B., Stewart, M.H., Wolfe, R.L. and Olson, B.H. (1991) Comparative study of commercial 4-methylumbelliferyl-,-D-glucuronide preparations with the *standard methods* membrane filtration fecal coliform test for the detection of *Escherichia coli* in water samples. *Appl. Environ. Microbiol.* **57**, 1528–1534.

Committee on Water Supply (1930) Bacterial aftergrowths in distribution systems. *Am. J. Public Health* **20**, 485–491.

Craun, G.F., Berger, P.S. and Calderon, R.L. (1997) Coliform bacteria and waterborne disease outbreaks. *J. AWWA* **89**, 96–104.

Davies, C.M., Long, J.A., Donald, M. and Ashbolt, N.J. (1995) Survival of fecal microorganisms in aquatic sediments of Sydney, Australia. *Appl. Environ. Microbiol.* **61**, 1888–1896.

Deetz, T.R., Smith, E.R., Goyal, S.M., Gerba, C.P., Vallet, J.V., Tsai, H.L., Dupont, H.L. and Keswick, B.H. (1984) Occurrence of rota and enteroviruses in drinking and environmental waters in a developing nation. *Wat. Res.* **18**, 572–577.

D'Herelle, F. (1926) *The Bacteriophage and Its Behavior* (English translation by G.H. Smith), Williams and Wilkins, Baltimore, MD.

Dufour, A.P. (1977) *Escherichia coli* : the fecal coliform. In *Bacterial Indicators/health Hazards Associated with Water* (eds A.W. Hoadley and B.J. Dutka), pp. 48–58, American Society for Testing and Materials, PA.

Durham, H.E. (1893) A simple method for demonstrating the production of gas by bacteria. *Brit. Med. J.* **1**, 1387 (cited by Hendricks (1978) p. 100).

Eckner, K.F. (1998) Comparison of membrane filtration and multiple-tube fermentation by the Colilert and Enterolert methods for detection of waterborne coliform bacteria, *Escherichia coli*, and enterococci used in drinking and bathing water quality monitoring in Southern Sweden. *Appl. Environ. Microbiol.* **64**, 3079–3083.

Edberg, S.C., Allen, M.J. and Smith, D.B. (1991) Defined substrate technology method for rapid and specific simultaneous enumeration of total coliforms and *Escherichia coli* from water: collaborative study. J. *Assoc. Off. Analy. Chem.* **74**, 526–529.

Edberg, S.C., Allen, M.J., Smith, D.B. and The National Collaborative Study (1988) National field evaluation of a defined substrate method for the simultaneous enumeration of total coliforms and *Escherichia coli* from drinking water: comparison with the standard multiple tube fermentation method. *Appl. Environ. Microbiol.* **54**, 1003–1008.

Edberg, S.C., Allen, M.J., Smith, D.B. and Kriz, N.J. (1990) Enumeration of total coliforms and *Escherichia coli* from source water by the defined substrate technology. *Appl. Environ. Microbiol.* **56**, 366–369.

Eggers, M.D., Balch, W.J., Mendoza, L.G., Gangadharan, R., Mallik, A.K., McMahon, M.G., Hogan, M.E., Xaio, D., Powdrill, T.R., Iverson, B., Fox, G.E., Willson, R.C., Maillard, K.I., Siefert, J.L. and Singh, N. (1997) Advanced approach to simultaneous monitoring of multiple bacteria in space. Chap. SAE Technical Series 972422. In *27th International Conference on Environmental Systems, Lake Tahoe, Nevada, 14–17 July*, pp. 1–8, The Engineering Society for Advancing Mobility Land Sea Air and Space, SAE International, Warrendale, PA.

EPA (1986) *Report of Task Force on Guide Standard and Protocol for Testing Microbiological Water Purifiers*. United States Environmental Protection Agency, CI, pp. 1–29.

Escherich, T. (1885) Die Darmbakterien des Neugeborenen und Säuglings. *Fortschr. Med.* **3**, 515–522, 547–554. (In German.)

EU (1998) Council Directive 98/83/EC of 3/11/98 on the quality of water intended for human consumption. *Off. J. Eur. Communit.* **L330**, 32–54.

Ferguson, C.M., Coote, B.G., Ashbolt, N.J. and Stevenson, I.M. (1996) Relationships between indicators, pathogens and water quality in an estuarine system. *Wat. Res.* **30**, 2045–2054.

Figureras, M.J., Polo, F., Inza, I. and Guarro, J. (1994) Poor specificity on m-Endo and m-FC culture media for the enumeration of coliform bacteria in sea water. *Lett. Appl. Microbiol.* **19**, 446–450.

Fleisher, J.M. (1990) The effects of measurement error on previously reported mathematical relationships between indicator organism density and swimming-associated illness: a quantitative estimate of the resulting bias. *Int. J. Epidemiol.* **19**, 1100–1106.

Fleisher, J.M. (1991) A re-analysis of data supporting US federal bacteriological water quality criteria governing marine recreational waters. *Res. J. Wat. Pollut. Contr. Fed.* **63**, 259–265.

Frahm, E., Heiber, I., Hoffmann, S., Koob, C., Meier, H., Ludwig, W., Amann, R., Schleifer, K.H. and Obst, U. (1998) Application of 23S rDNA-targeted oligonucleotide probes specific for enterococci to water hygiene control. *System. Appl. Microbiol.* **21**, 16–20.

Fricker, E.J., Illingworth, K.S. and Fricker, C.R. (1997) Use of two formulations of colilert and quantitray (TM) for assessment of the bacteriological quality of water. *Wat. Res.* **31**, 2495–2499.

Frost, F.J., Craun, G.F. and Calderon, R.L. (1996) Waterborne disease surveillance. *J. AWWA* **88,** 66–75.

Fuchs, B.M., Wallner, G., Beisker, W., Schwippl, I., Ludwig, W. and Amann, R. (1998) Flow-cytometric analysis of the *in situ* accessibility of *Escherichia coli* 16S rRNA for fluorescently labelled oligonucleotide probes. *Appl. Environ. Microbiol.* **64**, 4973–4982.

Furuse, K., Osawa, S., Kawashiro, J., Tanaka, R., Ozawa, Z., Sawamura, S., Yanagawa, Y., Nagao, T. and Watanabe, I. (1983) Bacteriophage distribution in human faeces: continuous survey of healthy subjects and patients with internal and leukemic diseases. *Journal of General Virology* **64**, 2039–2043.

Gauthier, M.J., Torregrossa, V.M., Balebona, M.C., Cornax, R. and Borrego, J.J. (1991) An intercalibration study of the use of 4-methylumbelliferyl-,-D-glucuronide for the specific enumeration of *Escherichia coli* in seawater and marine sediments. *System. Appl. Microbiol.* **14**, 183–189.

Geldreich, E.E. (1966) Sanitary Significance of Fecal Coliforms in the Environment. (Water Pollution Control Research Scrics, Publ. WP-20-3.) FWPCA, USDI, Cincinnati, OH.

Geldreich, E.E. (1976) Faecal coliform and faecal streptococcus density relationships in waste discharges and receiving waters. *Critical Rev. Environ. Contr.* **6**, 349–369.

Geldreich, E.E. (1978) Bacterial populations and indicator concepts in feces, sewage, stormwater and solid wastes. In *Indicators of Viruses in Water and Food* (ed. G. Berg), pp. 51–97, Ann Arbor Science, Ann Arbor, MI.

Geldreich, E.E. and Kenner, B.A. (1969) Concepts of faecal streptococci in stream pollution. *J. Wat. Pollut. Contr. Fed.* **41**, R336–R352.

George, I., Petit, M. and Servais, P. (2000) Use of enzymatic methods for rapid enumeration of coliforms in freshwaters. *J. Appl. Microbiol.* **88**, 404–413.

Gerba, C.P. (1987) Phage as indicators of faecal pollution. In *Phage Ecology* (eds S.M. Goyal, C.P. Gerba and G. Bitton), pp. 197–209, Wiley, New York.

Gleeson, C. and Gray, N. (1996) *The Coliform Index and Waterborne Disease*, E & FN Spon, London.

Goding, J.W. (1986). *Monoclonal Antibodies: Principles and Practice*, 2nd edn, Academic Press, London.

Grabow, W.O.K. (1986) Indicator systems for assessment of the virological safety of treated drinking water. *Wat. Sci. Technol.* **18**, 159–165.

Grabow, W.O.K. (1996) Waterborne diseases: Update on water quality assessment and control. *Water SA* **22**, 193–202.

Grabow, W.O.K., Gauss-Mller, V., Prozesky, O.W. and Deinhardt, F. (1983) Inactivation of Hepatitis A virus and indicator organisms in water by free chlorine residuals. *Applied and Environmental Microbiology* **46**, 619–624.

Grabow, W.O.K, Coubrough, P., Nupen, E.M. and Bateman, B.W. (1984) Evaluation of coliphages as indicators of the virological quality of sewage-polluted water. *Water SA* **10**, 7–14.

Grabow W.O.K., Holtzhausen C.S. and de Villiers J.C. (1993) *Research on Bacteriophages as Indicators of Water Quality.* WRC Report No 321/1/93, Water Research Commission, Pretoria.

Grabow, W.O.K., Neubrech, T.E., Holtzhausen, C.S. and Jofre, J. (1995) *Bacteroides fragilis* and *Escherichia coli* bacteriophages: excretion by humans and animals. *Wat. Sci. Technol.* **31**, 223–230.

Grabow, W.O.K., Botma, K.L., de Villiers, J.C., Clay, C.G. and Erasmus, B. (1999a) Assessment of cell culture and polymerase chain reaction procedures for the detection of polioviruses in wastewater. *Bulletin of the World Health Organization* **77**, 973–980.

Grabow, W.O.K., Clay, C.G., Dhaliwal, W., Vrey, M.A. and Müller, E.E. (1999b) Elimination of viruses, phages, bacteria and *Cryptosporidium* by a new generation *Aquaguard* point-of-use water treatment unit. *Zentralblatt für Hygiene und Umweltmedizin* **202**, 399–410.

Grabow, W.O.K., Taylor, M.B., Clay, C.G. and de Villiers, J.C. (2000) Molecular detection of viruses in drinking water: implications for safety and disinfection. Proceedings: Second Conference of the International Life Sciences Institute: *The Safety of Water Disinfection: Balancing Chemical and Microbial Risks.* Radisson Deauville Resort, Miami Beach, FL, 15–17 November.

Graczyk, T.K., M.R. Cranfield and R. Fayer (1996) Evaluation of commercial enzyme immunoassay (EIA) and immunofluorescent antibody (IFA) test kits for detection of *Cryptosporidium* oocysts of species other than *Cryptosporidium parvum. Am. J. Trop. Med. Hyg.* **54**, 274–279.

Guschin, D.Y., Mobarry, B.K., Proudnikov, D., Stahl, D.A., Rittmann, B.E. and Mirzabekov, A.D. (1997) Oligonucleotide microchips as genosensors for determinative and environmental studies in microbiology. *Appl. Environ. Microbiol.* **63**, 2397–2402.

Hardina, C.M. and Fujioka, R.S. (1991) Soil: The environmental source of *Escherichia coli* and enterococci in Hawaii's streams. *Environ. Toxicol. Wat. Qual.* **6**, 185–195.

Havelaar, A.H. (1993) The place of microbiological monitoring in the production of safe drinking water. In *Safety of Water Disinfection: Balancing Chemical and Microbial Risks* (ed. G.F. Craun), ILSI Press, Washington DC.

Hendricks, C.W. (1978) Exceptions to the coliform and the fecal coliform tests. In *Indicators of Viruses in Water and Food* (ed. G. Berg), pp. 99–145, Ann Arbor Science, Ann Arbor, MI.

HMSO (1934) *The Bacteriological Examination of Water Supplies.* Reports on Public Health and Medical Subjects, 1st edn, No. 71, HMSO, London.

HMSO (1969) *The Bacteriological Examination of Water Supplies.* Reports on Public Health and Medical Subjects, 4th edn, No. 71, HMSO, London.

HMSO (1989) Guidance on Safeguarding the Quality of Public Water Supplies. HMSO, London.

HMSO (1994) *The Microbiology of Water 1994: Part 1-Drinking Water. Reports on Public Health and Medical Subjects*, No. 71. Methods for the Examination of Water and Associated Materials. HMSO, London.

HMSO (1996) Method for the isolation and identification of *Escherichia coli* O157:H7 from waters. Methods for the Examination of Water and Associated Materials. HMSO, London.

Horrocks, W.H. (1901) *An Introduction to the Bacteriological Examination of Water*, J and C. Churchill, London.

Houston, A.C. (1900) On the value of examination of water for Streptococci and Staphylococci with a view to detection of its recent contamination with animal organic matter. In *Sup. 29th Ann. Report of the Local Government Board containing the Report of the Medical Officer for 1899–1900*, p. 548, London City Council, London.

Howell, J.M., Coyne, M.S. and Cornelius, P.L. (1995) Faecal bacteria in agricultural waters of the blue grass region of Kentucky. *J. Environ. Qual.* **24**, 411–419.

Hübner, I., Steinmetz, I., Obst, U., Giebel, D. and Bitter-Suermann, D. (1992) Rapid determination of members of the family *Enterobacteriaceae* in drinking water by an immunological assay using a monoclonal antibody against enterobacterial common antigen. *Appl. Environ. Microbiol.* **58**, 3187–3191.

Hutchinson, M. and Ridgway, J.W. (1977) *Microbiological Aspects of Drinking Water Supplies, p. 180,* Academic Press, London.

ISO (1975) Draft Report of SC4/WGS Meeting on Sulfite-Reducing Spore-Forming Anaerobes (Clostridia), 16 Januar.) International Standards Organization, Berlin.

ISO/FDIS (1998) 7899-1. Water Quality – Detection and enumeration of intestinal enterococci in surface and waste water – Part 1. Miniaturised method (Most Probable Number) by inoculation in liquid medium. International Standards Organization, Geneva.

ISO/FDIS (1999) 9308-3. Water Quality – Detection and enumeration of *Escherichia coli* and coliform bacteria in surface and waste water – Part 3. Miniaturised method (Most Probable Number) by inoculation in liquid medium. International Standards Organization, Geneva.

Jones, I.G. and Roworth, M. (1996) An outbreak of *Escherichia coli* O157 and campylobacteriosis associated with contamination of a drinking water supply. *Pub. Health* **110**, 277–282.

Keene, W.E., McAnulty, J.M., Hoesly, F.C., Williams, L.P., Hedber, K., Oxman, G.L., Barrett, T.J., Pfaller, M.A. and Fleming, D.W. (1994) A swimming-associated outbreak of hemorrhagic colitis caused by *Escherichia coli* 0157:H7 and *Shigella sonnei. New Engl. J. Med.* **331**, 579–584.

Kenner, B.A. (1978) Fecal streptococcal indicators. In *Indicators of Viruses in Water and Food* (ed. G. Berg), pp. 147–169, Ann Arbor Science, Ann Arbor, MI.

Klein, E. and Houston, A.C. (1899) Further report on bacteriological evidence of recent and therefore dangerous sewage pollution of elsewise potable waters. In *Supp. 28th Ann. Rept. of the Local Government Board Containing the Report of the Medical Officer for 1898–1899*, London City Council, London.

Kogure, K. and Ikemoto, E. (1997) Wide occurrence of enterohemorragic *Escherichia coli* O157 in natural freshwater environment. *Jap. J. Bacteriol.* **52**, 601–607. (In Japanese.)

Kott, Y., Roze, N., Sperber, S. and Betzer, N. (1974) Bacteriophages as viral pollution indicators. *Wat. Res.* **8**, 165–171.

Kramer, M.H., Herwaldt, B.L., Craun, G.F., Calderon, R.L. and Juranek, D.D. (1996) Surveillance for waterborne-disease outbreaks: United States, 1993–4. *Morbid. Mortal. Weekly Rep.* **45**, 1–33.

Kreader, C.A. (1995) Design and evaluation of *Bacteroides* DNA probes for the specific detection of human fecal pollution. *Appl. Environ. Microbiol.* **61**, 1171–1179.

Leclerc, H., Edberg, S., Pierzo, V. and Delattre, J.M. (2000) Bacteriophages as indicators of enteric viruses and public health risk in groundwaters. *J. Appl. Microbiol.* **88**, 5–21.

Leeming, R., Nichols, P.D. and Ashbolt, N.J. (1998) *Distinguishing Sources of Faecal Pollution in Australian Inland and Coastal Waters using Sterol Biomarkers and Microbial Faecal Indicators*. Research Report No. 204, Water Services Association of Australia, Melbourne.

Leeming, R., Ball, A., Ashbolt, N. and Nichols, P. (1996) Using faecal sterols from humans and animals to distinguish faecal pollution in receiving waters. *Wat. Res.* **30**, 2893–2900.

Levy, D.A., Bens, M.S., Craun, G.F., Calderon, R.L. and Herwaldt, B.L. (1998) Surveillance for waterborne-disease outbreaks: United States, 1995–6. *Morbid. Mortal. Weekly Rep.* **47**, 1–34.

Lisle, J.T., Broadway, S.C., Presscott, A.M., Pyle, B., Fricker, C. and McFeters, G.A. (1998) Effects of starvation on physiological activity and chlorine disinfection resistance in *Escherichia coli* O157:H7. *Appl. Environ. Microbiol.* **64**, 4658–4662.

Loesche, W.J. (1969) Oxygen sensitivity of various anaerobic bacteria. *Appl. Microbiol.* **18**, 723.

MacConkey, A.T. (1905) Lactose-fermenting bacteria in faeces. *J. Hyg.* **5**, 333.

MacConkey, A.T. (1908) Bile salt media and their advantages in some bacteriological examinations. *J. Hyg.* **8**, 322–334.

MacConkey, A.T. (1909) Further observations on the differentiation of lactose-fermenting bacilli with special reference to those of intestinal origin. *J. Hyg.* **9**, 86.

Mackenzie, E.F.W., Windle-Taylor, E. and Gilbert, W.E. (1948) Recent experiences in the rapid identification of *Bacterium coli* type 1. *J. Gen. Microbiol.* **2**, 197–204.

Marshall, M.M., Naumovitz, D., Ortega, Y. and Sterling, C.R. (1997) Waterborne protozoan pathogens. *Clin. Micro. Reviews* **10**(1), 67–85.

McCrady, H.M. (1915) The numerical interpretation of fermentation-tube results. *J. Inf. Diseases* **17**, 183–212.

Meier, H., Koob, C., Ludwig, W., Amann, R., Frahm, E., Hoffmann, S., Obst, U. and Schleifer, K.H. (1997) Detection of enterococci with rRNA targeted DNA probes and their use for hygienic drinking water control. *Wat. Sci. Tech.* **35**(11–12), 437–444.

Montgomery, J.M. (1982) Evaluation of treatment effectiveness for reducing trihalomethanes in drinking water. Final Report, US EPA, EPA-68-01-6292, Cincinnati, OH.

Morinigo, M.A., Wheeler, D., Berry, C., Jones, C., Munoz, M.A., Cornax, R. and Borrego, J.J. (1992) Evaluation of different bacteriophage groups as faecal indicators in contaminated natural waters in Southern England. *Wat. Res.* **26**, 267–271.

Mossel, D.A.A. (1958) The suitability of bifidobacteria as part of a more extended bacterial association, indicating faecal contamination of foods. In *Proc. 7th Internat. Congr. Microbiol.* Abstract of Papers, p. 440.

Naranjo, J.E., Chaidez, C.L., Quinonez, M., Gerba, C.P., Olson, J. and Dekko, J. (1997) Evaluation of a portable water purification system for the removal of enteric pathogens. *Wat. Sci. Technol.* **35**, 55–58.

Niemi, R.M., Niemelä, S.I., Lahti, K. and Niemi, J.S. (1997) Coliforms and *E. coli* in Finnish surface waters. In *Coliforms and E. coli. Problems or Solution?* (eds D. Kay and C. Fricker), pp. 112–119, The Royal Society of Chemistry, Cambridge.

Obst, U., Hübner, I., Steinmetz, I., Bitter-Suermann, D., Frahm, E. and Palmer, C. (1994) Experiences with immunological methods to detect Enterobacteriaceae and *Legionellaceae* in drinking water. AWWA-Proceedings 1993 WQTC, **Part I**, 879–897.

OECD (1999) Molecular Technologies for safe drinking water: The Interlaken workshop, OECD, Paris.

Ohno, A., Marui, A., Castrol, E.S., Reyes, A.A., Elio-Calvo, D., Kasitani, H., Ishii, Y. and Yamaguchi, K. (1997) Enteropathogenic bacteria in the La Paz River of Bolivia. *Amer. J. Trop. Med. Hyg.* **57**, 438–444.

Osawa, S., Furuse, K. and Watanabe, I. (1981) Distribution of ribonucleic acid coliphages in animals. *Appl. Environ. Microbiol.* **41**, 164–168.

Palmer, C.J., Tsai, Y.-L. and Lang, A.L. (1993) Evaluation of Colilert-marine water for detection of total coliforms and *Escherichia coli* in the marine environment. *Appl. Environ. Microbiol.* **59**, 786–790.

Parr, L.W. (1938) The occurrence and succession of coliform organisms in human feces. *Amer. J. Hyg.* **27**, 67.

Patel, R., Piper, K.E., Rouse, M.S., Steckelberg, J.M., Uhl, J.R., Kohner, P., Hopkins, M.K., Cockerill, F.R., III and Kline, B.C. (1998) Determination of 16S rRNA sequences of enterococci and application to species identification of nonmotile *Enterococcus gallinarum* isolates. *J. Clin. Microbiol.* **36**, 3399–3407.

Payment, P. and Franco, E. (1993) *Clostridium perfringens* and somatic coliphages as indicators of the efficiency of drinking water treatment for viruses and protozoan cysts. *Appl. Environ. Microbiol.* **59**, 2418–2424.

Pelczar, Jr M.J, Chan, E.C.S and Krieg, N.R. (1988) In *Microbiology*, 5th edn, p. 416, McGraw-Hill, Singapore.

Pommepuy, M., Butin, M., Derrien, A., Gourmelon, M., Colwell, R.R. and Cormier, M. (1996) Retention of enteropathogenicity by viable but nonculturable *Escherichia coli* exposed to seawater and sunlight. *Appl. Environ. Microbiol.* **62**, 4621–4626.

Puig, A., Queralt, N., Jofre, J. and Araujo, R. (1999) Diversity of *Bacteroides fragilis* strains in their capacity to recover phages from human and animal wastes and from fecally polluted wastewater. *Appl. Environ. Microbiol.* **65**, 1772–1776.

Pyle, B.H., Broadaway, S.C. and McFeters, G.A. (1995) A rapid, direct method for enumerating respiring enterohemorrhagic *Escherichia coli* 0157:H7 in water. *Appl. Environ. Microbiol.* **61**, 2614–2619.

Rhodes, M.W. and Kator, H. (1999) Sorbitol-fermenting bifidobacteria as indicators of diffuse human faecal pollution in estuarine watersheds. *J. Appl. Microbiol.* **87**, 528–535.

Rice, E.W., Fox, K.R., Miltner, R.J., Lytle, D.A. and Johnson, C.H. (1996) Evaluating plant performance with endospores. *J. AWWA* **88**, 122–130.

Rochelle, P.A., De Leon, R., Johnson, A., Stewart, M.H. and Wolfe, R.L. (1999) Evaluation of immunomagnetic separation for recovery of infectious *Cryptosporidium parvum* oocysts from environmental samples. *Appl. Environ. Microbiol.* **65**, 841–845.

Rosebury, T. (1962) *Microorganisms Indigenous to Man*, pp. 87–90 and 332–335, McGraw-Hill, New York.

Safarik, I., Safariková, M. and Forsythe, S.J. (1995) The application of magnetic separations in applied microbiology. *J. Appl. Bacteriol.* **78**, 575–585.

Sartory, D. and Watkins, J. (1999) Conventional culture for water quality assessment: is there a future? *J. Appl. Bact. Symp. Supp.* **85**, 225S–233S.

Scarpino, P.V. (1978) Bacteriophage indicators. In *Indicators of Viruses in Water and Food* (ed. G. Berg), pp. 201–227, Ann Arbor Science Publishers, Ann Arbor, MI.

Seeley, N.D. and Primrose, S.B. (1982) The isolation of bacteriophages from the environment. *Journal of Applied Bacteriology* **53**, 1–17.

Sheridan, G.E.C., Masters, C.I., Shallcross, J.A. and Mackey, B.M. (1998) Detection of mRNA by reverse transcription-PCR as an indicator of viability in *Escherichia coli* cells. *Appl. Environ. Microbiol.* **64**, 1313–1318.

Shin, G.A. and Sobsey, M.D. (1998) Reduction of Norwalk virus, poliovirus 1 and coliphage MS2 by monochloramine disinfection of water. *Wat. Sci. Tech.* **38**(12), 151–154.

Simkova, A. and Cervenka, J. (1981) Coliphages as ecological indicators of enteroviruses in various water systems. *Bulletin of the World Health Organization* **59**, 611–618.

Sinton, L.W., Finlay, R.K. and Lynch, P.A. (1999) Sunlight inactivation of fecal bacteriophages and bacteria in sewage-polluted seawater. *Appl. Environ. Microbiol.* **65**, 3605–3613.

Slanetz, L.W. and Bartley, C.H. (1957) Numbers of enterococci in water, sewage and feces determined by the membrane filter technique with an improved medium. *J. Bacteriol.* **74**, 591–595.

Smith, H.V. and Rose, J.B. (1998) Waterborne cryptosporidiosis – current status [review]. *Parasitology Today* **14**, 14–22.

Solo-Gabriele, H.M., Wolfert, M.A., Desmarais, T.R. and Palmer, C.J. (2000) Sources of *Escherichia coli* in a coastal subtropical environment. *Appl. Environ. Microbiol.* **66**, 230–237.

Sorensen, D.L., Eberl, S.G. and Diksa, R.A. (1989) *Clostridium perfringens* as a point source indicator in non-point-polluted streams. *Wat. Res.* **23**, 191–197.

Steinert, M., Emody, L., Amann, R. and Hacker, J. (1997) Resuscitation of viable but non-culturable *Legionella pneumophila* Philadelphia JR32 by *Acanthamoeba castellanii. Appl. Environ. Microbiol.* **63**, 2047–2053.

Steinmetz, I., Reinheimer, C. and Bitter-Suermann, D. (1992) Rapid identification of Legionellae by a colony blot assay based on a genus-specific monoclonal antibody. *J. Clin. Microbiol.* **30**, 1016–1018.

Stetler, R E. (1984) Coliphages as indicators of enteroviruses. *Applied and Environmental Microbiology* **48**, 668–670.

Tissier, M.H. (1889) La reaction chromophile d'Escherich et le *Bacterium coli. Comptes Rendus de l Academie des Sciences Serie III–Sciences de la Vie-Life Sciences* **51**, 943. (In French.)

Torrance, L. (1999) Immunological Detection and Quantification Methods. In *Proceedings of the OECD Workshop Interlaken.* (see http://www.eawag.ch/publications e/proceedings/oecd/proceedings/Torrance.pdf)

Toze, S (1999) PCR and the detection of microbial pathogens in water and wastewater. *Wat. Res.* **33**, 3545–3556.

Tyrrell, S.A., Rippey, S.R., Watkins, W.D. and Marcotte Chief, A.L. (1995) Inactivation of bacterial and viral indicators in secondary sewage effluents, using chlorine and ozone. *Wat. Res.* **29**, 2483–2490.

Vahey, M., Nau, M.E., Barrick, S., Cooley, J.D., Sawyer, R., Sleeker, A.A., Vickerman, P., Bloor, S., Larder, B., Michael, N.L. and Wegner, S.A. (1999) Performance of the Affymetrix GeneChip HIV PRT 440 platform for antiretroviral drug resistance genotyping of human immunodeficiency virus type 1 clades and viral isolates with length polymorphisms. *J. Clin. Microbiol.* **37**, 2533–2537.

Vaughn, J.M. and Metcalf, T.G. (1975) Coliphages as indicators of enteric viruses in shellfish and shellfish raising estuarine waters. *Wat. Res.* **9**, 613–616.

Venczel, L.V., Arrowood, M., Hurd, M. and Sobsey, M.D. (1997) Inactivation of *Cryptosporidium parvum* oocysts and *Clostridium perfringens* spores by a mixed-oxidant disinfectant and by free chlorine. *Appl. Environ. Microbiol.* **63**, 1598–1601 with erratum **63**(11), 4625.

Vivian, C.M.G. (1986) Tracers of sewage sludge in the marine environment: A review. *The Science of the Total Environment* **53**(1), 5–40.

Voelker, R. (1996) Lake-associated outbreak of *Escherichia coli* O157:H7 in Illinois, 1995. *J. Am. Med. Assoc.* **275**, 1872–1873.

Waite, W.M. (1985) A critical appraisal of the coliform test. *JIWSDI* **39**, 341–357.

Waite, W.M. (1997) Drinking water quality regulation – a European perspective. In *Coliforms and E. coli. Problem or solution?* (eds D. Kay and C. Fricker), pp. 208–217, The Royal Society of Chemistry, Cambridge.

Wang, J., Cai, X.H., Rivas, G., Shiraishi, H. and Dontha, N. (1997a) Nucleic-acid immobilization, recognition and detection at chronopotentiometric DNA chips. *Biosensors Bioelectron.* **12**, 587–599.

Wang, J., Rivas, G., Parrado, C., *et al.* (1997b) Electrochemical biosensor for detecting DNA sequences from the pathogenic protozoan *Cryptosporidium parvum. Talanta* **44**, 2003–2010.

Watkins, J. and Xiangrong, J. (1997) Cultural methods of detection for microorganisms: recent advances and successes. In *The Microbiological Quality of Water* (ed. D.W. Sutcliffe), pp. 19–27, Freshwater Biological Association, Ambleside, UK.

WHO (1997) Guidelines for Drinking-Water Quality Vol **3**: Surveillance and control of community supplies (Second Edition). World Health Organization, Geneva.

WHO (1998) Guidelines for Safe Recreational-Water Environments. Vol. 1. Coastal and Fresh-Waters. (Draft for consultation. WHO/EOS/98.14.) World Health Organization, Geneva.

Winslow, C.E.A. and Hunnewell, M.P. (1902) Streptococci characteristic of sewage and sewage-polluted waters. *Science* **15**, 827.

Winslow, C.E.A. and Walker, L.T. (1907) *Science* **26**, 797.

Wolf, H.W. (1972) The coliform count as a measure of water quality. In *Water Pollution Microbiology* (ed. R. Mitchell), pp. 333–345, Wiley-Interscience, New York.

Yates, M.V., Gerba, C.P. and Kelley, L.M. (1985) Virus persistence in groundwater. *Appl. Environ. Microbiol.* **49**, 778–781.

Yu, H. and Bruno, J.G. (1996) Immunomagnetic-electrochemiluminescent detection of *Escherichia coli* O157 and *Salmonella typhimurium* in foods and environmental water samples. *Appl. Environ. Microbiol.* **62**, 587–592.

Zhao, T., Clavero, M.R.S., Doyle, M.P. and Beuchat, L.R. (1997) Health relevance of the presence of fecal coliforms in iced tea and leaf tea. *J. Food Prot.* **60**, 215–218.

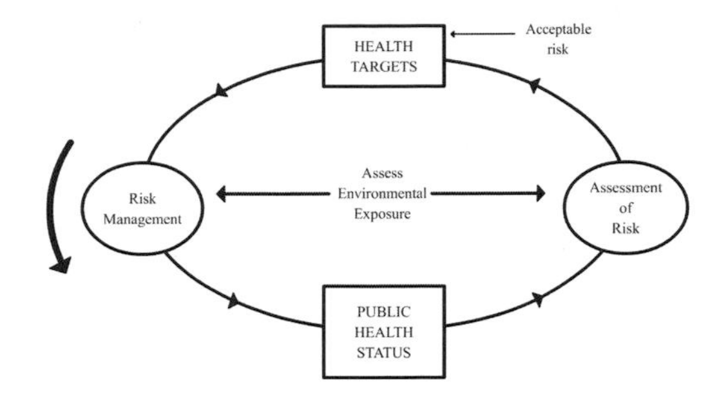

14

Risk communication

Sue Lang, Lorna Fewtrell and Jamie Bartram

There is an increasing number of factors affecting water supplies for which responsible agencies should have a risk communication programme in place. These factors might include chemical as well as microbiological hazards. In addition, there is a growing realisation that for risk communication to be effective it should be a continual and evolving process and not simply a crisis management measure.

This chapter considers some elements of effective risk communication that are applicable to the fields of recreational water and wastewater reuse as well as drinking water (from which most examples are drawn).

14.1 RISK COMMUNICATION

Risk communication is any purposeful exchange of information about risks between interested parties. More specifically in the context of this book, risk communication is the act of conveying or transmitting information between parties about a range of areas including:

 Water Quality: Guidelines, Standards and Health. Edited by Lorna Fewtrell and Jamie Bartram. Published by IWA Publishing, London, UK. ISBN: 1 900222 28 0

- levels of health or environmental risks
- the significance or meaning of health or environmental risks
- decisions, actions or policies aimed at managing or controlling health or environmental risks.

Interested parties include government, agencies, corporations and industry groups, unions, the media, scientists, professional organisations, interested groups, and individual citizens (Covello *et al.* 1991).

All too often it has been the case, with regard to policy making, that there was an emphasis on 'public misperceptions' with a tendency to treat all deviations from expert estimates as products of ignorance or stupidity (Bennett 1999), hardly an ideal basis for meaningful communication! Fortunately this stance is gradually changing, to acknowledge that public reactions to risk often have a rationality of their own, and that 'expert' and 'lay' perspectives should inform each other as part of a two-way process (Bennett 1999).

The necessity of the two-way process has been highlighted by the FAO/WHO:

> Ongoing reciprocal communication among all interested parties is an integral part of the risk management process. Risk communication is more than the dissemination of information, and a major function is the process by which information and opinion essential to effective risk management is incorporated into the decision. (Bennett and Calman 1999)

The days when it was possible to take a 'we know best' approach, simply informing the public that a risk has been identified, telling people not to worry, and stating what was intended to do about it, have in most cases long gone (Coles 1999). The public today no longer automatically acquiesce to authority and now demand a greater role in decision-making (McKechnie and Davies 1999). This, while opening up a route for better decision-making and stakeholder involvement, is no small undertaking and involves some major challenges (McCallum and Anderson 1991), including:

- Provision of information when science is uncertain.
- Explanation of the risk assessment process.
- Incorporation of the differing ways that various groups interpret the science into risk communication strategies.
- Accounting for differing concepts of an 'acceptable' level of risk.
- Provision of information that assists in personal decisions and informs opinions on policy.
- In terms of incident management, maximising appropriate public responses and minimising inappropriate public responses.

It is no accident that risk management, which was traditionally depicted as a linear process, is now generally viewed as a cyclic process with risk communication at its heart (Figure 14.1).

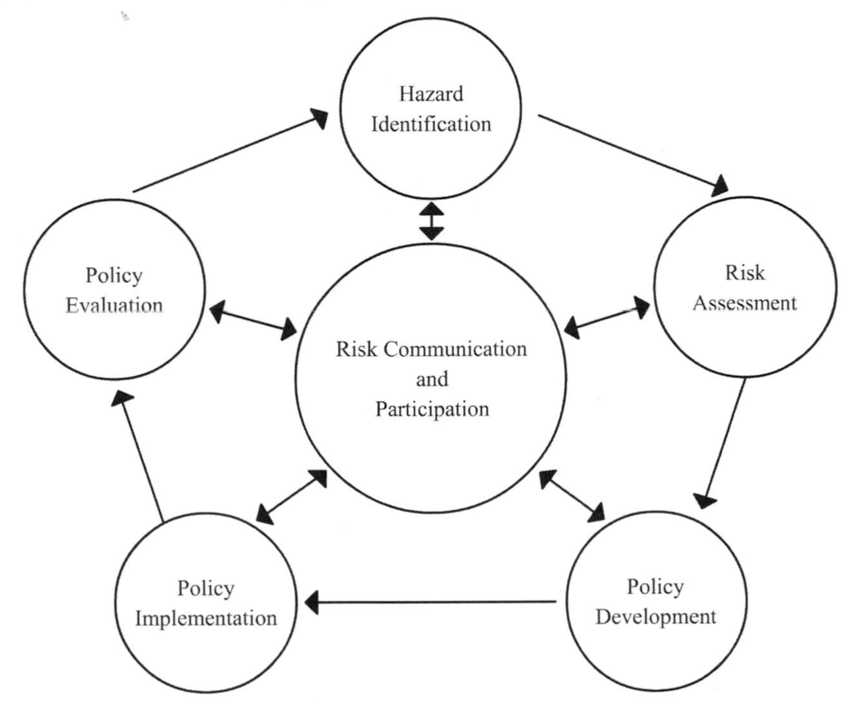

Figure 14.1. The risk management cycle (adapted from Chorus and Bartram 1999).

Responsible water management agencies should adopt a risk management philosophy through which the relevant agency is able to preserve its shareholder value, reputation and credibility, and market share (if appropriate) in the event of a health or environmental risk. An essential component of this philosophy is a risk audit process, which will assist to identify likely issues requiring risk communication strategies, with the central circle of Figure 14.1 being made up of numerous different audiences.

There are a number of functions that a risk communication programme might seek to fulfil (Renn and Levine 1991) including:

- Enlightenment role (aiming to improve risk understanding among target groups).
- Right-to-know (designed to disclose information about hazards to those who may be exposed).
- Attitude modification role (to legitimise risk-related decisions, to improve the acceptance of a specific risk source, or to challenge such decisions and reject specific risk sources).
- Legitimate function (to explain and justify risk management routines with a view to enhancing the trust in the competence and fairness of the management process).
- Risk reduction role (to enhance public protection through information about individual risk reduction measures).
- Behavioural change role (to encourage protective behaviour or supportive actions towards the communicating agency).
- Emergency readiness role (to provide guidelines or behavioural advice for emergency situations).
- Public involvement role (aiming at educating decision-makers about public concerns and perceptions).
- Participation role (to assist in reconciling conflicts about risk-related controversies).

Clearly, given these different possibilities it is important to have a defined objective (what is the aim of the risk communication?) before proceeding. As noted by Corvello (1998), however, the overall goal of risk communication should not be to diffuse public concerns but should be to produce an informed public that is involved, interested, reasonable, thoughtful, solution-orientated and collaborative.

A key consideration of risk communication is that the target will rarely be a single audience, but usually a variety of audiences, and as such messages must be tailored to consider the different audiences that are likely to have different interests, values, levels of intelligence, education and understanding. Audience types might include water consumers (which will encompass the old, young, mothers etc.), water-sports enthusiasts, shareholders, environmental groups, businesses using water, special needs consumers, hospitals and nursing homes, politicians, policy makers and so on.

Risk communication should not be restricted to negative messages and warnings but should include positive 'educational messages'. Whatever the topic, preparation is the key, as illustrated by the following list which attempts to characterise a local community in a developing country before putting out messages about the positive benefits of increased personal hygiene (WHO 1997). Determine:

- Local beliefs and attitudes regarding water, sanitation and health.
- Traditional water use and defecation habits and excreta disposal practices.
- Current levels of knowledge about disease transmission, especially among community leaders and other influential individuals.
- The priority given to improvements in water supply and sanitation in relation to other community needs.
- Existing channels of communication in the community including books, newspapers, and magazines, radio or television, tradition drama, songs and story-telling.
- Members of the community and field workers from other agencies who might be involved in spreading a similar message.

Such preparation will result in a far higher success rate as it will be more likely to engage the target audience in an appropriate and informed manner.

14.2 SITUATION MANAGEMENT

As disclosure and freedom of information laws are more common in many parts of the world, responsible agencies are increasingly focusing on how to communicate risk. At what point is the decision taken to make the public aware that there is an issue?

The responsible agency's risk management philosophy will, in some measure, dictate at what point the issue will be raised. The decision may relate to possibility/potential, combined with non-scientific evidence and field expertise in the absence of scientific evidence. Where the lines are unclear, independent advice may be sought from health departments (Chief Health Officer) or a scientific expert. As risks to health and the environment cannot be eliminated, value judgements are required.

A good risk communication programme will ensure that factual information is provided quickly, through an authoritative, accessible source with a clear, understandable message. Research has shown that organisations with strong relationships with key stakeholders will benefit from those relationships during a crisis. As crises magnify poor or non-existent relationships, investment in pre-crisis communications is a cost-effective strategy to minimise damage to an organisation during a crisis. Marra (1998) notes that six characteristics appear consistently in management and communication literature as a measure of a relationship:

(1) Trust
(2) Understanding
(3) Credibility
(4) Satisfaction
(5) Co-operation
(6) Agreement.

It is therefore important to have a crisis communications plan in place as a part of any organisation's risk communication programme. This allows accurate information to be provided in a timely fashion if an issue arises. A lack of available information leads to conjecture and seeking of information from less credible sources. Misinformation becomes news. Additionally, crisis conditions almost always reduce the likelihood of effective decision making, having effective procedures in place in advance should alleviate this problem (at least to some degree). Pre-planning should also reduce internal co-ordination problems and the possibility of confusing and contradictory messages which, unsurprisingly, can lead to external credibility problems.

Speedy provision of information and explanations that go beyond the basic information in media stories are likely to be viewed as an attempt to be open and address the situation. This is critical if the organisation is to maintain credibility and trust which is paramount in health-related issues. Examples of water-related issues include:

- Outbreak of illness linked to drinking water
- Microbiological contamination of bathing water
- Urban pollution (stormwater, sewage) of beaches
- Vegetables contaminated through irrigation with wastewater.

There are indications in some countries that media coverage of technical issues, including water, has become increasingly negative over the last 20 years, while objective indicators show either an improvement or no decline in quality. This increasing negativity may well be due to the perceived proliferation of health and environmental hazards resulting from new technologies (e.g. genetic modification of food) coupled with a corresponding push by lobby groups to focus on the possible impacts of these technologies. There has also been a feeling, for example within the UK, that the responsible agencies would sooner keep the public in the dark, or are too quick to provide unsupportable reassurances leading to a lack of trust, decreased credibility and an 'expect the worst' public attitude.

As media portrayals can have a significant impact on public attitudes, it makes sense to attempt to include the media as an ally in communication, rather than an audience. This can be done through invitations to the media to assist in conveying warnings and instructions to target audiences, reassuring the public, defusing inaccurate rumours, assisting in the response effort and soliciting assistance from the public as required. However, this may not always be possible, in which case it may be helpful to have an eye on a number of media 'triggers' (Table 14.1).

Table 14.1. Media triggers (adapted from Bennett 1999)

Triggers	
A possible risk to public health is more likely to become a major story if the following are prominent or can readily be made to become so:	
1	Questions of blame
2	Alleged secrets and attempted cover-ups
3	Human interest through identifiable heroes, villains, dupes etc. (as well as victims)
4	Links with existing high-profile issues or personalities
5	Conflict
6	Signal value: the story as a portent of further ills ('*what next*?')
7	Many people exposed to the risk, even if at low levels ('*it could be you*!')
8	Strong visual impact (e.g. pictures of suffering)
9	Links to sex and/or crime

With the possible exception of links to sex, it is not too difficult to imagine water-related scenarios that could hit all of these triggers!

14.2.1 Audience-focused communication

Once it is determined that public communication about a water quality issue is necessary, an audience-centred approach to communicating that risk is vital. According to Maibach and Parrott (1995), an individual's risk experiences and perceptions can affect their risk-related worry and eventual seeking of further information.

Predetermination of specific audiences requiring specially crafted messages can be extremely useful. It is helpful if health messages are designed to respond to the needs and situation of the target audience, rather than those of the responsible agency. It is suggested that a useful approach is to identify likely target audiences (e.g. families with young children, food processing businesses, dialysis patients, hospitals and nursing homes, water-sports enthusiasts), and be familiar with their preferred method of information extraction. The preparation of material in advance to address specific audience needs is of value in terms of being able to provide a rapid response.

It is important to bear in mind, however, that risk communication may work selectively, and often reaches those who are already better informed (Langford *et al.* 1999). This can be illustrated by a survey examining willingness to pay for clean bathing water (Georgiou *et al.* 1998). In this study attitudes regarding clean bathing water were canvassed among locals, day-trippers and holiday-makers at two sites, Lowestoft and Great Yarmouth in eastern England. Lowestoft has a beach that regularly passes the EC bathing water directive, while the beach at Great Yarmouth does not. In Lowestoft, 61% of people canvassed knew of its 'clean beach' status, i.e. people were well informed and many day-trippers had chosen the resort because of its clean beach. By contrast, only 12% of survey respondents at Great Yarmouth were aware that the beach failed to meet EC standards. Interestingly, those who actually bathed at Great Yarmouth had a significantly lower willingness to pay for improved water quality than those who didn't bathe, suggesting that bathers at Great Yarmouth were denying there was a possible health threat and just wanted to get on with their holiday! This study, therefore, also illustrates the problem of 'optimistic bias' or 'unreal optimism' (Weinstein 1980), where people tend to believe they are less at risk from a given hazard relative to an 'average' member of society.

14.2.2 Managing negative feedback and outrage

Risk communication experts in the US and Europe point to a risk comparison approach in determining the risk perception and evaluation. Sandman *et al.* (1993) points out that *outrage* (the relationship between the agency and the neighbourhood) affects the perceived seriousness of the situation by a factor of five relative to the 'actual' seriousness. He concludes that when people are outraged, they tend to think that the hazard is serious. Therefore, it is important to look at the factors which affect risk perception and evaluation and are thus likely to affect public concern (Table 14.2).

Given that a number of factors relating to water quality are likely to fall into the 'increase public concern' category, attempts to be trustworthy and to make the message understandable are likely to be well-received. The key is to control the message, not the messengers. As mentioned earlier, it is useful to provide all possible alternative information sources with the relevant simple facts and analogies. An invitation to include members of the public on advisory/consultative committees is also likely to gain favour and demonstrate openness.

Table 14.2. Risk perception (adapted from Covello 1998)

Factor	Increase public concern	Decrease public concern
Catastrophic potential	Fatalities and injuries grouped in time and space	Fatalities and injuries scattered and random
Controllability (personal)	Uncontrollable	Controllable
Manifestation of effects	Delayed effects	Immediate effects
Effects on children	Children specifically at risk	Children not specifically at risk
Familiarity	Unfamiliar	Familiar
Media attention	Much media attention	Little media attention
Origin	Caused by human actions or failures	Caused by 'Acts of God'
Reversibility	Effects irreversible	Effects reversible
Trust in institutions	Lack of trust in responsible institutions	Trust in responsible institutions
Uncertainty	Risks unknown	Risks known
Understanding	Mechanisms or processes not understood	Mechanisms and processes understood
Voluntariness of exposure	Involuntary	Voluntary

14.2.3 Anticipating concerns

Clearly, as part of the preparation process it is useful to be able to anticipate audience concerns. There are a number of approaches that can be taken to determine likely response:

- Researching the concerns raised in similar situations within your particular country if available (different cultures and societies are likely to have different concerns);
- Market research in the form of focus groups to determine the concerns of specific audience segments; and
- Monitoring throughout an active issue to ensure ongoing needs assessment – are there unanticipated audiences that require information? Are different issues arising within a recognised audience?

14.2.4 The choice of messenger – who people trust

Studies show that people, in general, get more information about risk and hazard from the media than from their own doctors, friends or relatives (Shaw 1994). Various polls taken in the US indicate that the public overwhelmingly relies on the mass media for information from which they will form their attitudes on water supply and health risks (Geldreich 1996). If this is taken to be the case, then the importance of using the media as an ally, rather than an audience, is even more pronounced.

A study conducted in the UK by the Consumers' Association (McKechnie and Davies 1999) surveyed over 2000 adults about whom they considered to be trustworthy sources of impartial advice (Table 14.3).

Table 14.3. Trusted sources of impartial advice in the UK (adapted from McKechnie and Davies 1999)

Source	Most trustworthy (%)	Least trustworthy (%)
Health professionals (e.g. GPs, health visitors)	36	3
Consumer organisations (e.g. National Consumer Council, Consumers' Association)	27	4
Scientists specialising in food safety	20	5
Government departments	5	49
The food industry	5	30

Although the survey had a food bias, the results make interesting reading. These estimates are unlikely to be static and will probably vary according to current news stories and other factors.

14.3 LONG-TERM TRUST

Although this chapter has largely been aimed at risk communication during situation management, many of the messages will be the same whether the risk communication is part of an ongoing process or a crisis situation. However, long-term trust is clearly an area that cannot be put in place in the context of situation management but, nonetheless, is likely to play an important role should a crisis occur. It is wise, therefore, to build a 'reservoir of goodwill' against which to 'borrow' if necessary. Although confidence and trust comprise goodwill and are often used interchangeably, confidence in a source can be distinguished as an enduring experience of trustworthiness over time. Trust can be broken down into perceived competence, objectivity, fairness, consistency and faith. Confidence is based on a good past record of trust-building communication (Kasperson and Stallen 1991).

People are unlikely to change their behaviour or attitudes if they distrust the source of risk information. Lack of credibility is often linked to incompetence, poor performance, incomplete or dishonest information, withholding of information, obscure or hidden decision-making processes, denial of obvious problems and denial of vested interests.

Credibility, however, can be reinforced by good performance, fast responses to public requests for information, consonance with highly esteemed social values, availability for communication with outsiders, unequivocal and highly focused information transfer, flexibility to respond to crisis situations or new public demands, and demonstration of public control over performance and money allocation. Overreacting to public requests for information never hurts.

14.4 COMMUNICATION TECHNIQUES

The amount of effort people use to process a message is important as it can affect what they remember, their attitudes, and their intent to comply with the message. Monahan (1995) concludes that negative messages foster the use of more elaborate, detail-oriented and analytical processing strategies, informing the audience that the current situation is problematic. Positively phrased messages inform the audience that the current situation is non-threatening and that a higher degree of attention is unnecessary. According to Holtgrave *et al.* (1995), arbitrary choices of wording can have a profound impact in terms of the decisions and behaviours they elicit from the audience.

There are seven key aspects to consider when communicating to an audience (Cutlip *et al.* 1985), namely:

(1) Credibility. The audience must have confidence in the agency and high regard for the agency's competence on the subject.
(2) Context. The communications programme must acknowledge the realities of its environment. The context must confirm, not contradict, the message. Effective communications require a supportive social environment, one largely set by the news media – hence the importance of using the media as a communication ally.
(3) Content. The message must have meaning for the audience and compatibility with the audience's value system. It should have relevance to the audience's situation. In general, people select the elements of the information that promise them the most reward. The content determines the audience.
(4) Clarity. Simple terms are most appropriate and it is important to ensure that the message means the same to the audience as it does to the communicating agency. Complex issues should be compressed into themes, analogies or stereotypes that are clear and simple. The further a message has to travel, the simpler it must be.

(5) Continuity and consistency. Communication requires repetition to achieve penetration. Repetition, with variation, contributes to both factual and attitudinal learning. The story should be consistent.
(6) Channels. Established channels that the audience uses and respects should be utilised. Different channels are required to reach different target audiences. People associate specific values with specific channels of communication, and this, too, should be kept in mind.
(7) Capability of audience. The capability of the audience should be addressed. Communications are most effective when they require the least effort on the part of the audience. This involves factors of availability, habits, reading ability and audience knowledge.

14.4.1 Empathy

There is no disadvantage in expressing concern and a willingness to take responsibility to address/rectify the situation. Indeed, it is likely to be a vital prerequisite to effective risk communication, especially if dealing with an outraged audience.

Another factor to consider is that any message that is heavily science-based is likely to be a barrier to public understanding and engagement. This, coupled with delivery of scientific results, which tend to be couched in dry unemotional language, is likely to alienate the audience with scientists coming across as distant and uncaring (Burke 1999).

14.4.2 Uncertainty

This area was raised earlier in the chapter as a major challenge (see Chapter 9), and while it may be difficult to acknowledge uncertainty (and indeed may go against demands for certainty from the public and policy-makers alike) failure to do so is likely to lead to greater problems in the long term (Bennett *et al.* 1999). In many countries, the public has become tired of false reassurances of safety and decisions presented as being conclusive when this is far from the case (McKechnie and Davies 1999). Such proclamations will drain trust as it becomes clear that the situation wasn't as cut and dried as originally presented. A related issue is that of presenting evidence: scientists will reject suggested causal links for which there is no positive evidence; however, the public will require strong proof against a link that looks intuitively plausible (Bennett *et al.* 1999). It has been suggested that '*there is no evidence that X causes a risk of Y*' be abandoned and the following, more constructive, approach be adopted:

(1) Acknowledge the initial plausibility of the link.
(2) Explain what evidence would be expected if such a link existed.
(3) Show that serious, well-conducted investigation has not found such evidence.

As Bennett *et al.* (1999) point out, if (2) or (3) cannot be provided, then 'no evidence' is a dubious reassurance!

14.4.3 Silence

If an organisation fails to communicate a risk issue (i.e. it is silent) the public are quick to judge that the organisation (or representative individuals) either doesn't have the requisite knowledge or information, is guilty and trying to 'cover-up', or is just plain arrogant, or possibly a combination of all three. If there is little information available, it is preferable to indicate what information is known and when further information is expected to be available.

14.5 EVALUATION

In any risk communication approach, especially in terms of crisis management, evaluation is important, both as part of the two-way process and checking assumptions about audiences. O'Donnell *et al.* (2000) recently examined the effectiveness of a 'boil water notice' issued in response to a drinking water pollution incident. The notice was issued to 878 households following possible sewage contamination of drinking water supplies. The notice was brightly coloured, included a telephone helpline number and provided the following simple advice (translated into several languages on the back of the notice):

- Boil water before use.
- Do not drink your tap water without first bringing it to the boil and letting it cool.
- Do not use unboiled water for preparing food, cleaning your teeth, or washing wounds.
- Remember your pets – they should not drink unboiled water either.
- You can still use tap water for washing and bathing without having to boil it.
- You can still use tap water for general household purposes and toilet flushing.

O'Donnell and her colleagues canvassed 350, randomly selected households by postal questionnaire about risk behaviour in light of the notice. Despite timely delivery of the notice, and the general feeling that the notice was easy to understand, 81% of households surveyed engaged in behaviour likely to increase the risk of waterborne infection. Most respondents said that they would appreciate more information about the nature of the incident and a description of possible health effects. More day-by-day information on the state of repairs and likelihood of the notice being lifted was also considered desirable.

14.6 RISK COMMUNICATION AND GUIDELINES

Risk communication plays an important role in the guidelines approach. WHO's water-related normative work attempts to provide a scientific basis to support individual countries in developing national (or potentially local or regional) risk management strategies – including the development of standards. The emphasis on providing a common worldwide scientific underpinning requires that the guidelines are orientated specifically towards health hazards and that aspects likely to vary widely between countries and regions are generally unsuitable for direct inclusion. For this reason the outputs are referred to as guidelines rather than standards to reflect the fact that they are intended to be adapted by countries to reflect their social/cultural, economic and environmental circumstances. The Guidelines for Drinking-water Quality (WHO 1993), for example, specifically advocate that a risk-benefit approach be adopted in developing overall strategy.

Figure 14.1 illustrates that risk communication is a circular process requiring two-way communication at all stages. As such the 'scientific' and 'rational' elements (which are typically the domain of environmental health administrations) cannot be isolated from other elements. WHO guidelines, therefore, typically recognise that factors such as societal values vary widely between cultures and therefore specific approaches and indeed standards themselves may vary between countries and cultures. This was one of the reasons behind the change from the earlier WHO 'International Standards for Drinking-water Quality' to 'Guidelines for Drinking-water Quality'.

The guidelines, however, are not limited to simple descriptions of what is safe in terms of the composition of water suitable for different purposes. Some (such as the Guidelines for Safe Use of Wastewater and Excreta in Agriculture and Aquaculture) place considerable emphasis on good practice, i.e. practices that would tend to prevent exposures that would be hazardous to human health. Most, either implicitly or explicitly, recognise the importance of individual behaviours in risk avoidance and, therefore, the need for an educated public provided with timely and appropriate information to enable them to interpret and act upon information available to them (from whatever source).

The area of risk communication is developing rapidly and, at present, there are great disparities across countries and regions in policy and practice. At the country level developments are likely to be influenced by parallel developments in the field of human rights and in relational to international trade. In the former, slow steps have been made towards the recognition of water and sanitation as 'human needs' and they are implicit as 'human rights' in a number of legal instruments. In the latter the involvement of international companies in service provision may lead to increasing pressure towards internal standardisation.

A risk communication strategy is very important in the process of adapting international guidelines to national policy. Regulators tend to be defensive and, thus, tend to exclude the public. This is the opposite of what is required and tends to be counterproductive. Engaging in risk communication creates an aware and informed public who should be allowed to have the right sort of input to the regulatory process.

14.7 REFERENCES

Bennett, P. (1999) Understanding responses to risk: some basic findings. In *Risk Communication and Public Health* (eds P. Bennett and K. Calman), pp. 3–19, Oxford University Press, Oxford.

Bennett, P., Coles, D. and McDonald, A. (1999) Risk communication as a decision process. In *Risk Communication and Public Health* (eds P. Bennett and K. Calman), pp. 207–221, Oxford University Press, Oxford.

Burke, D. (1999) The recent excitement over genetically modified foods. In *Risk Communication and Public Health* (eds P. Bennett and K. Calman), pp. 140–151, Oxford University Press, Oxford.

Chorus, I. and Bartram, J. (eds) (1999) *Toxic Cyanobacteria in Water. A Guide to Their Public Health Consequences, Monitoring and Management*, E & FN Spon, London.

Coles, D. (1999) The identification and management of risk: opening up the process. In *Risk Communication and Public Health*, (eds P. Bennett and K. Calman), pp. 195–204, Oxford University Press, Oxford.

Corvello, V.T. (1991) Risk comparison and risk communication: issues and problems in comparing health and environmental risk. In *Communicating Risks to the Public* (eds R.E. Kasperson and P.J.M. Stallen), pp. 79–118, Kluwer, Dordrecht.

Corvello, V.T. (1998) Risk communication. In *Handbook of Environmental Risk Assessment and Management* (ed. P. Callow), pp. 520–541, Blackwell Science, Oxford.

Cutlip, S.M., Center, A.H. and Broom, G.M. (1985) *Effective Public Relations*, Prentice-Hall, Englewood Cliffs, New Jersey.

Bennett, P. and Calman, K. (1999) Pulling the threads together. In *Risk Communication and Public Health* (eds P. Bennett and K. Calman), pp 205–206, Oxford University Press, Oxford.

Geldreich, E. (1996) *Microbial Quality of Water Supply in Distribution Systems*, CRC Press, Boca Raton, FL.

Georgiou, S., Langford, I.H., Bateman, I.J. and Turner, R.K. (1988) Determinants of willingness to pay for reductions in environmental health risks: a case study of bathing water quality. *Environment and Planning* **A**30, 577–594.

Holtgrave, Tinsley and Kay (1995) Encouraging risk reduction: A decision-making approach to message design. In *Designing Health Messages: Approaches from Communication, Theory and Public Health Practice* (eds E. Maibach and R.L. Parrott), pp. 24–40, Sage Publications, Thousand Oaks, CA.

Kasperson, R.E. and Stallen, P.J.M (1991) Chapter in *Communicating Risks to the Public*, (eds R.E. Kasperson and P.J.M. Stallen), Kluwer, Dordrecht.

Langford, I.H., Marris, C. and O'Riordan, T. (1999) Public reactions to risk: social structures, images of science, and the role of trust. In *Risk Communication and Public Health* (eds P. Bennett and K. Calman), pp. 33–50, Oxford University Press, Oxford.

Maibach, E. and Parrott, R.L. (1995) *Designing Health Messages: Approaches from Communication, Theory and Public Health Practice*, Sage Publications, Thousand Oaks, CA.

Marra, F.J. (1998) The importance of communication in excellent crisis management. *Australian Journal of Emergency Management* **13**(3), 7.

McCallum, D.B. and Anderson, L. (1991) Communicating about pesticides in the water. In *Communicating Risks to the Public* (eds R.E. Kasperson and P.J.M. Stallen), pp. 237–285, Kluwer, Dordrecht.

McKechnie, S. and Davies, S. (1999) Consumers and risk. In *Risk Communication and Public Health* (eds P. Bennett and K. Calman), pp. 170–182, Oxford University Press, Oxford.

Monahan, J. (1995) Thinking positively: Using positive affect when designing health messages. In *Designing Health Messages: Approaches from Communication, Theory and Public Health Practice* (eds E. Maibach and R.L. Parrott), pp. 81–98, Sage Publications, Thousand Oaks, CA.

O'Donnell, M., Platt, C. and Aston, R. (2000) Effect of a boil water notice on behaviour in the management of a water contamination incident. *Communicable Disease and Public Health* **3**(1), 56–59.

Renn, O. and Levine, D. (1991) Credibility and trust in risk communication. In *Communicating Risks to the Public* (eds R.E. Kasperson and P.J.M. Stallen) pp. 175–214, Kluwer, Dordrecht.

Sandman, P.M., Miller, P.M., Johnson, B.B. and Weinstein, N.D. (1993) Agency communication, community outrage and perception of risk: three simulation experiments. *Risk Analysis* **13**(6), 585–598.

Shaw, D. (1994) Cry Wolf Stories Permeate Coverage of Health Stories. *Los Angeles Times*, 12 September.

Weinstein, N.D. (1980) Unrealistic optimism about future life events. *Journal of Personality and Social Psychology* **39**, 806–820.

WHO (1993) Guidelines for drinking water quality. Volume 1. Recommendations. World Health Organization, Geneva.

WHO (1997) Guidelines for drinking-water quality. Volume 3. Surveillance and control of community supplies. World Health Organization, Geneva.

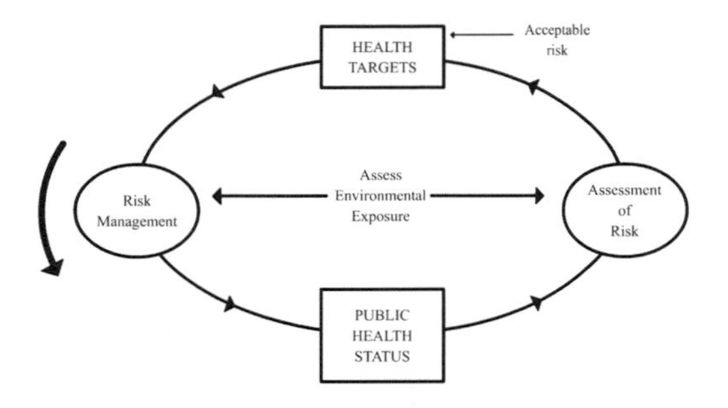

15

Economic evaluation and priority setting in water and sanitation interventions

Guy Hutton

There is always a need to weigh up costs against benefits and in doing so one of the more difficult problems is to come up with a monetary measure for different benefits. This chapter examines some of the instruments available to guide priority-setting and their use in assessing water and sanitation interventions.

15.1 INTRODUCTION

The discipline of economics essentially deals with the allocation of scarce resources amongst competing alternatives, with the aim of maximising an outcome of interest (e.g. profit, health or social welfare). In the health arena, policy makers and programme managers are constantly faced with economic decisions: how to spend a limited budget and have the biggest impact on health? The technique of economic evaluation can contribute to these decisions by providing information on the costs and benefits of alternative interventions, summarising information in a cost-effectiveness or cost-benefit ratio. In addition to the information it provides economic evaluation helps to bring elements of transparency and objectivity to policy making.

Water and sanitation interventions provide an interesting but challenging application of economic principles to resource allocation issues. The challenge is partly that economic evaluation guidelines were developed to evaluate health interventions provided by core health services, with a focus on health sector costs and benefits. However, like many environmental interventions aimed at improving or sustaining health, water and sanitation interventions are different in that:

- they are more likely to be regulatory in nature (such as the meeting of quality criteria)
- they involve cross-sector collaboration and are often financed by non-health agencies (Varley *et al.* 1998)
- they provide large non-health benefits (such as time saving, increasing amenity etc.) which are important to consider (Hutton 2000)
- they are less amenable to controlled trials to evaluate effectiveness (due to confounding factors, for example Blum and Feachem 1983)
- different studies have reported wide ranges of effect (Esrey *et al.* 1985) leading to difficulties in generalising results between different settings.

The implication of these aspects is that appropriate methods for evaluating water and sanitation interventions have remained underdeveloped, and subsequently there are few published studies that have dealt with the economics of water and sanitation interventions in a comprehensive or satisfactory way (Hutton 2000).

Another particular challenge faced in implementing water and sanitation interventions in developing countries is that the expenditure patterns required to meet current guidelines and standards are unrealistic in many developing countries (WHO 1997). This requires many resource-poor countries to make choices over which quality standards they should meet using a risk-benefit or economic evaluation approach, since meeting some quality standards may be less expensive and/or have a larger health effect than others. However, again, there is remarkably limited evidence on the cost-effectiveness of water and sanitation interventions to make these choices (Hutton 2000).

Therefore, the overall aim of this chapter is to assess the applicability of a recently developed and widely supported economic evaluation framework to appraise alternative water and sanitation interventions, and make recommendations for those wishing to conduct economic evaluations in this area.

15.2 ECONOMIC EVALUATION FRAMEWORK

15.2.1 Outline of economic evaluation framework

The past two decades have experienced a proliferation in published economic evaluations in the medical literature, reflecting the increasing importance of economic aspects of medical interventions in resource allocation decisions (Elixhauser *et al.* 1993, 1998; HEED 2000; Walker and Fox-Rushby 2000). Economic evaluation guidelines such as those put forward by Drummond *et al.* (1997) have played an important part in improving the quality of such evaluations, albeit gradually (Adams *et al.* 1992; Baladi *et al.* 1998). More recently, the use of the economic evaluation framework recommended by guidelines has been linked both formally and informally to the peer review process for publishing articles, and has been used by funding bodies of health-care research in allocating research funds. For example, the *British Medical Journal* (BMJ) commissioned an Economic Evaluation Working Party to put together a series of critical elements for journal reviewers and editors to use when deciding whether to publish economic evaluations (Drummond and Jefferson 1996). Also, the *New England Journal of Medicine* published a series of articles with recommendations produced by the United States Panel on Cost-Effectiveness in Health and Medicine (Gold *et al.* 1996; Weinstein *et al.* 1996). Currently the World Health Organization (WHO) is designing its own cost-effectiveness guidelines (Murray *et al.* 2000).

The main purposes of these economic evaluation guidelines are to increase consistency and to allow comparison of the results of different studies, as well as to clarify the methodological choices that can be made at various stages of the evaluation for those less familiar with the economic evaluation framework. The economic evaluation framework recommended by the BMJ is summarised in Box 15.1 (Drummond and Jefferson 1996) and consists of three main stages, namely: study design, data collection, and data analysis and interpretation of results.

Box 15.1. Summary of BMJ economic evaluation guidelines (reproduced, with permission, from Drummond and Jefferson 1996)

Study design

(1) Study Question

–The economic importance of the research question should be outlined.

–The hypothesis being tested, or question being addressed, in the economic evaluation should be clearly stated.

–The viewpoint(s) – for example, health care system, society – for the analysis should be clearly stated and justified.

(2) Selection of alternatives

–The rationale for choice of the alternative programmes or interventions for comparison should be given.

–The alternative interventions should be described in sufficient detail to enable the reader to assess the relevance to his or her setting – that is, who did what, to whom, where, and how often.

(3) Form of evaluation

–The form(s) of evaluation used – for example, cost minimisation analysis, cost-effectiveness analysis (CEA) – should be stated.

–A clear justification should be given for the form(s) of evaluation chosen in relation to the question(s) being addressed.

Data collection

(4) Effectiveness data

–If the economic evaluation is based on a single effectiveness study – for example, a clinical trial – details of the design and results of that study should be given – for example, selection of study population, method of allocation of subjects, whether analysed by intention to treat or evaluable cohort, effect size with confidence intervals.

–If the economic evaluation is based on an overview of a number of effectiveness studies, details should be given of the method of synthesis or meta-analysis of evidence – for example, search strategy, criteria for inclusion of studies in the overview.

(5) Benefit measurement and valuation

–The primary outcome measure(s) for the economic evaluation should be clearly stated – for example, cases detected, life years, quality-adjusted life years (QALYs), willingness to pay.

–If health benefits have been valued, details should be given of the methods used – for example, time trade off, standard gamble, contingent valuation – and the subjects for whom valuations were obtained – for example, patients, members of the general public, health care professionals.

–If changes in productivity (indirect benefits) are included they should be reported separately and their relevance to the study question discussed.

(6) Cost data
-Quantities of resources should be reported separately from the prices (unit costs) of those resources.
-Methods for the estimation of both quantities and prices (unit costs) should be given.
-The currency and price date should be recorded and details of any adjustment for inflation, or currency conversion, given.

(7) Modelling
–Details should be given of any modelling used in the economic study – for example, decision tree model, epidemiology model, regression model.
–Justification should be given of the choice of the model and the key parameters.

Analysis and interpretation of results
(8) Adjustment for timing and costs of benefits
–The time horizon over which costs and benefits are considered should be given.
–The discount rate(s) should be given and the choice of rate(s) justified.
–If costs or benefits are not discounted an explanation should be given.

(9) Allowance for uncertainty
–When stochastic data are reported details should be given of the statistical tests performed and the confidence intervals around the main variables.
–When a sensitivity analysis is performed details should be given of the approach used – for example, multivariate, univariate, threshold analysis – and justification given for the choice of variables for sensitivity analysis and the ranges over which they are varied.

(10) Presentation of results
–An incremental analysis – for example, incremental cost per life year gained – should be reported, comparing the relevant alternatives.
–Major outcomes – for example, impact on quality of life – should be presented in a disaggregated as well as an aggregated form.
–Any comparison with other health care interventions – for example, in terms of relative cost-effectiveness – should be made only when close similarity in study methods and settings can be demonstrated.
–The answer to the original study question should be given; any conclusions should follow clearly from the data reported and should be accompanied by appropriate qualifications or reservations.

Current evaluation guidelines recommend presentation of incremental cost-effectiveness (i.e. current care versus the best alternative). However, there is increasing support for presentation of average cost-effectiveness ratios as well (Murray *et al.* 2000), where alternatives are compared with the costs and consequences of a do nothing alternative (i.e. no intervention). Economic evaluation guidelines distinguish between studies relating to decisions of programme managers in the short term, which should use marginal costs, and those relating to national policy, which should use average costs (Drummond and Jefferson 1996).

15.2.2 Implications of economic evaluation framework for water and sanitation interventions

As already mentioned, when applied to water and sanitation interventions, there is a risk that the current economic evaluation guidelines are not comprehensive in scope, as they may be confined to include only those interventions typically delivered by core health services, with an emphasis on curative treatment. As outlined earlier, environmental health interventions differ from core health services. It is for these reasons that selective primary health-care interventions, such as those suggested in the influential article by Walsh and Warren (1980) contain limited environmental health interventions, and those included appeared much less cost-effective than most curative measures. Attempts have recently been made to formulate essential national packages of services in developing countries; however, few contained environmental health interventions. Exceptions included that proposed by Jha *et al.* (1998) which included the construction of pit latrines and safe water provision as part of a package of 40 health interventions in Guinea, although this intervention turned out to be considerably less cost-effective than the treatment of diarrhoea. However, Varley *et al.* (1998) argued that environmental health interventions to prevent diarrhoea can compete with other means of controlling diarrhoea, such as oral rehydration therapy, once the non-health benefits are taken into consideration.

Therefore, the special nature of environmental health interventions in general, and water and sanitation interventions in particular, should be considered when applying current economic evaluation guidelines to estimate cost-effectiveness. Before these issues raised above are discussed further, a review of literature on economic aspects of water and sanitation interventions is presented and discussed briefly, to act as a backdrop to the discussion of issues in the following section.

15.3 ECONOMIC STUDIES IN WATER AND SANITATION

A brief search was made of electronic databases using key words and researchers in this area contacted to identify articles on the economics of water supply and sanitation. The purpose of the search and review was not to compare the actual cost-effectiveness of alternative water and sanitation interventions, but instead a more important first step was to summarise the range of studies conducted to date, and comment on methodological approaches and study quality. Twenty-four studies were located on the economics of water and sanitation interventions, and these are summarised in Table 15.1. Three main types of study are classified in the table: those evaluating cost-effectiveness or cost-of-illness of water and sanitation interventions; those measuring willingness to pay (WTP) for water and sanitation interventions; and those measuring WTP, cost and cost-effectiveness of water quality improvement.

None of the studies estimated cost-effectiveness of water and sanitation interventions using primary data from a single setting, and only four studies considered both the costs and consequences of at least two alternatives, thus meeting the criteria for a full 'economic evaluation' (see Box 15.1). The most comprehensive study was that by Varley *et al.* (1998) who modelled the cost-effectiveness of water and sanitation interventions in a hypothetical city in a developing country, using secondary data from a variety of sources and a number of assumptions. Phillips (1993) discussed the potential cost-effectiveness of hand-washing to prevent diarrhoea, and used published studies of effectiveness data to build a plausible picture of procedures and resource use, and hence of cost-effectiveness. Briscoe (1984) discussed methodological issues in evaluating the cost-effectiveness of water and sanitation interventions, presenting data to support the hypothesis that water and sanitation interventions can compete with oral rehydration in terms of cost-effectiveness in reducing the incidence of diarrhoeal diseases. Very few published studies measured the actual costs of water and sanitation services, and those that did (such as Varley *et al.* 1998) used available data that were often transferred from other countries.

Generally the studies assessing the water market focused on the demand side, measuring willingness to pay of actual or potential customers and identifying options for cost recovery. Most willingness to pay studies measured either (a) the value to consumers of improvements in the availability and quality of drinking water or (b) the value to consumers of improvements in the quality of surface water (rivers, lakes or coastal waters) for amenity uses. Most WTP studies used the contingent valuation method (see later) to identify the potential demand curve for improved water supply and quality, and many of these also identified current water markets and compared them with WTP (Whittington *et al.* 1990a,b). Franceys (1997) also discussed several options for private sector participation in the provision of water and sanitation facilities, using case studies taken from both developed and developing countries.

In conclusion, the literature reviewed covered several economic aspects of water supply, water quality and sanitation interventions, including costs, cost-effectiveness, willingness to pay, and cost-of-illness. However, few studies measured the costs and benefits of alternative interventions to provide policy makers with the information to choose the most efficient intervention from the viewpoint of society or the health sector. Generally, it would seem that there has been inadequate attention to economic issues in water and sanitation interventions.

Table 15.1. Cost-effectiveness, cost-of-illness or willingness to pay studies on water and sanitation services

Reference	Study aim and country	Costs included	Benefits included
Cost-effectiveness or cost-of-illness studies			
Briscoe (1984)	Review of cost-effectiveness of water supply	R: HS	R: MOR
Harrington *et al.* (1989)	Costs of a waterborne disease outbreak (USA)	P: HS, PT	P: COI
Paul and Mauskopf (1991)	Methodology for cost-of-illness studies	None	R: COI
Philips (1993)	Review of diarrhoea control (LDCs)	S: HS	S: CDA
WASH (1993)	COI of cholera epidemic (Peru)	None	P: COI
Varley *et al* (1998)	CE of WS interventions (LDCs)	S: HW/SW	S: CDA, DALY
Willingness to pay (WTP) studies on water supply and sanitation services			
Boadu (1992)	WTP for water piped to households (Ghana)	None	P: WTP
Whittington *et al.* (1990a)	WTP for water from village standposts (Haiti)	None	P: WTP
Whittington *et al.* (1990b)	WTP for water piped to households (Nigeria)	S: PIP	P: WTP
Whittington *et al.* (1990c)	WTP for water – vendor/kiosk/wells (Kenya)	None	P: WTP
Whittington *et al.* (1991)	WTP for improved piped water supply (Nigeria)	P: VE, HW	P: WTP
Darling *et al.* (1992)	WTP for sewerage facilities (Caribbean)	None	P: WTP
Whittington *et al.* (1993)	Time to think in WTP valuations (Nigeria)	None	P: WTP
Hanley (1989)	WTP for reducing nitrate level of water (UK)	None	P: WTP
North and Griffin (1993)	Water supply and house prices (Philippines)	None	P: WTP
Whittington *et al.* (1993)	WTP for improved WS services (Ghana)	P: HW	P: WTP

Reference	Study aim and country	Costs included	Benefits included
Willingness to pay, cost and cost-effectiveness studies on water quality improvement			
Dixon *et al.* (1986)	Industrial waste water disposal (Philippines)	S: IND	None
Hanley (1989)	Costs of reducing nitrate pollution (UK)	P: IND	None
Hanley and Spash (1993)	Review of CB of controlling nitrate pollution	R: PC	R: WTP, CAV
Kwak and Russell (1994)	WTP to stop contaminating river water (Korea)	None	P: WTP
WHO (1994)	Review of cost recovery approaches for WS	S: GOV	None
Giorgiou *et al.* (1996)	WTP to improve bathing water quality (UK)	None	P: WTP
Day and Mourato (1998)	WTP to improve river water quality (China)	None	P: WTP
Machado and Mourato (1999)	WTP to improve bathing water quality (Portugal)	None	P: WTP

Abbreviations: CE – cost-effectiveness; WS – water and sanitation; WTP – willingness to pay; LDCs – less developed countries; CB – cost-benefit. ***Data type***: P – primary data; R – review; S – secondary data. ***Costs included***: HS – health service; PT – patient; PC – pollution control; GOV – government; VE – private vendors; IND – industry; HW – hardware; SW – software. ***Benefits included***: MOR – morbidity and mortality; COI – cost-of-illness; CAV – costs averted; CDA – cases and deaths averted; DALY – disability-adjusted life years saved.

This highlights the need for an economic framework that is specific to water and sanitation interventions, but which still allows comparison of economic efficiency with other health interventions. Therefore, the rest of this chapter focuses on issues where greater clarification and agreement are needed.

15.4 ISSUES IN APPLYING THE ECONOMIC EVALUATION FRAMEWORK TO WATER AND SANITATION INTERVENTIONS

This section examines the issues arising through application of the economic evaluation guidelines to the water and sanitation field.

15.4.1 Study viewpoint: benefit inclusion

Berman (1982) points out that cost-effectiveness comparisons tend to undervalue interventions that provide important outcomes other than the one being considered, and are thus particularly inappropriate where programmes produce a broad mix of benefits. Water and sanitation interventions are a good example of health-related programmes with a broad mix of benefits. For example, WASH (1993) argued that:

> ...benefits analysis related to water supply and sanitation projects should include measurement of direct economic benefits, such as increased time availability when water is more conveniently located, commercial benefits (reflected in infrastructure improvement leading to increased investment and other opportunities) and health benefits, both direct in terms of avoided medical expenses and indirect in terms of productivity gains due to reduced morbidity.

Several categories of potential benefit arising from water and sanitation interventions are identifiable; Postle 1997). These are summarised in Table 15.2.

There are two main questions that follow from identifying the beneficiaries of water and sanitation interventions. The first question concerns which benefits should be included in the economic evaluation. The second question concerns identifying the main beneficiaries of an intervention and whether they would be willing to contribute to cost recovery.

15.4.1.1 Deciding which benefits to include

The answer to the first question invariably depends on the viewpoint of the policy maker (or those conducting research), whether representing a single ministry or government department, consumers, industry, or society as a whole.

While economic evaluation guidelines support a societal perspective (a view endorsed by Philips (1993) for water and sanitation interventions) the division of budgets between different government ministries or departments means that there are few incentives to estimate an overall 'societal' cost-effectiveness ratio, unless government departments work together in implementing and funding water and sanitation interventions. For example, in a purely health sector analysis, only the health gains and impact on medical costs would be included. This means that many of the non-health benefits are likely to be of less interest to the health ministry, despite empirical evidence to demonstrate the many benefits of water supply (for example, Briscoe 1984; Whittington 1990a,b,c, 1991).

There has also been some discussion surrounding which health benefits to include in the evaluation of water and sanitation interventions. Citing examples from Berman (1982) and Briscoe (1984), Feachem (1986) writes that:

> ...special difficulties are inherent in applying cost-effectiveness analysis to interventions having multiple benefits, and water and sanitation interventions present these difficulties in an extreme form. In addition to their impact on diarrhoea rates among young children, these interventions may avert diarrhoea in other age groups, reduce the incidence of other infectious diseases and have a variety of benefits unrelated to health.

The implication of this view is that other studies (for example Varley *et al.* 1998 who modelled the cost-effectiveness of water and sanitation interventions on diarrhoeal incidence in under-fives) will have underestimated the overall health benefits and thus the true cost-effectiveness of water and sanitation interventions.

Another influence on benefit inclusion is the availability of data and ease of data collection or benefit valuation (see later). At the planning stage of the study, some idea is required of where the greatest data deficiencies or uncertainties lie, and which should first be addressed. Many of the data listed in Table 15.2 may already be available from routine sources such as government records. Other data, such as information on individual productivity, avertive expenditures and time saved, and recreational use or non-use values, however, will need special collection efforts. While it is recognised that data cannot necessarily be collected on all the beneficiaries in Table 15.2, lack of data should not be used to justify the exclusion of important benefits from the cost-effectiveness ratio.

Table 15.2. Categorisation of benefits to society of water and sanitation interventions

Benefit to	Type of benefit	Code
Health sector	Reduction in current costs due to health intervention: materials such as oral rehydration therapy and antibiotics, staff time	Med-cost-avert
	Savings in poison control centre costs	Med-cost-avert
	Reduction in future costs (fewer cases, less severe cases)	Med-cost-avert
Third party payer	Reduction in payouts to health-care providers	Med-cost-avert
Patient[1]	Reduced morbidity and mortality	Health benefit
	Increased life expectancy	Health benefit
	Increased health-related quality of life	Health benefit
	Reduced direct costs of attending care (out-of-pocket expenses)	Med-cost-avert
	Reduced future medical or social care costs	Med-cost-avert
	Increased productivity or capital formation activities such as less time off work and school and increased efficiency while at work or school	Prod -loss-saved
	Reduced risk avertive expenditures such as money cost (capital, recurrent) and time input	Avert-exp-saved
Family or carers of patient	Reduced time caring (back to work)	Prod -loss-saved
	Reduced out-of-pocket payments for medical care	Med-cost-avert
	Reduced risk avertive expenditures (see above)	Avert-exp-saved
Industry	Direct economic value of high quality water such as irrigation water for crops, fishery production, and sea ecosystems	Other-not-health
	Reduced sick leave of employees (paid sick leave, lost production)	Other-pay-avert
	Reduced medical expenses	Med-cost-avert
	Reduced avertive expenditures	Avert-exp-saved
Other government ministries	Reduced running costs or maintenance	Other-pay-avert
	Reduced avertive expenditures	Avert-exp-saved
All consumers	Reduced running costs or maintenance	Other-pay-avert
	Non-health benefits such as increased convenience of a good water supply, increased amenity (laundry, recreational uses), and non-use values (option, existence, bequest)	Other-not-health

Codes: Med-cost-avert: medical costs averted; Avert-exp-saved: avertive expenditure saved; Prod-loss-saved: production loss saved; Other-pay-avert: other payments averted; Other-not-health: other benefits not related to health impact.

[1] The patient is the person who would have been ill in the absence of environmental health intervention.

15.4.1.2 Identifying beneficiaries for cost recovery purposes

In addressing the second question, that of identifying means of cost recovery, there has been considerable work and advancement of methods in this area. For example, Whittington and others have shown that even poor people are willing to pay significant amounts for improved water supply. Also, Franceys (1997) presented examples showing how the private sector can become involved in water and sanitation provision.

Table 15.2 shows that there are many different agencies that may be willing to pay for the identified health services that avoid either tangible (real health and economic losses suffered) or intangible (non-use value) costs. One approach, the 'cost of illness' (COI) approach, has been found to be useful in identifying the size of the tangible costs, which approximates the overall willingness to pay to avoid the illness. As Mills (1991) notes, the inclusion of cost-of-illness aspects has tended to blur the distinction between cost-effectiveness analysis and cost benefit analysis. However, questions may be raised over the relevance of identifying WTP if it is not technically correct to include non-health benefits in the cost-effectiveness ratio. On the other hand, if benefits can be measured and used as evidence that there are options for cost recovery, cross-sector collaboration may be facilitated for the reason that interested parties are less concerned about projects being under-funded.

The importance of averting the indirect economic impacts that result from poor water supply and sanitation was illustrated in a study by Paul and Mauskopf (1991) on the impact of the cholera epidemic in Peru. In this epidemic, it was estimated that three-quarters of the economic costs were from indirect productivity losses due to morbidity (US$2.6m) and mortality (US$93.9m), as well as the macroeconomic impact of loss of exports (US$8.1m), tourism (US$15.4m) and domestic production (US$26.9m). Out of a total economic loss of US$200m, it was estimated that only US$53m was met by the health sector in terms of treatment of cholera cases and public education campaigns. Therefore, these data suggest that other beneficiaries, such as consumers, industry and other government departments, would be willing to pay to prevent such an outbreak from happening again.

In addition to these short-term costs associated with illness, WASH (1993) recognised longer term impacts of water and sanitation interventions, such as changes in population pressures through decreased mortality, and changes in physical capital formation through savings rates and school attendance.

One of the problems of cost recovery is that often not all the benefits are realised instantaneously, whereas significant costs may need to be recovered in the short term. This budget constraint means that while many agents may show willingness to pay for water and sanitation interventions in hypothetical surveys, few may be willing to pay for the benefits before they occur.

15.4.2 Study viewpoint: cost inclusion

This section discusses which environmental health intervention costs should be included in economic evaluations under a variety of viewpoints. Table 15.3 lists the range of agencies that may incur costs in relation to water and sanitation interventions. Several questions are raised in relation to cost inclusion, although not all are discussed here:

- What do water and sanitation facilities cost in different settings?
- Which costs do economic evaluation guidelines recommend to include?
- What proportion of costs fall within and outside the health sector?
- What is the possible impact of the inclusion/exclusion of non-health sector costs on the cost-effectiveness ratio?
- To what extent should the health sector be interested in funding non-health sector costs? Conversely, to what extent should the other agents be interested in funding health sector costs?
- Given the range of agencies funding water and sanitation interventions, which costs should be included in the cost-effectiveness ratio?

Without access to primary data sources, the first question is particularly difficult to answer, due to the lack of cost information provided in the medical literature (see Table 15.1), despite the WHO booklet outlining issues in financial management of water supply and sanitation (WHO 1994). While Varley *et al.* (1998) estimated 'hardware costs' of US$72 per household per year, and 'software costs' of US$3 per household per year, there was no indication of how these costs may vary with bulk purchase, location or quality. However, these data do suggest that a high proportion of water and sanitation cost consists of hardware costs, which are traditionally not paid for by the health sector.

Regarding the second question, costs included in the cost-effectiveness ratio in a purely health sector analysis should be costs met by the health sector. This view is supported by Varley *et al.* (1998) who recommended that the cost of water and sanitation interventions should be included in the health programme budget. This approach is justified in that it yields results that are useful for allocating health programme resources. On the other hand, Briscoe (1984) suggests including the full costs of water and sanitation activities, but subtracting from this figure the amount that users are willing to pay (thus giving the net cost to the providing agency).

Table 15.3. Categorisation of costs of water and sanitation health interventions

Cost borne by	Type of cost[1]
Health sector	Health education outreach and media
	Research costs such as epidemiological study and economic evaluation
	Monitoring and surveillance
Industry	Compliance with emissions regulations[2]
Agriculture	Change in land use following water management
Local council	Waste disposal services
	Water treatment activities
Other government/ public sector	Check compliance with regulations
	Providing clean water and water quality maintenance (e.g. finding new sources)
	Laying water and sewerage pipes (pipes, latrines, digging equipment, labour)
	Repairs to hardware
	Water treatment activities
	Education activities
Consumers	Compliance with waste disposal regulations
	Increased prices passed on by industry
	Charges for sewerage and water facilities

[1] Costs related as well as unrelated to the water and sanitation intervention (in both the initial and extended life span). Note that some of the costs attributed to certain agencies may in fact be met by other agencies in the first column.
[2] Regulations can have two principal types of effect. The first is when a regulation applies at the local level only. In this case it imposes costs on a producer, causing it either to earn less profit, to pay lower wages, or to go out of business. The second is when a regulation applies at the entire industry level. In this case costs are passed on to consumers in a higher price, or companies attempt to cut costs to keep prices stable.

The problem with the approach recommended by current economic evaluation guidelines is that it implicitly assumes a zero cost for non-health programme resources used for water and sanitation interventions. Thus it is unlikely to make optimal use of society's resources allocated to these interventions. For this reason, WASH (1993) states that:

...comprehensive analysis of the economic effects of water supply and sanitation services have to include cost analysis components, such as construction costs, costs related to community organisation and participation, training, and ongoing operations and maintenance.

15.4.3 Valuation of benefits in monetary units

Economic evaluation guidelines recommend the use of economic value, wherever possible. Market prices are usually a good measure of economic value. However, the two problems faced in many economic evaluations are:

- Market values do not represent economic value, because there are some distortions present in the market such as monopoly, subsidy or taxes. The presence of any of these means that prices do not reflect the 'true' market rate. If the divergence of price with economic value is suspected to be substantial, then adjustments are recommended. For example, the profit element in medical charges could be subtracted if profit margins are known.
- The market values are not available to represent economic value. This is more of a serious problem, as it requires the use of other methods to value willingness to pay, and considerable controversy remains over optimal valuation methods.

Therefore, this section aims to identify the strengths and weaknesses of the different methods for valuing different types of benefit using the willingness to pay (WTP) method. The four methods for valuing willingness to pay identified in the economic literature (Hanley and Spash 1993; Postle 1997) are:

- market price
- household production function
- revealed preference
- contingent valuation.

The approaches are described briefly, and advantages and disadvantages discussed, in the following sections, while Table 15.4 summarises the different methods according to the benefits examined.

15.4.3.1 Market price of goods and activities

Market prices are used to value the costs or benefits associated with changes in environmental quality. This includes the 'cost of illness' approach discussed earlier, or the 'replacement costs' approach which values the damage to assets using market prices. This approach assumes that market price represents the economic value, and that there are no taxes and subsidies.

Essentially, market prices can be used for those changes in activity where markets exist. For example, changes in medical costs can be estimated by aggregating the unit costs of those services for the numbers of services saved.

Market prices are also used in the 'human capital' approach, where human life and time spent ill or recovering from illness are valued using future expected earnings. The calculation uses approximations of the value of the increased productivity of individuals through fewer days lost from work or days with restricted activity. For a person who dies prematurely, the lost productivity estimate is often given as the stream of earnings that the person would have earned if he or she had not died. The human capital approach can also be applied to time saving not associated with health, such as the economic value of reduced water collection time.

Table 15.4. Recommended methods of valuation for benefits of environmental health interventions

	Method of measuring willingness to pay[1]			
Type of benefit	Market value	Household production	Revealed preference	Contingent valuation
Health-related benefits				
Improved quality of life				4
Improved life expectancy				4
Medical costs avoided	4			(4)
Reduced time spent in care	4			(4)
Reduced travel expenses to care	4			(4)
Reduced avertive expenditure		4	(4)	(4)
Increased productivity	4			(4)
Reduced sick leave	4			(4)
Benefits not related to health				
Increased competitiveness	4			(4)
Reduced running costs	4			(4)
Reduced emergency services	4			(4)
Increased convenience	4		(4)	4
Increased amenity	(4)	(4)	(4)	4
Non-use option value[2]			(4)	4
Non-use existence value[2]			(4)	4
Non-use bequest value[2]			(4)	4

[1] See text for description of methods; 4 = preferred method(s); (4) = second best method(s); blank = no method available. [2] See text for explanation.

The human capital approach is perhaps the most difficult and controversial aspect of valuing health effects (Freeman III 1993). The most serious shortcoming of the human capital approach is that it does not provide information about what the individual would be willing to pay to obtain a given reduction in the probability of loss of life (Fisher 1981). Also, it does not measure the net contribution to society, it ignores non-market activities important to individuals, and the loss of leisure time or activities. Also, there is considerable uncertainty about the number of days or years that individuals actually take off work (Hanley and Spash 1993). Therefore, if used, this method must be applied with caution, and interpreted appropriately.

15.4.3.2 Household production function

The production function method may be applied either to private sector companies producing goods or services, or to households producing services that generate positive utility. For example, a household may react to water contamination by either purchasing water treatment equipment or by boiling water, both of which involve changes in expenditure patterns and the use of time. This behaviour is called mitigative behaviour, or avertive expenditure. The value of an improvement in water quality can be inferred directly from reductions in averting expenditure (Courant and Porter 1981). However, avertive expenditure may not capture all aspects of a benefit, it may overstate the benefit, and it is not widely applicable, but is instead specific to occasions when individuals change their activities to prevent an outcome. It requires surveys of individual behaviour, and results are likely to be highly setting-specific due to the many contextual factors that affect human behaviour (e.g. norms, income, risk perception etc.).

Another approach, the ‘travel cost’ method, has also been shown to be a useful method for measuring the value associated with environmental benefits, such as recreational benefits of water, although it has not been used to value health benefits. The travel cost method also suffers from weaknesses, such as whether a journey is made for reasons other than simply the environmental benefit.

15.4.3.3 Revealed preferences

The revealed preference method (also called ‘hedonic pricing') seeks to find a relationship between the levels of environmental services (such as a water supply), and the prices of the marketed goods (houses). Most studies found in the literature used regression analysis to identify this relationship. Several problems exist with this method, including large sample size requirements, omitted variable bias, multi-colinearity, wrong choice of functional form, not recognising market segmentation, not accounting for impact of expected environmental goods, and not meeting restrictive assumptions of the model (Hanley and Spash 1993).

Another application of the revealed preference method is the valuation of incremental morbidity or mortality risks by identifying wage differentials due to risk differences. The theory is that workers have to be paid a premium to undertake jobs that are inherently risky (or disagreeable) and this information can be used to estimate the implicit value individuals place on sickness or premature death associated with the job. Thus it measures, albeit inaccurately, an implicit willingness to pay for reductions in risk of death, or willingness to accept increases in the risk of death. However, it has limited applicability to water and sanitation interventions, and has several weaknesses (Hanley and Spash 1993).

15.4.3.4 Contingent valuation

In this method, the public is asked to value non-market goods within a hypothetical market. The contingent valuation method enables economic values to be estimated for a wide range of commodities not traded in markets, such as health and public goods (for example, clean air and scenery). The technique is now widely accepted by resource economists, following a great deal of empirical and theoretical refinements in the 1970s and 1980s (Hanley and Spash 1993). The contingent valuation method works directly by soliciting from a sample of consumers their willingness to pay for an improvement in the level of environmental service flows (or willingness to accept compensation) in a carefully structured hypothetical market. Bids are then obtained from the consumers, bid curves estimated, and the data aggregated to estimate the market demand curve.

There are several advantages of the contingent valuation method over other valuation techniques:

- It can take into account non-use values, such as the utility individuals derive from the existence of environmental goods, even if they do not use it. Non-use is divided into option value (the possibility that the person may want to use it in the future), existence value (the person values the fact that the environmental good exists, irrespective of use), and bequest value (the person wants future generations to enjoy it).
- It can be designed to include only the variables or characteristics of the market relevant to the objective of the study. For example, it can be designed to include only willingness to pay for health effects, or it can include productivity effects, expenditure averted, etc.
- It allows individuals to consider the true cost to themselves of a particular injury or illness. Results have been shown to be repeatable, both in terms of similarity in results across different settings, but also using a test-retest methodology. Whittington *et al.* (1990a) have found contingent valuation methods to be an appropriate instrument to elicit valuations in a very poor, illiterate population in Haiti, where reasonable, consistent answers were provided.

There are also several potential problem areas associated with this method, including bias, protest bids, the lack of verification procedures, and research cost.

15.4.3.5 Conclusions

In conclusion, market valuations are the best valuation method (if available) because they use existing prices and behaviour and are therefore generally valid. However, when markets do not exist, market behaviour must be extracted from surrogate or proxy markets, or from questionnaires. In general, the contingent valuation method is preferable to revealed preference as it is more reliable and questionnaires can be adapted to answer primary objectives. The household production approach is the least applicable, in that it only values averting expenditures or activities in relation to their health benefits.

While the general concept of willingness to pay is widely accepted by economists, there still exist several methodological problems associated with conducting these studies, whichever elicitation method is used:

- They assume rational individuals.
- They assume people are well informed about the choices they make.
- They assume a well functioning market.
- Through aggregating values, the preferences of the many are remorselessly outweighing the preferences of the few. This is especially problematic when the majority of people are ill informed.
- Under the cost-benefit analysis system, intrinsic value exists only in humans and not in animals, plants and other natural resources. Therefore cost-benefit analysis is anthropocentric, and only 'values' non-human entities when humans themselves value them. Put another way, an environmental good that does not enter at least one person's utility function or at least one private company's production function will have no economic value (Hanley and Spash 1993). Field (1997) suggests a 'stewardship value', related to the desire to maintain the environment for the continued use of all living organisms.

15.4.4 Discounting future costs and benefits

The economic evaluation guidelines state that the time horizon of the economic evaluation should be long enough to capture all the differential effects of the alternative options, and recommend discounting of future costs and benefits occurring during different time periods to their present value. The rationale for discounting is based on the observation that individuals discount the future, because:

- they may expect to be richer in the future
- there is risk attached to investment
- people prefer present to future consumption.

Economic evaluation guidelines also argue that monetary values and health outcomes should be discounted at the same rate. According to Weinstein *et al.* (1996) future health effects should be discounted at the same rate as future costs because people have opportunities to exchange money for health, and vice versa, throughout their lives. Therefore, failure to discount health effects will lead to inconsistent choices over time.

Table 15.5 shows how the net present value of future income streams is reduced with higher discount rates. For example, for the age group 20–24 years, use of a 10% discount rate reduces future income streams to a third of those at 2.5%. The table also shows how the net present value of income of children is very small at higher discount rates, as they will not become productive for many years. These data therefore illustrate the potential impact on future events of discount rates, whether they are costs or benefits (costs saved, health gain), and has implications for the relative cost-effectiveness of health interventions that have costs and benefits in different time periods.

Table 15.5. Net present value of future income streams (no specified currency) for different age groups and discount rates (from Landefield and Seskin 1982)

	Discount rate		
Age group	2.5%	6%	10%
1–4 years	405,802	109,368	31,918
20–24 years	515,741	285,165	170,707
40–44 years	333,533	242,600	180,352
65–69 years	25,331	21,807	18,825

Therefore, how should the discount rate be chosen, and should the same discount rate be applied to all health interventions, and to costs and benefits? Weinstein *et al.* (1996) suggest that a convention is needed for choosing the discount rate in order to achieve consistency across analyses. They argue that theoretical considerations suggest that the real discount rate should be based on time preference, the difference in value people assign to events occurring in the present versus the future. This is reflected in the rate of return on riskless, long-term securities, such as government bonds, which empirical evidence shows to be in the vicinity of 3% per annum. Rates of between 0% and 7% are recommended in the sensitivity analysis.

Discussion surrounding the discount rate and its role in economic evaluation has been given considerable attention by those working in environmental projects, and is of key interest in projects both related and unrelated to health. Baldwin (1983) argued in the context of rural water supply projects, 'the process of discounting removes from consideration a higher and higher proportion of values that fall in the future'. Therefore, the comparison of cost-effectiveness ratios of water and sanitation interventions with curative interventions leaves the former disadvantaged because curative interventions have more immediate effect. Also, the bulk of the costs of water and sanitation interventions are incurred early in the life of the project. Therefore, a positive discount rate reduces the relative costs of low technology curative interventions.

A number of solutions have been suggested but it is clear that there is no single alternative solution for the choice of discount rate that would not attract severe criticisms. Therefore, analysts and decision makers should be aware of the extent to which discount rates make environmental health interventions with high short-term costs and high long-term effects less cost-effective compared with other health interventions.

15.4.5 Dealing with uncertainty

The issue of uncertainty and how to deal with it plays an important role in cost-effectiveness analysis, particularly for water and sanitation interventions. Uncertainty stems from a lack of information about the consequences of a given action (data uncertainty), a lack of agreement in methods (model uncertainty), or uncertainty in the degree to which data can be transferred across settings (generalisability) (Briggs *et al.* 1994). Data uncertainty can include uncertainties in measurement, future values, scientific uncertainties (e.g. cause-effect relation), or the timeframe over which costs and benefits occur (Postle 1997). Model uncertainty can include methods for measuring economic value, the discount rate, and which costs and benefits are included, and have already been discussed in detail. Uncertainty associated with generalisability involves whether cost-effectiveness values from one setting (whether at the village, town, or country level) are applicable in another setting and, if not, whether adjustments can be made to make better predictions. These issues are discussed below for both effectiveness and costs.

15.4.5.1 Effectiveness of water and sanitation interventions

Briscoe (1984) argues that, 'an assessment of the likely impact of water supply and sanitation programmes on health is far more problematic than the assessment of other components of primary health care which operate more directly on the causes of disease'. There are several reasons why uncertainty in the effectiveness of water and sanitation interventions may be greater than for many other types of health intervention. First, evaluating health effects from a change in the natural or human environment is more difficult to do using controlled experiments such as the randomised controlled trial (Luken 1985), and therefore many assumptions are usually required in estimating health benefit. Blum and Feachem (1983) list methodological problems of previous epidemiological studies in measuring the impact of water and sanitation investments on diarrhoeal diseases. These included: lack of adequate control; sample size of one in cluster randomisation studies; confounding variables not controlled for; health indicator recall bias; poor health indicator definition; failure to analyse by age; failure to record facility usage; and failure to analyse by season. Subsequently larger confidence intervals exist around health effects for water and sanitation interventions than for curative activities, which tend to have more high-quality studies of effect performed (as evidenced by the weight of evidence in reviews of epidemiological evidence such as collected by the Cochrane Collaboration).

Second, there is substantial variability in dose–response relationship, and therefore the effectiveness of water and sanitation interventions. Machado and Mourato (1999) discussed the problems in identifying a dose–response relationship when considering the health risks of different levels of coliforms and streptococci, due to variability in levels between location, different weather conditions/times of day, and characteristics of person (gender, age, health condition, hygiene), all of which affect vulnerability to polluted water. This raises the need for subgroup analysis, to better understand dose–response relationships for specified conditions.

Third, due to the lack of evidence on causes of variability in dose–response relationships, it makes generalisations of effectiveness data between settings a highly uncertain process. For example, Hanley and Spash (1993) argue that the benefit of controlling nitrate pollution depends on percolation rates through groundwater, which are highly locale-specific. This raises serious questions about the appropriateness of taking effectiveness data from reviews of studies. For example, Varley *et al.* (1998) used a review of 65 studies to generate a plausible range for the minimum effectiveness of water and sanitation interventions in a hypothetical city in the developing world. However, actual effect may fall anywhere within that range. In this case, it may be better to use the results of the best quality study that was conducted in similar conditions to the setting of interest, thus reducing the range of effectiveness and therefore cost-effectiveness.

15.4.5.2 Costs of water and sanitation interventions

As argued earlier, there is limited primary data in the published medical literature on the costs of water and sanitation interventions. The published cost-effectiveness studies identified have largely used secondary data sources, thus increasing the degree of uncertainty in cost-effectiveness ratios. The implications are that researchers and policy makers using these cost-effectiveness studies to plan services are using outdated or inappropriate cost data.

Therefore, two measures are recommended. The first measure is the use of sensitivity analysis to quantify the impact of uncertainty in costs on the overall cost-effectiveness ratio. Thus the cost-effectiveness ratio is presented as a range as opposed to a point estimate. For example, Luken (1985) suggests the use of worst case and best case scenarios for estimating the costs of compliance with regulations. However, this approach does not attach probabilities to different outcomes, which may be important for policy makers to weigh up the risks of taking certain actions.

The second measure is improving access for researchers and policy makers to cost information, via the internet, local and international organisations, and published cost data in the medical literature. These costs should be both comprehensive (i.e. include all aspects of water and sanitation interventions) and detailed, thus providing data on the costs of different types and specifications of the required materials and equipment.

15.5 CONCLUSIONS

The economic evaluation guidelines (Drummond and Jefferson 1996) presented in this chapter are recognised to be important in that they reflect consensus among mainstream health economists and they increase consistency and comparability between cost-effectiveness ratios for a wide range of health interventions. However, several limitations or disadvantages were discussed in this chapter when applying these guidelines to water and sanitation interventions. These included uncertainty about which costs and benefits to include in the cost-effectiveness ratio, and the choice of discount rate for future costs and effects. Also, the advantages and disadvantages of different benefit valuation methods need to be understood fully by those undertaking such research, and this chapter provided a brief discussion of issues. There are a number of characteristics of water and sanitation interventions that make them particularly difficult to estimate cost-effectiveness with any degree of certainty, including the lack or poor quality of current evidence on costs and effects, and uncertainty associated with generalising cost-effectiveness across settings.

This chapter highlights the problems associated with placing an economic value on water-related interventions. Clearly, however, in terms of adapting international guidelines to national regulations such a valuation should play an important role in the process if standards are to be cost-effective and appropriate to local circumstances. It is, perhaps, the role of future guidelines to provide standardisation and guidance on how such a valuation should be achieved.

15.6 REFERENCES

Adams, M.E., McGall, N.T. *et al.* (1992) Economic analysis in randomized control trials. *Medical Care* **30**(3), 231–243.

Baladi, J., Menon, D. *et al.* (1998) Use of economic evaluation guidelines: Two years' experience in Canada. *Health Economics* **7**(3), 221–227.

Baldwin, G. (1983) Why present value calculations should not be used in choosing rural water supply technology. *World Development* **11**, 12.

Berman, P.A. (1982) Selective primary health care: Is efficient sufficient? *Social Science and Medicine* **16**, 1054–1059.

Blum, D. and Feachem, R. (1983) Measuring the impact of water supply and sanitation investments on diarrhoeal diseases: problems of methodology. *International Journal of Epidemiology* **12**, 357–365.

Boadu, F. (1992) Contingent valuation for household water in rural Ghana. *Journal of Agricultural Economics* **43**, 458–465.

Briggs, A., Sculpher, M. and Buxton, M. (1994) Uncertainty in the economic evaluation of health care technologies: The role of sensitivity analysis. *Health Economics* **3**, 95–104.

Briscoe, J. (1984) Water supply and health in developing countries: Selective primary health care revisited. *American Journal of Public Health* **74**(9), 1009–1013.

Courant, P. and Porter, R. (1981) Averting expenditures and the costs of pollution. *Journal of Environmental Economics and Management* **8**, 321–329.

Darling, A.H., Gomez, C., *et al.* (1993) The question of a public sewerage system in a Caribbean country: A case study. Environmental economics and natural resource management in developing countries. World Bank, Washington DC.

Day, B. and Mourato, S. (1998) Willingness to pay for water quality maintenance in Chinese rivers. CSERGE Working Paper WM 98-02.

Dixon, J.A., Scura, L F. *et al.* (1986) *Economic Analysis of Environmental Impacts*, Earthscan, London.

Drummond, M.F. and Jefferson, T.O. (1996) Guidelines for authors and peer reviewers of economic submissions to the BMJ. *British Medical Journal* **313**, 275–283.

Drummond, M.F., O'Brien, B., Stoddart, G.L. and Torrance, G.W. (1997) *Methods for the Economic Evaluation of Health Care Programmes*, 2nd edn, Oxford University Press, Oxford.

Elixhauser, A., Luce, B.R., Taylor, W.R. and Reblando, J. (1993). Health care CBA/CEA: an update on the growth and composition of the literature. *Medical Care* **31**(7 suppl), JS1-11, JS18–149.

Elixhauser, A., Halpern, M., Schmier, J. and Luce, B.R. (1998) Health care CBA and CEA from 1991 to 1996: an updated bibliography. *Medical Care* **36**(5 suppl), MS1-9, MS18–147.

Esrey, S., Feachem, R. and Hughes, J.M. (1985) Interventions for the control of diarrhoeal diseases among young children: Improving water supplies and excreta disposal facilities. *Bulletin of the World Health Organization* **63**, 757–772.

Feachem, R. (1986) Preventing diarrhoea: What are the policy options? *Health Policy and Planning* **1**(2), 109–117.

Field, B.C. (1997) *Environmental Economics*, McGraw-Hill, UK.

Fisher, A.C. (1981) *Resource and Environmental Economics*, Cambridge University Press, Cambridge.

Francеys, R. (1997) Private sector participation in the water and sanitation sector. UK Department for International Development Occasional Paper No. 3, London.
Freeman III, M. (1993) The measurement of environmental and resource values. Theory and methods. Resources for the Future, Washington DC.
Georgiou, S., Langford, I. *et al.* (1996) Determinants of individuals' willingness to pay for reduction in environmental health risks: A case study of bathing water quality. CSERGE Working Paper GEC 96-14.
Gold, M.R., Siegel, J.E., Russell, L.B. and Weinstein, M.C. (1996) *Cost-effectiveness in Health and Medicine*, Oxford University Press, Oxford.
Hanley N. (1989) Problems in valuing environmental improvements resulting from agricultural policy changes: The case of nitrate pollution. Discussion paper no. 89/1, Economics Dept, University of Stirling, UK.
Hanley, N. and Spash, C.L. (1993) *Cost-benefit Analysis and the Environment*, Edward Elgar, Cheltenham, UK.
Harrington, W., Krupnick, A.J. *et al.* (1989) The economic losses of a waterborne disease outbreak. *Journal of Urban Economics* **25**, 116–137.
HEED (2000) Health Economic Evaluations Database. Office of Health Economics, UK. See www.ohe/org/HEED.htm
Hutton, G. (2000) Contribution to WHO guidelines on cost-effectiveness analysis: considerations in evaluating environmental health interventions. Unpublished working document, Cluster of Sustainable Development and Healthy Environments, WHO.
Jha, P., Bangoura, O. and Ranson, K. (1998) The cost-effectiveness of 40 health interventions in Guinea. *Health Policy and Planning* **13**(3), 249–262.
Kwak, S.J. and Russell, C.S. (1994) Contingent valuation in Korean environmental planning: A pilot application for the protection of drinking water in Seoul. *Environmental and Resource Economics* **4**, 511–526.
Landefield, S. and Seskin, E. (1982) The economic value of life: Linking theory to practice. *American Journal of Public Health* **72**, 6, 555–566.
Luken, R.A. (1985) The emerging role of benefit-cost analysis in the regulatory process at EPA. *Environmental Health Perspectives* **62**, 373–379.
Machado, F. and Mourato, S. (1999) Improving the assessment of water-related health impacts: Evidence from coastal waters in Portugal. CSERGE Working Paper GEC 99-09.
Mills, A. (1991) The economics of malaria control. Waiting for the vaccine, Wiley, Chichester.
Murray, C.J., Evans, D.B., *et al.* (2000) Development of WHO guidelines on generalised cost-effectiveness analysis. *Health Economics* **9**(3), 235–252.
North, J. and Griffin, C. (1993) Water source as a housing characteristic: Hedonic property valuation and willingness to pay for water. *Water Resources Research* **29**(7), 1923–1929.
Paul, M. and Mauskopf, J. (1991) Cost-of-illness methodologies for water-related diseases in developing countries. Water and Sanitation Health Project, US AID.
Phillips, M. (1993) Setting global priorities for strategies to control diarrhoeal disease: the contribution of cost-effectiveness analysis. In *Health Economics Research in Developing Countries* (eds A. Mills and K. Lee), Oxford University Press, Oxford.
Postle, M. (1997) Cost-benefit analysis in chemical risk management. Risk and Policy Analysts Ltd, International Council for Metals and the Environment, Ottawa, Canada.
Varley, R., Tarvid, J. *et al.* (1998) A reassessment of the cost effectiveness of water and sanitation interventions in programmes for controlling childhood diarrhoea. *Bulletin of the World Health Organization* **76**(6), 617–631.

Walker, D. and Fox-Rushby, J. (2000) Economic evaluation and parasitic diseases: A critique of the internal and external validity of published studies. *Tropical Medicine and International Health* **5**, 4, 237–249.

Walsh, J.A. and Warren, K.S. (1980) Selective primary health-care: An interim strategy for disease control in developing countries. *Social Science and Medicine* **14C**, 145–163.

WASH (1993) The economic impact of the cholera epidemic in Peru: An application of the cost-of-illness methodology. Water and Sanitation for Health Project, Field Report 415.

Weinstein, M.C., Siegel, J.E. *et al.* (1996) Recommendations of the panel on cost-effectiveness in health and medicine. *Journal of the American Medical Association* **276**(15), 1253–1258.

Whittington, D., Mu, X. *et al.* (1990a) Calculating the value of time spent collecting water: Some estimates for Ukunda, Kenya. *World Development* **18**, 2.

Whittington, D., Briscoe, J. *et al.* (1990b) Estimating the willingness to pay for water services in developing countries: A case study of the use of contingent valuation surveys in Southern Haiti. *Economic Development and Cultural Change*, 293–311.

Whittington, D., Okorafor, A. *et al.* (1990c) Strategy for cost recovery in the rural water sector: A case study of Nsukka district, Anambra state, Nigeria. *Water Resources Research* **26**(9), 1899–1913.

Whittington, D., Lauria, D.T. *et al.* (1991) A study of water vending and willingness to pay for water in Onitsha, Nigeria. *World Development* **19**(2/3), 179–198.

Whittington, D., Smith, V.K. *et al.* (1992) Giving respondents time to think in contingent valuation studies: A developing country application. *Journal of Environmental Economics and Management* **22**, 205–225.

Whittington, D., Lauria, D.T. *et al.* (1993) Household demand for improved sanitation services in Camas, Ghana: A contingent valuation study. *Water Resources Research* **29**(6), 1539–1560.

WHO (1994) Financial management of water supply and sanitation. A handbook. World Health Organization, Geneva.

WHO (1997) Health and environment in sustainable development. Five years after the summit. World Health Organization (WHO/EHG/97.8), Geneva.

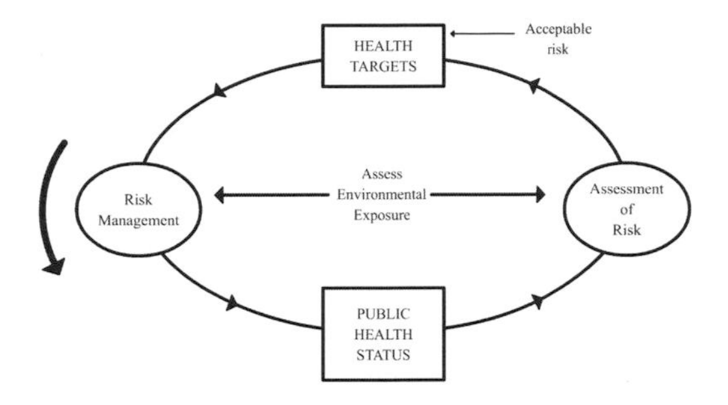

16

Implementation of guidelines: some practical aspects

Marcos von Sperling and Badri Fattal

The setting up of adequate legislation for the protection of the quality of water resources is an essential point in the environmental development of all countries. The transfer of guidelines into practicable standards, which are used not merely for enforcement, but as an integral part of public health and environmental protection policy, has been a challenge for most countries. This chapter examines that process, with an emphasis on the developing country situation.

16.1 INTRODUCTION

One of the main stages of guideline implementation is the conversion and adaptation of the philosophy, guidance and numeric values of the general guidelines, such as those set by the World Health Organization (WHO), into quality standards, defined by each individual country. WHO guidelines are generic by nature, aimed at protecting public health on a worldwide basis. National standards are defined by each country, have legal status and are based on the specific conditions of the country itself. Depending on the political structure of the country, regional standards may also be developed. Economic, social and cultural aspects, prevailing diseases, environmental circumstances, acceptable risks and technological development are all particular to each country or region, and are better taken into account by the country or region itself when converting the WHO guidelines into national/regional standards. This adaptation is crucial: adequate consideration of the guidelines prior to the adoption of standards may be an invaluable tool in the health and environmental development of a country, whereas inadequate consideration may lead to discredit, frustration, unnecessary monetary expenditure, unsustainable systems and other problems. The setting of standards should be based on sound, logical, scientific grounds and should be aimed at achieving a measured or estimated benefit or minimising a given risk for a known cost (Johnstone and Horan 1994).

16.2 COMPARISON BETWEEN DEVELOPED AND DEVELOPING COUNTRIES

It is very difficult to make comparisons and generalisations regarding developed and developing countries. There are large disparities within countries as well as between countries. The aim of the present section is to highlight some aspects that are important in terms of the implementation of guidelines in developing countries and to demonstrate the need for specific approaches.

Developed nations have, to some degree, overcome the basic stages of water pollution problems, although there are still numerous problems and little room for complacency. Developing nations, however, are under pressure from two sides: on the one hand, observing or attempting to follow the international trends of reducing standard concentration levels and, on the other, being unable to reverse the trend of environmental degradation. In many countries the increase in sanitary infrastructure can barely cope with the net population growth. The implementation of water and sanitary regulations depends to a large extent on political will and, even when this is present, financial constraints are often the final barrier, which undermines the necessary steps towards environmental restoration and public health maintenance. Time passes, and the distance between desirable and achievable, between laws and reality, continues to grow.

Figure 16.1 presents a comparison between the current status of developed and developing countries in terms of microbiological drinking-water quality. In this example, the microbiological standard is assumed to be the same for both developed and developing countries. In developed countries, compliance is achieved most of the time, and the main concern is related to occasional episodes of non-compliance. However, in developing nations pollutant levels are still very high, and efforts are directed towards reducing the gap between existing values and the prescribed standards with a view to eventually achieving compliance.

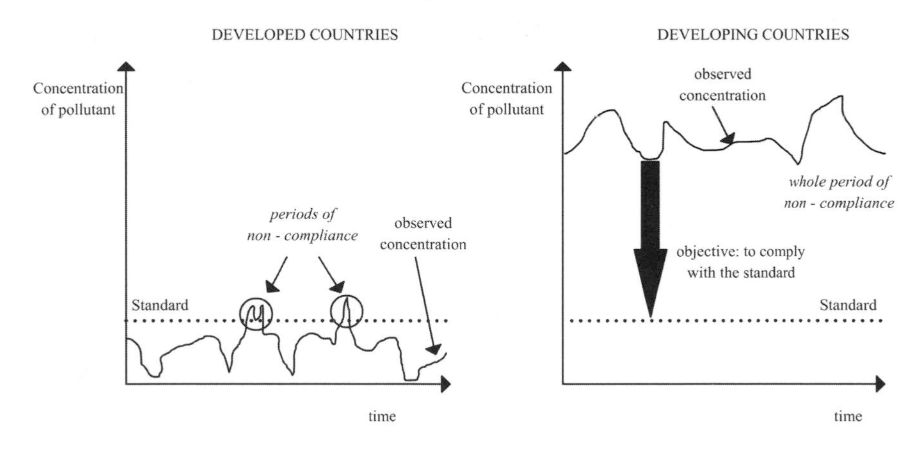

Figure 16.1. Comparison between developed and developing countries in terms of compliance with standards.

The implementation of national water quality standards is intimately linked to the adoption of adequate technologies for the treatment of water and wastewater. There is a wide variety of systems that can be used for wastewater treatment. This, in addition to the diversity of standards encountered in the different countries, will influence the choice of technology. The cost component and the operational requirements, while important in developed countries, play a much more decisive role in developing countries. A further aspect in developing countries is the marked contrast often seen between urban areas, periurban and rural areas. All of these factors make the preliminary selection of the most appropriate system for the intended application a critical step. An additional factor in developing countries may be the influence of foreign expertise. Foreign consultancies may advise according to standards and conditions with which they are familiar, rather than the ones that may be appropriate or those that prevail in the country in question.

Figure 16.2 presents a comparison of important aspects in the selection of water and wastewater treatment systems, analysed in terms of developed and developing countries.

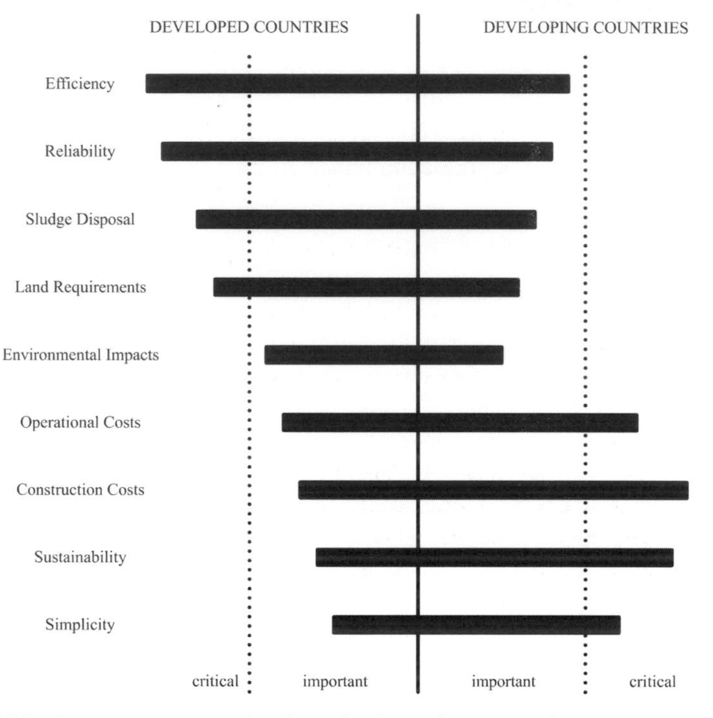

Figure 16.2. Important aspects in the selection of water and wastewater treatment systems: a comparison between developed and developing countries.

The comparison is necessarily general, due to the specificities of individual countries and the contrasts seen within the developing countries. The items are organised in descending order of importance for the developed countries. In these countries, critical items are usually efficiency, reliability, sludge disposal aspects and land requirements. In developing countries, these aspects follow the same pattern of decreasing importance but are less important than in developed countries. In contrast to developed countries the factors of over-riding importance (von Sperling 1996) for developing countries are:

- construction costs
- sustainability
- simplicity
- operational costs.

16.3 TYPICAL PROBLEMS WITH SETTING UP AND IMPLEMENTING STANDARDS IN DEVELOPING COUNTRIES

Several researchers have discussed the inadequacies and difficulties in setting up discharge standards for developing countries. Johnstone and Horan (1994, 1996) presented some interesting papers in which they analysed institutional aspects of standards and river quality and compared different scenarios for the UK and other developed and developing countries. Von Sperling and Nascimento have conducted a detailed analysis of the Brazilian legislation (von Sperling 1998), covering aspects such as comparisons between the limit concentrations in the standards with quality criteria for different water uses (Nascimento and von Sperling 1998), standards for coliforms, sensitivity of laboratory techniques (Nascimento and von Sperling 1999) and requirements for dilution ratios (river flow/effluent flow) in order to match the compliance of water and discharge standards (von Sperling 2000).

Table 16.1 (over) presents a selected list of common problems associated with setting up and implementing standards, especially in developing countries. A further issue relates to international trade and the globalisation of services. Increasingly, companies operate in both developing and developed countries and the acceptability of offering different levels of service (based on different standards) has to be questioned.

It is clear from this table that there is no substitute for adequate examination of guidelines according to prevailing conditions and the adoption of standards based upon realistic expectations.

16.4 STEPWISE IMPLEMENTATION OF STANDARDS

Usually, the stepwise implementation of a water supply or sewerage system is through the physical expansion of the size or number of units. A plant can have, for example, two tanks built in the first stage, and another tank built in the second stage, after it has been verified that the influent load has increased (through, for example, population growth). This stepwise implementation is essential in order to reduce the initial construction costs.

Table 16.1. Common problems associated with establishing and implementing standards, especially in developing countries

Problem	Ideal situation	Frequent outcome
Guidelines are directly taken as national standards.	Guidelines are general worldwide values. Each country should adapt the guidelines, based on local conditions, and derive individual national standards.	In many cases the adaptation is not carried out in developing countries, and the worldwide guidelines are directly taken as national standards, without recognising the country's specific characteristics.
Guideline values are treated as absolute values, and not as target values.	Guideline values should be treated as target values, to be attained in the short, medium or long term, depending on the country's technological, institutional or financial conditions.	Guideline values are treated as absolute rigid values, leading to simple 'pass' or 'fail' interpretations, without recognising the current difficulty of many countries to comply with them.
Protection measures that do not lead to immediate compliance with the standards do not obtain licensing or financing.	Control agencies and financial institutes should license and fund control measures (e.g. wastewater treatment plants) which allow for stepwise improvement of water quality, even though the standards are not immediately achieved.	Agencies or financial institutions do not support control measures which, based on their design, do not lead to immediate compliance with the standards. Without licensing or financing, intermediate measures are not implemented. The ideal solution, even though approved, is also not implemented, because of lack of funds. As a result, no control measures are implemented.
Some standards are excessively stringent or excessively relaxed.	Standards should reflect water quality criteria and objectives, based on the intended water uses.	In most cases, standards are excessively stringent, more than is necessary to guarantee the safe use of water. In this case, they are frequently not achieved. Designers may also want to use additional safety factors in the design, thus increasing the costs. In other cases, standards are too relaxed, and do not guarantee the safe intended uses of the water.

Table 16.1 (cont'd)

Problem	Ideal situation	Frequent outcome
Discharge standards are not compatible with water quality standards.	In terms of pollution control, the objective is the preservation of the quality of the water bodies. However, discharge standards should be compatible with water quality standards, assuming a certain dilution or assimilation capacity of the water bodies.	Even if water quality standards are well set up, based on water quality objectives, discharge standards may not be compatible with them. The aim of protecting the water bodies is thus not guaranteed.
There is no affordable technology to lead to compliance of standards.	Control technologies should be within the countries' financial conditions. The use of appropriate technology should be always pursued.	Existing technologies are in many cases too expensive for developing countries. Either because the technology is inappropriate, or because there is no political will or the countries' priorities are different, control measures are not implemented.
Monitoring requirements are undefined or inadequate.	Monitoring requirements and frequency of sampling should be defined, in order to allow proper statistical interpretation of results. The cost implications for monitoring need to be taken into account in the overall regulatory framework.	In many cases, monitoring requirements are not specified, leading to difficulty in the interpretation of the results.
Required percentage of compliance is not defined.	It should be clear how to interpret the monitoring results and the related compliance with the standards (e.g. mean values, maximum values, absolute values, percentiles or other criteria).	Lack of specification regarding the treatment of monitoring results may lead to different interpretations, which may result in diverging positions as to whether compliance has been achieved.
There is no institutional development to support and regulate the implementation of standards.	The efficient implementation of standards requires an adequate infrastructure and institutional capacity to license, guide and control polluting activities and enforce standards.	In many countries appropriate institutions are not adequately structured or sufficiently equipped, leading to poor control of the various activities associated with the implementation of standards.

However, another use for stepwise implementation that should be considered, especially in developing countries, is the gradual improvement of the quality of the water or wastewater. It should be possible, in a large number of situations, to implement an initial stage that is not optimally efficient (or a process that does not remove all pollutants), graduating at a later stage (as funds become available) to a system that is more efficient or more wide-reaching in terms of pollutants. If the planning is well structured, with a well-defined timetable, it may be possible for allowances to be made permitting a temporary standards violation in the first stage. Naturally a great deal of care must be exercised to prevent a temporary situation from becoming permanent (a common occurrence in developing countries). This use of a stepwise development of water or wastewater quality is undoubtedly much more desirable than a large violation of the standards, the solution to which is often unpredictable over time.

Figure 16.3 presents two alternatives in wastewater treatment implementation. If a country decides to utilise treatment plants that can potentially lead to immediate compliance with the standards, this is likely to require a large and concentrated effort, since the baseline water quality is probably very poor (especially in developing countries). This effort is naturally associated with a high cost, which most developing countries will be unable to afford, the result being that the plant construction is postponed and may never be put into effect. On the other hand, if the country decides to implement only partial treatment, financial resources may be available. A certain improvement in the water quality is achieved and health and environmental risks are reduced, even though the standards have not been satisfied. In this case, the standards are treated as target values, to be achieved whenever possible. The environmental agency is a partner in solving the problem, and establishes a programme for future improvements. After some time additional funds are available and the standards are eventually satisfied. In this case, compliance with the standards is likely to be obtained before the alternative without stepwise implementation.

In developing countries it is not only water and wastewater systems that should expand on a stepwise basis, but also the national water quality standards. The following situations may be encountered:

- If the legislation in a developing country explicitly states that the standards are to be considered a target, then the national standards could have the same values as in the guidelines. Stepwise implementation, however, is complex and requires the provision that if a target value is achieved there should be no slipping back to the previous level.

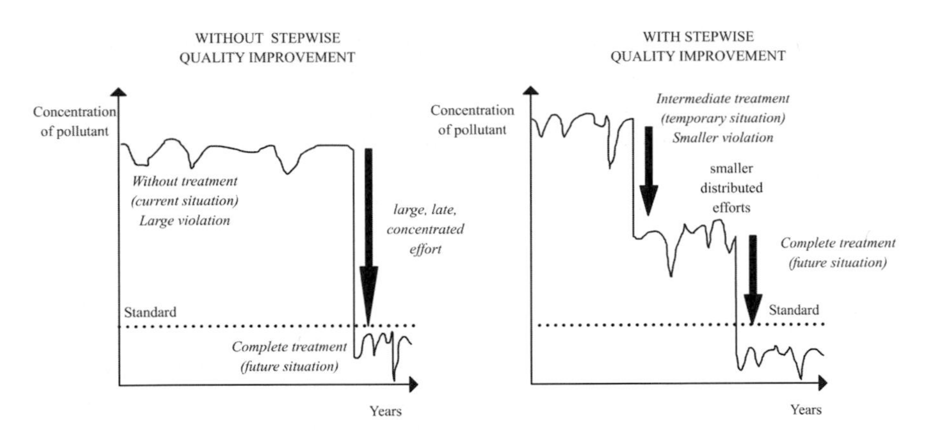

Figure 16.3. Concept of the stepwise improvement of water quality.

- If the concept of targets is not clear in the legislation, then the numerical values of the limit concentrations could progress in a stepwise fashion towards increasing stringency. The standards should be adapted periodically, eventually reaching the same values as those in the guidelines. Ideally the timetable for progressive implementation should be defined, and adequate/ appropriate lead time should be allowed.
- If there are specific conditions in a particular country then the related standards may not necessarily need to converge with the guideline values.

Further advantages of stepwise implementation of standards and sanitary infrastructure are discussed in Table 16.2.

An important issue in the stepwise approach is how to guarantee that the second, subsequent stages of improvement will be implemented, not interrupted after the first stage. Due to financial restrictions, there is always a risk that the subsequent stages will be indefinitely postponed, using the argument that the priority has now shifted to systems that have not yet been implemented in the first stage. Even though this might well be justifiable, it cannot be converted into a commonly used excuse. The control agency or responsible institution must set up a series of intervention targets with the body responsible for the required improvements. These should include the minimum intervention associated with the first stage and subsequent specifications, including required measures, benefits, costs and timetable. The formalisation of the commitment also helps in ensuring continuation of water quality improvement.

Table 16.2. Advantages of stepwise implementation of standards and sanitary infrastructure

Advantage	Comment
Polluters are more likely to afford gradual investment for control measures.	Polluters and/or water authorities will find it much more feasible to divide investments into different steps than to make a large and, in many cases, unaffordable investment.
The present value of construction costs is reduced.	The division of construction costs into different stages leads to a lower present value than a single, large, initial cost. This aspect is most relevant in countries in which (due to inflation) interest rates are high.
The cost-benefit of the first stage is likely to be more favourable than the subsequent stages.	In the first stage, when environmental conditions are poor, a large benefit is usually achieved at a comparatively low cost. In the subsequent stages, the increase in benefit is not so substantial, but the associated costs are high (i.e. there are diminishing returns).
Actual water or wastewater characteristics can be determined.	Operation of the system will involve monitoring, which will result in familiarity with the water or wastewater characteristics. The design of the second or subsequent stages can, therefore, be based on first-hand experience and not on generic values taken from the literature.
There is the opportunity to optimise operation, without necessarily requiring physical expansion.	Experience in operating the system will lead to a good knowledge of its behaviour. This will allow, in some cases, optimisation of the process (improvement of efficiency or capacity), without necessarily requiring physical expansion of the system. The first stage will be analogous to a pilot plant.
There is time and opportunity to implement, in the second stage, new techniques or more developed processes.	The availability of new or more efficient processes for water and wastewater treatment increases with time. Second or subsequent steps can make use of these better and/or cheaper technologies, and realise benefits that would not have been possible with a single step.
The country has more time to develop its own standards.	As time passes, the experience gained in operating the system and evaluating its positive and negative implications in terms of water quality, health status and environmental conditions will lead to the establishment of standards that are really appropriate to the local conditions.
The country has more time and better conditions for developing a suitable regulatory framework and institutional capacity.	Experience gained in operating the system and in setting up the required infrastructure and institutional capacity for regulation and enforcement will also improve progressively, as the system expands in the second and subsequent stages.

16.5 THE PRINCIPLE OF EQUITY

The principle of equity is well rooted within the ethos of the World Health Organization, in that all peoples, irrespective of race, culture, religion, geographic position or economic status are entitled to the same life expectancy and quality of life. Broadly speaking, the reasons for a lower quality of life are associated with environmental conditions. If these improve the quality of life is expected to increase accordingly. On this basis, there is no justification for accepting different environmental guideline values between developed and developing countries.

If guideline values are treated as absolute values, then only developed countries are likely to achieve them, and developing nations will probably not be able to afford the required investments. However, if guideline values are treated as targets, then all countries should eventually be able to achieve them, some on a short-, some on a medium- and others only on a long-term basis.

16.6 COST IMPLICATIONS

Any analysis of guidelines and standards is incomplete and merely an academic exercise if cost implications are not taken into account. Ideally, a cost-benefit analysis should be undertaken when implementing a system of standards or sanitary infrastructure system, although it should be noted that there may be a host of non-health benefits that are difficult to account for (see also Chapter 15). However, in many cases, even though the cost-benefit analysis may prove to be entirely favourable, in developing countries financial resources may not be available to cover the required costs, and the system will remain unimplemented. This point reinforces the need for stepwise implementation and the consideration of guidelines as target values.

16.7 CASE STUDY

The need for defensible standards, both in terms of the degree of protection offered and cost-effectiveness, is a global requirement but one that takes on even greater significance in cash-strapped developing countries. Adopting the wrong approach has led to hundreds of cities in the developing world not being able to afford to meet the standards that they had innocently copied from elsewhere, and thus taking no action. This is a classic tragedy of where insisting on the *very best* prevented achievement of the *good*. This case study examines the level of protection afforded by existing microbiological guidelines for the reuse of wastewater in agriculture in light of acceptable levels of risk and comments on justification of standards on a cost basis. It is based on the publication by Shuval *et al.* (1997) and is revised and reproduced here with the permission of the authors.

16.7.1 Background

In 1982, the World Bank and the World Health Organization embarked upon a broad spectrum, multi-institutional scientific study in order to provide a rational health basis for the revaluation of microbial guidelines for wastewater irrigation. This involved three teams of independent scientists reviewing the epidemiological and technological evidence available concerning health risks associated with wastewater irrigation (Feachem *et al.* 1983; Shuval *et al.* 1986; Strauss and Blumenthal 1989). These studies resulted in the WHO Health Guidelines for the Use of Wastewater in Agriculture and Aquaculture (WHO (1989); reviewed in detail in Chapter 2) which recommended a mean of 1000 faecal coliforms (FC)/100ml and less than one helminth egg per litre of effluent for the irrigation of vegetables eaten raw. These new guidelines have become widely accepted by international agencies including the FAO, UNDP, UNEP and the World Bank, and have been adopted by the French health authorities and the governments of a number of developing as well as developed countries.

In 1992, the US EPA together with the US Agency for International Development (US AID) published their own Guidelines for Water Reuse intended both for internal use in the US and for use by the USAID missions working in developing countries (US EPA/USAID 1992). These new guidelines, for irrigation of crops eaten uncooked are extremely strict and, in microbiological terms, call for no detectable FC/100ml – essentially a drinking water standard.

16.7.2 Methodology

For the purposes of this case study (funded by USAID) the risk assessment model, estimating the risk of infection and disease from ingesting micro-organisms in drinking water, developed by Haas *et al.* (1993) has been used (see Chapter 8), adapted to estimate the risk of infection associated with eating vegetables irrigated with wastewater of various microbial qualities.

Estimates of pathogen levels ingested from eating selected wastewater-irrigated vegetables were made from laboratory experiments which determined the amount of water that might cling to the irrigated vegetables, and then by estimating the concentration of indicator organisms and pathogens that might remain on such irrigated vegetables. A worst-case scenario was chosen by assuming that any micro-organisms contained in the residual wastewater retained on the irrigated vegetables would cling to the vegetables even after the wastewater evaporated.

Based on the laboratory determinations it was estimated that the amount of wastewater that would cling to the outside of irrigated cucumbers would be 0.36ml/100g (or one large cucumber) and 10.8ml/100g on long-leaf lettuce (about three leaves). To estimate the risk of infection and illness from ingesting selected wastewater-irrigated vegetables a numbers of assumptions were made, namely:

- Raw wastewater has a FC concentration of 10^7/100ml.
- The enteric virus:faecal coliform ratio in wastewater is 1:10^5 (Schwartzbrod 1995).
- The degree of pathogen reduction, between irrigation and consumption, is 3 logs.
- All of the enteric viruses are a single pathogen such as infectious hepatitis or polio (allowing assumptions to be made about median infectious dose and infection to morbidity ratios).
- An infection to disease ratio of 50%, i.e. $P_{D:I}$ = 0.5.
- N_{50} values range between 5.6 to 10^4 (see Table 16.3).
- = 0.2 (assuming = 0.5 decreases the risk by about 1 log).
- Individuals eat 100g of either cucumber or long-leaf lettuce (unwashed) per day. For an annual estimate of risk, the same level of daily consumption takes place for 150 days of the year.

16.7.3 Results

A total of four pathogens were examined; two enteric viruses (rotavirus and hepatitis A) and two enteric bacteria (*V. cholerae* and *S. typhi*), all of which have a clear epidemiological record indicating environmental and waterborne transmission (Schwartzbrod 1995). Table 16.3 shows the estimated risk of infection and illness from eating lettuce (which carries a higher risk than cucumbers) irrigated with either raw wastewater or wastewater complying with WHO guidelines.

Comparison of the hypothetical examples with data obtained from an outbreak of cholera in Jerusalem in 1970 allowed the validation of some of the assumptions used in Table 16.3 (Fattal *et al.* 1986).

Table 16.3. Risk of infection and disease from eating 100 grams (3 leaves) of long-leaf lettuce irrigated with raw- and WHO guideline compliant-wastewater effluent

Pathogen	N_{50}	One time risk of eating lettuce (100g)		Annual risk of eating lettuce (100g/d for 150 days)	
		P_I	P_D	P_I	P_D
Raw wastewater					
Rotavirus*	5.6	2.7×10^{-3}	1.3×10^{-3}	4.0×10^{-1}	1.0×10^{-1}
Hepatitis A**	30	1.3×10^{-3}	6.5×10^{-4}	1.7×10^{-1}	4.4×10^{-2}
*V. cholerae***	10^3	6.2×10^{-3}	3.1×10^{-3}	6.0×10^{-1}	1.5×10^{-1}
*S. typhi***	10^4	6.2×10^{-3}	3.1×10^{-3}	6.0×10^{-1}	1.5×10^{-1}
WHO compliant wastewater effluent					
Rotavirus*	5.6	2.7×10^{-7}	1.3×10^{-7}	4.0×10^{-5}	1.0×10^{-5}
Hepatitis A**	30	1.3×10^{-7}	6.5×10^{-8}	1.7×10^{-5}	4.7×10^{-6}
*V. cholerae***	10^3	6.2×10^{-7}	3.1×10^{-7}	9.2×10^{-5}	2.3×10^{-5}
*S. typhi***	10^4	6.2×10^{-7}	3.1×10^{-7}	9.2×10^{-5}	2.3×10^{-5}

P_I = Risk of infection; P_D = Risk of developing clinical disease
N_{50} number of pathogens required to infect 50% of the exposed population
* α=0.265 ** α=0.20 where α= a slope parameter (ratio between N_{50} and P_I)

16.7.4 Case study conclusions

The US EPA has determined that microbial guidelines for drinking water should be designed to ensure that human populations are not subjected to a risk of infection by enteric disease greater than 10^{-4} (or 1 case per 10,000 persons/year, Regli *et al.* 1991). Thus, compared with this US EPA level of acceptable risk the WHO Wastewater Reuse Guidelines, based upon the outlined calculations, appear to be some one or two orders of magnitude more rigorous in terms of protecting consumers.

It is questionable, therefore, whether additional expenditure to provide further treatment to comply with more rigorous standards (such as those proposed by US EPA/USAID, which are 1000-fold more stringent) could be justified in terms of consumer protection. This risk assessment, however, does not account for the risks that may be run by agricultural workers using the wastewater, nor does it take into consideration other benefits that may derive from installing additional infrastructure.

16.8 IMPLICATIONS FOR INTERNATIONAL GUIDELINES AND NATIONAL REGULATIONS

This chapter highlights the complex nature of adopting standards at national level based on guidelines and details a range of factors that need to be considered. Developed countries have generally undergone an implicit stepwise implementation of standards as regulations have become progressively more stringent. Many developing countries are now faced with trying to comply with these stringent levels, but are far from meeting them. For this reason, the concept of stepwise implementation needs to be explicit and it has been recommended that specific guidance on this issue be included in future guidelines.

16.9 REFERENCES

Fattal, B., Yekutiel, P. and Shuval, H.I. (1986) Cholera outbreak in Jerusalem 1970 revisited: The case for transmission by wastewater irrigated vegetables. In *Environmental Epidemiology,* CRC Press, Boca Raton, FL.

Feachem, R.G., Bradley, D.H., Garelick, H. and Mara, D.D. (1983) *Sanitation and Disease: Health Aspects of Excreta and Wastewater Management*, Wiley, New York.

Haas, C.N., Rose, J.B., Gerba, C. and Regli, S. (1993) Risk assessment of virus in drinking water. *Risk Analysis* **13**, 545–552.

Johnstone, D.W.M. and Horan, N.J. (1994) Standards, costs and benefits: an international perspective. *J. IWEM* **8**, 450–458.

Johnstone, D.W.M. and Horan, N.J. (1996) Institutional developments, standards and river quality: A UK history and some lessons for industrialising countries. *Wat. Sci. Tech.* **33**(3), 211–222.

Nascimento, L.V. and von Sperling, M. (1998) Comparação entre padrões de qualidade das águas e critérios para proteção da vida aquática e da saúde humana e animal. In *Anais, XXVI Congreso Interamericano de Ingenieria Sanitaria y Ambiental*, AIDIS, Lima, 1–6 November. (In Portuguese.)

Nascimento, L.V. and von Sperling, M. (1999) Comparação entre os limites de detecção dos métodos analíticos e os padrões de qualidade das águas e de lançamento de efluentes da Resolução CONAMA 20/86. In *Congresso Brasileiro de Engenharia Sanitária e Ambiental*, **20**, Rio de Janeiro, 10–14 May, pp. 2407–2412. (In Portuguese.)

Regli S., Rose, J.B., Haas, C.N. and Gerba, C.P. (1991) Modelling risk for pathogens in drinking water. *Jour. Am. Water Works Assoc.* **83**(11), 76–84.

Schwartzbrod, L. (1995). *Effect of Human Viruses on Public Health Associated with the Use of Wastewater and Sewage Sludge in Agriculture and Aquaculture.* World Health Organization-WHO/EOs/95/.19, Geneva.

Shuval, H.I, Lampert, Y. and Fattal, B. (1997) Development of a risk assessment approach for evaluating wastewater reuse standards for agriculture. *Wat. Sci. Tech.* **25**,15–20.

Shuval, H.I., Adin, A., Fattal, B., Rawitz, E. and Yekutiel, P. (1986) *Wastewater Irrigation in Developing Countries: Health Effects and Technical Solutions.* World Bank Technical Paper Number 51, World Bank, Washington, DC.

Strauss, M. and Blumenthal, U.J. (1989) *Health aspects of human waste use in agriculture and aquaculture-utilization practices and health perspectives.* Report No. 08/88, International Reference Center For Waste Disposal, Dubendorf.

US EPA/USAID (1992) *Guidelines for Water Reuse.* United States Environmental Protection Agency, Washington (Wash Technical Report No. 81, September 1992).

von Sperling, M. (1996) Comparison among the most frequently used systems for wastewater treatment in developing countries. *Wat. Sci. Tech.* **33**(3), 59–72.

von Sperling, M. (1998) Análise dos padrões brasileiros de qualidade de corpos d'água e de lançamento de efluentes líquidos. *Revista Brasileira de Recursos Hídricos* **3**(1), 111–132. (In Portuguese.)

von Sperling, M. (2000) Wastewater discharges and water quality standards in Brazil – Implications on the selection of wastewater treatment technologies. In *Water, Sanitation and Health* (eds I. Chorus, U. Ringelband, G. Schlag and O. Schmoll), pp. 141–146, IWA Publishing, WHO Series, London.

WHO (1989) Health guidelines for the use of wastewater in agriculture and aquaculture – Report of a WHO Scientific Group. Technical Report Series 778, WHO, Geneva.

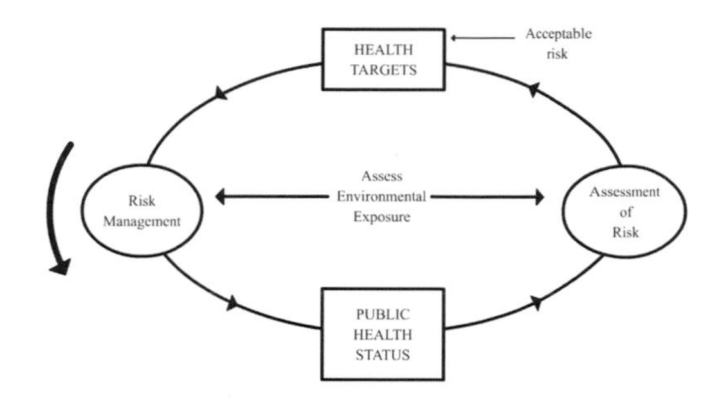

17

Regulation of microbiological quality in the water cycle

Guy Howard, Jamie Bartram, Stephen Schaub, Dan Deere and Mike Waite

Regulation focused on the control of microbiological hazards is important in reducing the incidence of infectious disease. Controls are required throughout the water and waste cycle and many stages are inter-related. The regulator should, therefore, take a 'whole of system' view of microbiological risks and ensure that essential and cost-effective interventions are promoted. This chapter provides an overview of the regulatory issues related to the proposed harmonised framework.

17.1 INTRODUCTION

Regulation is driven by two important objectives:

 Water Quality: Guidelines, Standards and Health. Edited by Lorna Fewtrell and Jamie Bartram. Published by IWA Publishing, London, UK. ISBN: 1 900222 28 0

(1) The protection of the public interest against sub-standard services that raise the risks of health impairment; and
(2) The provision of a transparent system of management where roles, responsibility and liability are clearly defined.

Historically, the development of standards and regulations related to microbiological quality in the water and waste cycle focused on drinking water supplies. In Europe and North America in the nineteenth century this was primarily driven by the need to address epidemics of infectious disease. During the early development of standards, the importance of faecal contamination of water was recognised and the roles of filtration and, later, disinfection were emphasised as critical to the control of drinking water quality. A strong emphasis was placed on sanitary integrity of water sources and supplies and on the use of sanitary inspection in monitoring water supplies. This was associated with the development of the concept of (faecal) indicator bacteria combined with simple methods to test for their presence as a means for assessing the potential presence of pathogens (Chapter 13 and Helmer *et al.* 1999).

Within the water sector, WHO has continued to advocate this approach with respect to control of drinking-water quality (WHO 1993), recreational water quality (Bartram and Rees 2000) and the quality of wastewater reused in agriculture and aquaculture (Mara and Cairncross 1989). Inspection, protection and treatment measures are given a high priority. Indicators to determine the acceptability of water quality include faecal indicator bacteria, turbidity, pH and free residual chlorine for drinking waters and intestinal helminth counts and trematode eggs for wastewater reuse. In the latter case, revised numerical values were recently proposed, taking into account epidemiological evidence for health risks (Blumental *et al.* 1999).

By contrast, national standards have tended to place the greatest importance on indicator bacteria rather than the many other indicators of system integrity and water quality noted above. In some countries, turbidity limits have recently been targeted as a treatment standard in response to risks of *Cryptosporidium* breakthrough. Risk assessment is being used in a number of countries to determine drinking water treatment requirements based on the risk of infection from reference pathogens (Regli *et al.* 1991, 1993).

Despite recent moves to expand the scope of regulations, in most jurisdictions, regulatory enforcement (i.e. action resulting from infringement) remains primarily based on the performance of the supply as determined by indicator bacteria. This almost exclusive reliance on indicator bacteria makes application of risk-based cost-benefit approaches difficult and inhibits the refinement of the definition of tolerable health risks. It also fails to address the breadth of interventions required to reduce disease burdens. It may unintentionally result, for example, in an over-emphasis on water quality control in piped drinking water supply where investment in improving access and reliability of the water supply or improved excreta disposal might yield greater overall health benefits. The sole use of numerical limit values for indicator bacteria also mitigates against the process of incremental improvements and innovation in water supply that are frequently required (Briscoe 1996; Kalbermatten and Middleton 1999; see Chapter 16).

There is little doubt that water supplies which consistently meet standards set for indicator bacteria represent reduced risks to public health. The detection and control of indicator bacteria has proven effective in reducing the frequency of epidemics of bacterial pathogens. However, the value of these indicators to predict the presence of non-bacterial pathogens is limited (see Chapter 13) and there is increasing evidence of infections in populations consuming water that meets current indicator-based standards for drinking water quality (Payment *et al.* 1991; see Chapters 4 and 7).

Interpretation of the results of indicator bacteria analyses may not be straightforward. Typically, indicator bacteria are discrete in water and generally have a non-random distribution in water (Lightfoot *et al.* 1994). They are more likely to be found in clumps following treatment, rather than uniformly spread throughout the water (Gale *et al.* 1997). The volume of water actually analysed by taking occasional 100 ml samples from a large water supply is often less than one millionth of 1% of that produced. Therefore, the absence of indicator bacteria in these small samples may not reflect their true density in water (Gale 1996). Current approaches with a heavy reliance on the use of indicator bacteria are simplistic and are not based on a holistic understanding of the actual health risk derived from exposure. However, because penalties are linked to exceedance of the numerical value for the indicator, the achievement of the standard for the indicator inevitably assumes greater importance than the production of water that is of a quality suitable to protect public health. Thus the tool of monitoring and regulation has become, in many circumstances, the objective of treatment. Further difficulties can arise when the role or applicability of indicators is confused – often the very important difference between total coliforms and faecal coliforms is not clear to non-microbiologists. This leads to the application of excessive disinfection (and the production of disinfection by-products) to reduce total coliform levels even where there is no evidence of faecal contamination.

To address these concerns, revised health protection approaches need to be explored (such as the development of the proposed harmonised framework), and a more process-driven approach considered. Potential implications for future regulation include changes to the indicators of performance with reduced reliance on microbiological parameters and greater reliance on process and system management. These should, in turn, be derived from evidence-based assessment of efficacy to develop a closer link to health outcomes. The use of risk assessment, process control and system management offer advantages for regulatory bodies, enabling them to establish systems that will promote realistic standards that can be modified with changing conditions.

17.2 DEFINING HAZARDS AND ACCEPTABLE LEVELS OF RISK

The proposed harmonised framework translates into a series of activities that regulatory bodies would undertake. These include:

- Identification of hazards and their significance in the local context – types of pathogen, health consequences (diseases and severity), prevalence studies, and possible identification of vulnerable and sentinel groups.
- Identification of health impacts – costs to individuals, costs to society.
- Public consultation – define acceptable risk, tolerable disease burden, willingness to pay for improvement.
- Characterisation of waters with respect to the hazards of concern.

The first two activities are based on sound scientific and health evidence to determine likely prevalence of diseases and the overall impact on the health of the population at large and on sensitive sub-groups. The differentiation of sensitive sub-populations may be important, for instance pregnant women in South Africa are at much greater risk from hepatitis E virus than males or younger children (Grabow 1997). The costs to the individual and society may be less easy to calculate and should take into account non-monetary costs and benefits from improved water and waste services.

Information from international reference sources such as Guidelines documents (see Chapter 2) provide much of the required information and can form the basis for public consultation. Consultation will only produce useful outputs if it is based on a thorough understanding (by all stakeholders) of the major issues and a balanced dialogue can be maintained (see Chapter 14). This is particularly important, as establishing an acceptable risk or tolerable disease burden is effectively asking people to define the level of ill-health they are willing to tolerate (see Chapter 10). In order to do this, the implications of different levels of protection and/or treatment of water on costs to the consumer must be clear and understood by the general population.

In addition to targeted public consultation within this process, there are broader considerations. These include consideration of the proportion of costs that should be absorbed by the service provider without direct consequence on tariff. A further consideration is the incorporation of the estimates of costs accrued to society as a whole rather than to individuals or communities. This includes, for instance, the financial costs of medical treatment in epidemics. Furthermore, there are a broad range of engineering options for developing water and wastewater treatment and use strategies. For example, point-of-use treatment and/or provision of small volumes of specially treated drinking water to high vulnerability groups (e.g. those that are immune-suppressed) can reduce the treatment requirements for the bulk water supply and be a more cost-effective approach in some systems.

Externalities, such as regional agreements or protocols that must be adhered to, must also be addressed and in some cases will be the principal consideration. Such externalities should also encompass potentials for lost earnings either from the presence of unacceptable microbiological hazards in key exports (e.g. shellfish, raw fruit and vegetables) or through lost income from reduced tourism due to poor international perceptions of safety. The latter point is increasingly important for some lower-income countries for which tourism is a rapidly growing sector, but where demands from the tourist population for health protection are high.

In all these stages the regulator would normally take the lead to ensure that the standards and norms established match current capacities and demands. In order to achieve this, a degree of consensus is required between the different stakeholders, and inter-agency collaboration is essential. In most cases, the standard-setting body would, by preference, be multi-sectoral in order to achieve this and would in particular ensure that health and social welfare concerns are adequately addressed in addition to technical and economic considerations.

As a first step, the available water resources should be characterised according to their use and quality requirements. For example, it is common for source waters that are destined to become drinking water to be protected through set-back distances, protection zones and discharge permit levels. The value of characterisation is that it may reduce the frequency with which detailed hazard assessment needs to be carried out and simpler techniques can be used to evaluate hazards in an approach based on sanitary survey/inspection. This characterisation will need periodic updating, but can form the basis of establishing the requirements of individual water and wastewater plants and the degree to which watercourses will require protection based on their use. It will also inform decision-making regarding allocation of different resources to different purposes.

17.3 RISK MANAGEMENT

From a regulator's viewpoint there are three broad approaches to risk management. While they are not entirely mutually exclusive, each has a very different focus, as summarised below.

- Specifying water quality requirements. This is the traditional approach outlined above in which indicators are used as the primary regulatory requirement. The problem of the target of monitoring becoming the treatment objective was introduced earlier. Where this type of approach is used, some division of monitoring between the supply agency responsible for 'quality control' and an independent agency responsible for 'surveillance' is typically encountered. Where a significant volume of testing is undertaken by the regulatory agency this may constitute a transfer of costs from the operator to the regulator.
- Direct regulation of processes. Specification of, for example, treatment processes to be applied to waters of differing qualities is commonly encountered in water quality legislation – both for drinking water supply and for wastewater treatment. This approach places additional burdens on the regulatory agency, which becomes responsible for the validity of the requirements made (i.e. the supplier is not responsible for public health but rather to put in place the treatment requirements specified by the regulations). For the purposes of surveillance, compliance is relatively easily assessed since the existence and operation of processes may be readily verified. Complementary measures such as the maintenance of records available for audit by the regulator may contribute to this.
- Requirement to demonstrate safe practice. In some instances, and more frequently in sectors other than water, the approach taken has been to require that the operator undertakes a risk assessment and puts in place adequate measures to protect public health. The assessment may require approval by a third party (which may have responsibility for public health) and would commonly specify elements such as definition of critical control points and validation and verification requirements (see

below). Such an approach places a burden on the operator to adequately assess and document safety. Where small operators dominate this may not be achievable unless a support system can be developed (through provision of model codes and operating practices for different types of facility, for example). For the regulator, such an approach provides a simple framework within which to apply an auditable approach that may be more effective and reduce costs in many circumstances.

It is the third of these approaches that was recommended at the Stockholm meeting as providing the basis for the harmonised framework. There were many reasons for this, with one of the most important being the adaptability of the approach, enabling different local circumstances to be taken into account. In addition, the third approach is adaptable enough to include appropriate elements of the first two. For example, there needs to be some form of process specification. This can be locally derived and can be site-specific or involve adoption of generic approaches. The *prima facie* verification programme may include indicator measurements.

When applying this type of approach the basic requirement is that an operator prepares a hazard analysis and risk management plan that would include, as a minimum, the following components:

- Baseline characterisation of the source water quality and its variability.
- A verified description of the system and processes.
- The description of water quality objectives appropriate for the specified water use.
- Hazard analysis including an assessment of the type and magnitude of risks.
- The identification of points at which hazards need to be controlled (control points – jargon terms include critical control points, points of attention, sanitary operating practices or preventative measures and choice depends on local preference).
- Monitoring of treatment efficiency indicators to pick up potentially problematic failures of the process at the control points.
- The setting of critical limit targets for monitored activities.
- A corrective action plan in case of failure to comply with the critical limit targets (which would normally distinguish between minor events and events of potentially major public health significance).

- Validating (proving) that the facility is *prima facie* capable of meeting the appropriate water quality targets or other regulatory requirements.
- Verification activities to provide *prima facie* evidence that water is meeting the requirements and that public health targets are being achieved (including record keeping).

This process is outlined in more detail in Chapter 12. However, some of these components are reviewed from a regulatory perspective below.

17.3.1 Critical control point identification

One of the key elements within risk assessment and management is the identification of points within the water or wastewater chain that will either:

- reduce pathogen presence;
- remove/inactivate pathogens, or
- prevent exposure to pathogens.

These are often termed the critical control points as they represent the parts of the cycle where definable action can be taken that will result in (often quantifiable) change in risk and which will provide protection against unacceptable microbiological quality. A range of terms have been used by different regulators for these critical control points depending on the level of criticality (i.e. 'control points' or 'points of attention'), or nature of the process (i.e. 'preventative measures' or 'sanitary standard operating procedures'). For simplicity, here we will use the term critical control points.

In order that critical control points have relevance for the regulatory regime (and especially enforcement), they should be in areas where specific action is required. For instance, while control of agricultural pollution in a catchment is a 'conceptual' critical control point, it is little practical value in terms of the regulator to measure compliance. This would need to be translated into a specific action. An example might be a seasonal restriction on the application of manure, or a restriction on feedlots within a distance specified on the basis of the potential for pathogen migration. A similar process would be seen in regard to treatment processes where the degree of specification would typically be expected to cover defined operational performance criteria (length of filter run, backwash efficiency or effluent retention time, for instance). Regulators have options to employ direct regulation of specific critical control point limits or indirect regulation by requiring that an operator demonstrate that their total system has adequate capacity to reach a defined water quality target, without specifying the processes through which this should be achieved.

As different microbiological hazards have different characteristics in terms of pathogenicity, occurrence and survival, different pathogens may require different critical control points. Some types of pathogens may represent particular hazards or challenges and it is sensible to define the critical control points with regard to such ‘reference’ pathogens. Clearly, different reference pathogens will be required within the various parts of the water cycle and the most appropriate micro-organism selected for each stage of risk mitigation (based on resistance, pathogenicity and nature/magnitude of exposure). Within this, due consideration should be given to average conditions, seasonal variations and extreme events. The latter may, for instance, take into account treatability of drinking water under extreme contamination due to floods or may be used to define effluent quality when flows in receiving water are very low.

17.3.2 Process adequacy (validation)

A key component underlying the process of critical control point identification and application is that there should be evidence of efficacy in terms of risk mitigation or reduction. In this way, the critical control point is related back to the hazard assessment.

During planning and commissioning it is essential that any facility be demonstrated to be capable of meeting the water quality targets or other regulatory targets assigned. During the design phase this may imply theoretical estimations or, in some cases, pilot plant work. For smaller facilities ‘standard designs’ may be adopted in some circumstances. There would be a requirement that plans are approved and performance certified during commissioning by a ‘competent authority’. These are relatively straightforward regulatory requirements to implement and costs largely accrue to the operator where the costs of the competent authority are of experts paid from operator funds rather than provided by the regulator itself. This does, however, raise the issue of shifting liability and also the need for guidelines or regulations on how to certify competence. The objective of this validation exercise is to provide objective evidence that the water quality for the designated use is unlikely to deviate beyond a stated target.

One of the consequences of treating the different components of the water and wastewater cycle separately has been to distort the control of risks derived from infectious diseases towards the production of drinking water, despite obvious comparative advantages in many cases of controlling risks closer to the source of contamination. The multiple barrier principle has long been applied to drinking water supply and the same principle can easily be applied at the broader water and waste cycle level.

For the regulator, the most important aspect at this stage is to define the appropriate intervention within the cycle. This approach tries to answer the following questions:

- Where is human exposure to the hazard within the water and waste cycle most likely to occur or be most significant?
- What is an acceptable risk within each stage of the water and waste cycle based on the hazard and nature of exposure for each pathogen?
- At which point in the cycle will action be most cost-effective?
- What will be the impact on downstream stages of the cycle of the application of an acceptable risk level at an upstream stage?

The next stage is to assess the efficacy of the critical control points in meeting the acceptable risk level. This requires an initial 'research' stage that assesses how effective different processes are in producing water or wastewater of acceptable quality and the operational boundaries that describe performance. The treatment of wastewater and drinking water typically utilises multiple stages of treatment (and in the case of drinking water, source protection measures), thus such boundaries should define not only the combined effect of the multiple stages, but the performance of each individual process.

In many cases, such research has already been undertaken through studies of inactivation rates of particular pathogens in unit processes and through treatment trains. In most cases, therefore, the 'research' component may be limited to literature-based assessments of efficacy. Experimental research may only be required where the level of acceptable risk is significantly lower than attainable by typical treatment performance reported by previous research, where ambient conditions are significantly different from those challenges reported from experimental treatment efficacy research, or where either a new hazard is defined or new process evaluated.

The critical control points within the treatment process can then be defined for the plant as a whole and for unit processes. These critical control points are effectively the key operational parameters that control the overall capability of the process to reduce the hazard to the acceptable risk level.

17.3.3 Monitoring to match the critical control points

For each critical control point, there should be some means of monitoring its effectiveness to ensure that performance targets are within critical limits. This monitoring system needs to provide a reliable assessment of whether the critical control point is being applied effectively and the residual risk is acceptable. Unless there is a simple means of monitoring, repeated and routine hazard assessment would be required, which would become unduly expensive and ultimately difficult to sustain.

Monitoring systems need to be:

- Specific - related to a particular critical control point and not to a broad set of inter-related factors.
- Measurable – it should be possible to translate the critical control point status into some form of quantifiable assessment, even if data collection is based on semi-quantitative or qualitative approaches.
- Accurate – providing an accurate reflection of the critical control point status and sensitive to changes that are of relevance to changes in exposure; they should also have fairly small and precise confidence and prediction intervals, to increase the value of the data they produce.
- Reliable – to give similar results each time it is measured; again this should be within precisely defined confidence and prediction intervals to allow the degree of uncertainty to be described within routine monitoring.
- Transparent – the process of selection of the monitoring variable, the method and frequency of measurement and the interpretation of the results should be transparent and accepted by all stakeholders.

For all components of the system, the choice of monitored variables should be evaluated and validated alongside the critical control point efficacy validation and testing in order that they can be calibrated against an acceptable risk of exposure. The subject of the monitoring should be relatively simple to measure and permit information to be collected frequently and cheaply. Any system of monitoring that becomes too complicated or expensive is unlikely to be effective.

17.3.4 Corrective actions

Where monitoring demonstrates that critical control points are likely to fail, based on the exceedance of a critical limit, corrective actions need to be taken. Regulators can ensure that the appropriate organisations have incident management plans to regain control. These can be generic plans that describe incident management protocols, lines of communications and strategies that can be applied to any incident. These generic plans describe the process by which an incident team will regain control of the situation. For reasonably foreseeable system failures it is better to develop and test specific plans to enable a rapid and effective response. Careful analysis of responses to system failures can be used to help organisations prepare better for subsequent system failures.

17.3.5 Verification and auditing

It should be stressed that it would be expected that some additional *prima facie* verification that water quality targets were being met would be required. Most likely, this would retain the use of indicator bacteria and public health verifications. However, the use and interpretation of such information would be incorporated into a multi-factorial assessment of risks to health. Additional verification activities would assess the adherence of operational systems and personnel to appropriate practice.

17.4 APPLICATION OF THE FRAMEWORK TO COMMUNITY DRINKING-WATER SUPPLY SYSTEMS

A large proportion of the world's population relies on services that are not utility-managed, but are managed by the users or community that they serve. This group represents special problems for regulation (Howard 2000). As regulation is based on a principle of protecting the public interest, it is effective when there is a clear organisational separation between the supplier of the water and the consumers. Where the consumers also operate the supply, the enforcement of standards becomes difficult unless the impact on health can be seen to be affecting people outside the immediate community, for instance in tourist locations or where water is used for food processing. Direct regulation, therefore, is often restricted to the design and construction phases rather than subsequent operation and maintenance.

However, while direct regulation may be problematic, there is a great need for surveillance as a supporting function that promotes improved public health and the ongoing (often incremental) improvement in services. This role is therefore often geared towards training and support to communities in an attempt to improve the quality of services. The application of the framework in these situations is discussed below.

Community-managed drinking water supplies range in size from single point sources, such as a borehole with handpump, to relatively sophisticated piped distribution systems that utilise multi-stage filtration and/or disinfection. Some of these serve single households, while others are designed for relatively large communities of several tens of thousands of people. While the majority of these supplies are found in developing countries, they also represent a significant proportion of supplies in the countries of Central and Eastern Europe and the newly independent states as well as in Western Europe and North America.

Community-managed supplies are not restricted to rural areas and small towns, but are common in many urban areas worldwide (Howard *et al.* 1999). Their use may be found in very large cities. For instance, in Dhaka, tubewells with handpumps are a highly significant source of drinking water given very low rates of access to piped water (Ahmed and Hossain 1997).

The microbiological quality of the water supplied to small and community-managed water supplies is a major concern worldwide. In developing countries, many supplies routinely show contamination whether in urban areas (Gelinas et al. 1996; Howard et al. 1999; Rahman et al. 1997) or in rural areas (Bartram 1998). In industrialised countries, similar problems are noted. For instance, an assessment of the quality of small supplies in the UK found that almost 50% of supplies failed to meet prevailing microbiological criteria and had increased problems (Fewtrell et al. 1998). Similar problems are noted in the US and in Germany.

Some of these problems relate to lack of technical capacity and expertise within the communities for undertaking water quality analysis. Few communities that manage their own supply have access to the equipment and skills to undertake routine water quality monitoring. As a result, monitoring necessarily becomes increasingly infrequent and must be done by an outside agency. In many cases, the methods adopted for such monitoring result in lengthy delays in reporting of results to users and managers of the supply. This inevitably compromises the usefulness of such data in implementing remedial actions, and the results may have limited value in more complex systems as water quality at the time of sampling may not reflect subsequent (often rapid) changes in quality.

Furthermore, where results are relayed to the community, there are often difficulties in their interpretation in relation to potential health risks and in the appropriate remedial actions that should be taken. This lack of understanding is frequently translated into a lack of action of behalf of the community, leading to frustration among staff from local environmental health and water supply sectors.

An important component in ensuring that communities take appropriate action is to ensure that there is effective management to direct operation and maintenance activities and to respond rapidly to failures in the water supply. In many countries, a water source committee would undertake the management of a community water supply. These committees are usually made up of between 6 and 12 members of the community, and are responsible for overall management of the source. As women tend to be the managers of water, such committees are often formed to ensure that women are adequately represented.

The role of the water source committee includes setting and collecting revenue from users and agreeing community contributions in kind to undertake routine maintenance and cleaning. It also liaises closely with the caretaker (who may also be a member of the committee) in agreeing the timing and resources required for maintenance and repair work.

When water source committees run effectively, the management of the water supply and the quality of water provided is usually good. Failures in management by the committee often translate into poor management and poor water quality. For example, in Uganda, a common feature in the failure of many small supplies (including point sources and public taps) was the absence of an active water source committee, many of which had become non-functional over time. Where such committees were reactivated, improvements in overall water supply quality were seen. In this case, an important factor in the promotion of better quality of drinking water could be an active and effective water source committee. Factors that would support the role of the committee include receiving adequate training in monitoring, maintenance and management of the supply and having access to appropriate tools and spare parts to carry out maintenance activities. These could be likened to critical control points and monitoring could be focused on the frequency and scope of training, use of specific maintenance and management tools. Another activity that could be thought of as a critical control point would be ongoing support through surveillance programmes.

17.5 APPLICATION OF THE FRAMEWORK TO WASTES MANAGEMENT

In terms of the management of wastes, an equal potential is noted for improved management of microbiological risks as noted for drinking-water supply. Local use of wastes is common in many parts of the world where excreta has traditionally been used as a fertiliser and effluent for irrigation. The use of untreated wastes in agriculture and aquaculture may also be common.

There may be specific issues that relate to small-scale waste reuse applications where excreta is used from pit latrines. In these cases, critical control points are usually based on storage of excreta, with the length of residence time of the excreta within the pit being the critical limit used as a surrogate measure of likely inactivation of *Ascaris* eggs. However, as this critical control point is a direct responsibility of the user, the technical component must be supported by training and guidance from agriculture and environmental health field staff. A similar situation is likely to be found where a household or community fish pond is supplied by human waste as a nutrient source. The technical basis for the critical control point may be easy to define (based on treatment of waste or retention within a container), but the educational component is likely to be the more critical focus in practice.

For all aspects of the control of microbiological quality in water and wastes, general environmental health protection will also be critical. This will have impacts on the quality of water used for drinking, the quality of wastes and wastewater reused and on water used for other purposes, including recreation and also domestic chores such as laundry and bathing. The promotion of sanitation, proper siting of excreta disposal facilities in relation to drinking water sources, fish ponds and natural water courses and good management of wastes and hygiene will all lead to reduced hazards. This will again require an interface between the technical and educational components of critical control points, with promotion of good practice at a community level being more important than external systems of verification.

The implications for water supply agencies and regulators in reducing risks to users of community-managed services is clear. In the assessment of whether these services are adequate, not only should the infrastructure critical control points be assessed but also the educational and management points. The absence of management structures such as committees should imply action is required by the sector regulator to ensure that agencies engaged in the delivery of services address this properly.

While the framework does not necessarily overcome the legal problems relating to the regulation of community-managed water and wastes systems, it does provide a mechanism by which reductions in health risks from water and wastes can be significantly enhanced. It also provides much greater potential for communities to be active players in managing risks and monitoring the changing levels of risk that they are exposed to and this, in the long term, should translate into improved sustainability.

17.6 IMPLICATIONS FOR INTERNATIONAL GUIDELINES AND NATIONAL REGULATIONS

The development of the harmonised framework has regulatory implications as it will require the regulator, in conjunction with water suppliers and other stakeholders, to establish standards that are acceptable and systems of verification that are reliable. The development of the framework will allow more realistic and effective control of health risks from infectious diseases. This more intelligent regulatory approach will require regulators and operational organisations to think more about the most effective and efficient way to protect public health. This may involve increasing the level of resources (including personnel) that are dedicated to water cycle management. The expectation is that this systematic and evidence-based approach to regulation will lead to better-targeted and, possibly, less costly engineered works while at the same time enhancing public health overall.

This development should not inhibit innovation by applying rigid standards and prescriptions that will limit the potential for new treatment technologies or distribution materials to be developed. One purpose of the regulatory regime is to ensure that 'consumers' get access to a product that is of an acceptable quality in the most cost-effective manner. The need to innovate is particularly acute in developing countries where the derivation of more realistic, evidence-based and balanced standards would contribute greatly to the broader need to address the challenges of providing adequate services to the whole population.

17.7 REFERENCES

Ahmed, F. and Hossain, M.D. (1997) The status of water supply and sanitation access in urban slums and fringes of Bangladesh. *Journal of Water Supply Research and Technology – Aqua* **46**(1), 14–19.

Bartram, J. (1998) Effective monitoring of small drinking-water supplies. In *Providing Safe Drinking Water in Small Systems. Proceedings of the NSF Conference on Small Water Systems, May 1998, Washington DC, USA*, (eds J.A. Cotruvo, G.F. Craun and N. Hearne), pp. 353–366, Lewis Publishers, Boca Raton, FL.

Bartram, J. and Rees, G. (2000) *Monitoring Bathing Waters: A Practical Guide to the Design and Implementation of Assessments and Monitoring Programmes*, E&F Spon, London.

Blumental, U.J., Peasey, A., Ruiz-Palacios, G. and Mara, D.D. (1999) Guidelines for wastewater reuse in agriculture and aquaculture: recommended revisions based on new research evidence. WELL study, London School of Hygiene & Tropical Medicine and WEDC, London.

Briscoe, J. (1996) Financing water and sanitation services: the old and new challenges. *Water Supply* **14**(3/4), 1–17.

Fewtrell, L., Kay, D. and Godfree, A. (1998) The microbiological quality of private water supplies. *Journal of the Chartered Institution of Water and Environmental Management* **12**, 45–47.

Gale, P. (1996) Coliforms in drinking-water supply: what information do the 0/100ml samples provide? *Journal of Water Supply Research and Technology – Aqua* **45**(4), 155–161.

Gale, P., van Dijk, P.A.H. and Stanfield, G. (1997) Drinking-water treatment increases micro-organism clustering: the implications for microbiological risk assessment. *Journal of Water Supply Research and Technology – Aqua* **46**(3), 117–126.

Gelinas, Y., Randall, H., Robidoux, L. and Schmit, J-P. (1996) Well water survey in two Districts of Conakry (republic of Guinea) and comparison with the piped city water. *Water Resources* **30**(9) 2017–2026.

Grabow, W.O.K. (1997) Hepatitis viruses in water: Update on risk and control. *Water SA* **23** (4), 379–386.

Helmer, R., Bartram, J. and Gala-Gorchev, H. (1999) Regulation of drinking water supplies. *Water Supply* **17**(3/4), 1–6.

Howard, G. (ed.) (2000) Urban water surveillance: a reference manual: Final draft document. WEDC/RCPEH, Loughborough.

Howard, G., Bartram, J.K. and Luyima, P.G. (1999) Small water supplies in urban areas of developing countries. In *Providing Safe Drinking Water in Small Systems: Technology, Operations and Economics* (eds J.A. Cotruvo, G.F. Craun and N. Hearne), pp. 83–93, Lewis Publishers, Washington, DC.

Kalbermatten, J.M. and Middleton, R.N. (1999) The need for innovation. *Water Supply* **17**(3/4), 389–395.

Lightfoot, N.F., Tillet, H.E., Boyd, P. and Eaton, S. (1994) Duplicate spilt samples for internal quality control in routine water microbiology. *Letters in Applied Microbiology* **19**, 321–324.

Mara, D.D. and Cairncross, S. (1989) Guidelines for the safe use of wastewater and excreta in agriculture and aquaculture: measures for public health protection. WHO, Geneva.

Payment, P., Richardson, L., Siemiatycki, J, Dewar, R, Edwardes, M and Franco, E. (1991) A randomised trial to evaluate the risk of gastrointestinal disease due to consumption of drinking water meeting current microbiological standards. *American Journal of Public Health* **81**(6), 703–708.

Rahman A., Lee H.K. and Khan M.A. (1997) Domestic water contamination in rapidly growing megacities of Asia: Case of Karachi, Pakistan. *Environmental Monitoring and Assessment* **44**(1–3), 339–360.

Regli, S. Rose, J.B., Haas, C.N. and Gerba, C.P. (1991) Modeling the risk from Giardia and viruses and drinking water. *Journal of the American Water Works Association* **83**(6), 76–84.

Regli, S., Berger, P., Macler, B. and Haas, C. (1993) Proposed decision tree management of risks in drinking water: consideration for health and socio-economic factors. In *Safety of Drinking Water: Balancing Chemical and Microbial Risks* (ed. G.F. Craun), pp. 39–80, ILSI Press, Washington, DC.

WHO (1993) Guidelines for drinking-water quality. Volume 1. Recommendations. World Health Organization, Geneva.

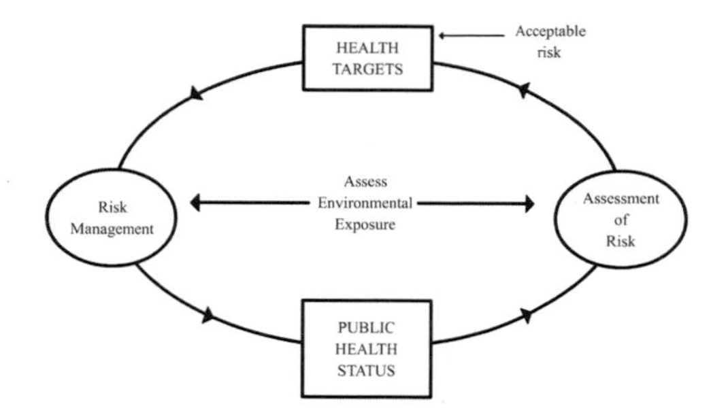

18

Framework for guidelines development in practice

David Kay, Dan Deere, Marcos von Sperling and Martin Strauss

This chapter outlines a series of hypothetical studies that demonstrate the use of the proposed framework for guidelines development in practice. Examples are taken from each of the water-related guideline areas, namely: drinking water, recreational water and wastewater use.

18.1 INTRODUCTION

The proposed harmonised framework for guidelines development, in terms of water-related microbiological risk, was developed during a five-day WHO workshop held in Stockholm in September 1999. Many of the chapters in this book have developed and expanded upon issues and concepts relating to the framework that arose during the meeting. During the workshop an initial attempt was made to 'trial' the framework by working through hypothetical examples for each of the guideline areas of interest. All the workshop participants were involved in this process. These examples are designed to be illustrative and, for the purposes of the exercise, a number of assumptions was made based upon the participants knowledge of each area. Clearly, for a 'proper' iteration of the framework it will be necessary to evaluate, and document, the relevant literature.

The trial studies are outlined, one by one, along the lines of the framework, although the specific approach taken within the bounds of the framework was determined by sub-group participants. Each group's starting point was a health target and acceptable risk level to be considered in relation to a specific pathogen. Health outcomes were expressed as acute gastrointestinal infection (AGI) equivalents. The following sections outline the trial studies and the reader is referred back to earlier chapters for specific details, if required.

18.2 DRINKING WATER

The drinking water group worked through the framework using *Campylobacter* sp. as their reference pathogen. The tolerable burden was considered to be one case of AGI per person during a 10-year period. Due to possible chronic sequelae *Campylobacter* infection was considered to be 2.5 times worse than AGI, resulting in a tolerable risk of one *Campylobacter* infection per person during a 25-year period. An additional requirement was that there should be an avoidance of a detectable rise in morbidity (i.e. an outbreak) due to *Campylobacter* arising from the water source.

18.2.1 Trial study setting

In order to direct the group's thinking a water supply scenario was defined. The water supply system was taken from a surface water river. This was collected into a reservoir, disinfected, and passed into a reticulated water supply that included balancing storages. One of these balancing storages was uncovered.

18.2.2 Assess environmental exposure

Human infectious *Campylobacter* were considered to be potentially present wherever warm-blooded animal faeces contaminate water. However, they were not considered to be free-living in typical environmental waters. Therefore, any surface water or storage basin that could be subject to bird, animal or human faecal contamination could have a potentially unacceptable exposure. Groundwater protected from recent faecal contamination should not pose a significant risk. Therefore, a sanitary survey was proposed as a means of identifying the water source and the potential for faecal impacts.

18.2.2.1 Predictive assessment

For the example, the water supply system was taken from a surface water source that was assumed, from the sanitary survey, to be subject to animal and human faecal contamination. This was collected into a reservoir and passed into a reticulated water supply that included balancing storages, one of which was uncovered and was, therefore, subject to faecal contamination from birds. In conclusion, the supply was subject to a potentially significant *Campylobacter* exposure and further analysis was required.

18.2.2.2 Measured assessment

Monitoring methods do enable quantification of *Campylobacter* for exposure assessment. However, many laboratories and jurisdictions would not have access to such tests, preventing the direct assessment of exposure. In addition, quality may vary widely, requiring high sample numbers in order to properly assess exposure. Since sources of *Campylobacter* are also sources of *E. coli* and enterococci, these were reasoned to be suitable measures as indicators of general faecal contamination levels. It was noted, however, that all monitoring methods are subject to limitations due to the potential for a variation in densities of *Campylobacter* relating to:

- variations in the presence of animal and human hosts and defecation patterns
- variations in the prevalence and nature of infections carried by hosts
- variations in the origins of water reaching the sampling point due to hydrological variation.

These three factors can vary greatly both temporally and spatially and, therefore, so can exposure. The sanitary survey was considered the most important part of the exposure assessment process. Monitoring was thought to be useful as a means of verification of the level of faecal contamination. A statistically valid long-term monitoring regime was recommended. *E. coli* or enterococci were considered to be preferable alternatives to *Campylobacter* for exposure assessment to support the sanitary survey. This is because:

- The presence of these indicators is less subject to host carriage rate variation.

- Indicators are easier to detect and are typically present in higher densities within hosts than pathogens.
- Indicators are useful because although their presence does not mean *Campylobacter* are present and that the supply is unsafe at a point in time, it implies that the water has the potential to become unsafe.

For this example, monitoring results taken weekly from the reservoir water and analysed for *E. coli* and enterococci were assumed to have demonstrated the frequent presence of these organisms at a concentration greater than 1 per 100 ml. This indicated the presence of faecal contamination from warm-blooded animals at densities of potential significance.

18.2.3 Assessment of risk

In practice, it is likely that different strains of *Campylobacter* will have different infection probabilities and also that the health outcome following infection may vary according to the population. Furthermore, the level of acceptable risk may be different in individual countries according to specific national circumstances. The following is illustrative, therefore, only for the trial study.

For waterborne exposures in relatively clean water, faecal contamination would be likely to be very dilute and pathogen densities low. Therefore, the relevant dose–response relationship is found by extrapolating from human feeding trials to the lower doses.

It was assumed that the probability of ingestion leading to both infection and AGI was 0.1%. The severity of symptoms following infection can be converted to Disability-Adjusted Life Years (DALYs) (see Chapter 3) to enable a more general comparison, although that has not been done here and health outcomes are enumerated as AGI equivalents. This was assumed to give a consequence of infection of 2.5 AGI equivalents per infection (average).

18.2.4 Acceptable risk and health targets

Health targets in this example were two-fold:

- One AGI per 10 years from *Campylobacter* for long-term exposure.
- The avoidance of a detectable rise in morbidity (an outbreak) resulting from the water source due to *Campylobacter* arising from an acute exposure period.

An acceptable long-term risk level for the case study community for the water supply was set at one AGI per person per 10 person-years (ppy) from

Campylobacter. This equated to one *Campylobacter* infection per 25 years or, based on the assumed infection rate of 0.1%, 1000 exposures per person per 25 years. On the assumption that people drink 2 litres (l) of water per day, over a 25-year period an individual would be exposed to over 18,000 litres of water. This can be translated into an acceptable exposure concentration of 1 *Campylobacter* per 18 l by equating to 1000 exposures over the same time period.

The acute exposure limit set for short-term occasional high exposures was calculated differently. In the trial study community considered, there was assumed to be a background rate of 0.4% per person-year of notified campylobacteriosis. The surveillance system in place was assumed to be capable of detecting a 10-fold increase in the rate of campylobacteriosis after one week duration as an outbreak. This would equate to a case rate of 4% ppy for a period of one week. It turns out that, in this example, this is equivalent to the acceptable rate for long-term exposure and thus the acute exposure limit for periods up to one week is also 1 *Campylobacter* per 18 l.

18.2.5 Risk management

18.2.5.1 Basic control approaches

There may be a number of immediate actions that can be taken to give rapid reductions in exposure. These may, for example, be infrastructural. In our hypothetical trial, vermin-proofing grills on water tanks were assumed to be damaged. However, these were thought to be quickly and cheaply repairable to reduce avian faecal contamination during reticulation. Other examples could include land-use issues. In our example, the sanitary survey suggested that livestock faeces was heaped in storage piles near a watercourse that ran into the reservoir. It was assumed to be a simple matter for the land manager to store this at the other extremity of the land area farthest away from the river. This would reduce the risk of direct runoff and increase the amelioration effects of sub-surface and overland flow. Taking these simple, and often highly effective, measures can lead to rapid reductions in risk. This illustrates the importance of implementing basic controls while the more detailed risk assessment and management cycle gets underway.

18.2.5.2 Water quality objectives

Water quality objectives are designed to describe the desirable water quality for exposure. For our example, 1 *Campylobacter* per 18 l provides an exposure that is consistent with short- and long-term health objectives. This requires faecal contamination to be very dilute since even 1 g of faeces from an infected host (which could contain millions of *Campylobacter*) could contaminate many megalitres of water to beyond this limit. Such water would also be expected to contain even greater concentrations of indicator bacteria. For our example we have assumed that *Campylobacter* would be present in concentrations at least 1000-fold lower than *E. coli*, and that it has a lesser environmental persistence (it is also more sensitive to disinfectants than indicator bacteria). Therefore, a water quality objective of an average of less than one *E. coli* per 100 ml was thought to represent an appropriate monitoring target both for long- and short-term (outbreak) exposures.

18.2.5.3 Other management objectives

The next step is to set system management objectives to ensure that the system meets the water quality objectives and any other objectives required of the system. These objectives might include technical system management aspects as well as training of staff and education and communication with stakeholders and customers. In our trial example, this involved ensuring water quality objectives were met under normal circumstances and preventing gross contamination during unusual events, such as system failures, to meet those same water quality objectives. This involved influencing a number of identifiable groups. Some groups have roles that are not related to water supply, such as those that manage land-uses that could impact on reservoir water quality. Education, guidelines and regulation are tools to influence such groups. Others are internal, such as utility staff for whom training and appropriate resourcing would be used to ensure the supply of the best quality reservoir water, management of the disinfection system and protection of the reticulation system.

18.2.5.4 Current condition

Measures and interventions to manage the risk involve assessing the environmental exposure by systematically analysing the system from contamination source(s) through to the point(s) of consumption. Possible points of entry of hazards (i.e. *Campylobacter* spp.) would be identified along with any points of removal or inactivation. In the study example, the group was interested in the possible points of entry of faecal contamination from warm-blooded animals and birds. It considered the storage of water for a prolonged period (as a removal and inactivation barrier) and a disinfection system (as an inactivation barrier).

18.2.5.5 Key risk points and audit procedures

The key preventative measures to minimise contamination and barriers to inactivate and control contamination are to be identified and their effectiveness assessed by audit.

In the study example, point sources of faecal contamination were thought to represent a high priority hazard that required audit to ensure that the best reasonable and practicable measures were being taken to prevent faecal material entering the water. For example, for faecal material from agricultural facilities, storage in heaps for elevated temperature composting, storage as far away from the water source as possible and the use of wastewater treatment systems were examples of preventative measures aimed at reducing contamination. The selective harvesting of water from the rivers to the reservoir was another area of potential intervention. The group concluded that since this system drew water from a river into a reservoir, harvesting of water would cease after events in which fresh contamination would run into the river, such as heavy storms or a notified wastewater treatment system failure upstream. The point from which water is drawn into supply from the reservoir could be positioned to maximise the quality of water withdrawn. Turbidity could be used as a surrogate for likely bacterial contamination. The disinfection barrier, if properly applied, was thought to represent the most significant critical control point as it was assumed to provide very large reductions in *Campylobacter* densities (residual disinfectant levels also provide protection within the distribution system). Effectively, the heavy reliance on the disinfection barrier results in a 'fragile' system and suggests the need for an automatic cut-off if the disinfection process fails.

Vermin and bird-proofing in the storages, systems to prevent backflow, careful attention to the location and maintenance of pumps and suction lines, and the use of maintained/continuous positive system pressures were further points of attention recommended to prevent recontamination in the reticulation system. A longer-term intervention that may be considered in the study system was the covering of the open storage.

18.2.5.6 Analytical verifications

The key analytical monitoring and verification procedures are focused on the major preventative measures and critical control points. In the case study example, monitoring points were proposed throughout the system. Point sources of faecal contamination could be inspected regularly to ensure that appropriate waste management practices were being adhered to. The frequency of inspection would be proportional to the likely rate of change of practices at the site as well as the level of risk presented and inspectors would need to be properly trained. Where effective management measures were not being adhered to, corrective actions would need to be taken, such as advising the landholders on appropriate management measures or enforcing regulations. The decisions on selective harvesting of water from the river to supply the reservoir could be linked to a water quality monitoring programme. Action could be triggered by results from water quality parameters monitored instantaneously, such as turbidity, or by rainfall itself. Long-term monitoring could be used to understand the relationship between those factors that can be measured early enough to use as cues for preventative action (such as rainfall) and those that cannot (such as bacterial indicator readings). A system for monitoring and notifying major treated wastewater discharges could be included to enable the avoidance of water harvesting after treatment system failures. Within the reservoir, the water quality could be monitored at a range of depths and at regular intervals to ensure that the point of offtake from the reservoir is optimal for water quality. The disinfection barrier could include an alarm to enable rapid corrective actions to be taken in response to malfunctions measured in terms of pH, chlorine or turbidity. In the short term, triggering of the alarm could result in an automatic redirection of the flow to waste. Corrective actions could include repair or engaging of backup disinfection systems. Engineered items such as tanks, pumps and suction lines and reticulation system backflow preventors need to be appropriately designed and could be carefully monitored at appropriate intervals with repairs made where problems are found. Where system pressures are found to have been lost, potentially leading to ingress, such as after the repair of bursts, an appropriate flushing regime could be used to remove contamination prior to resumption of supply.

A process for verification would be used to ensure that training is up to date and that people are performing their tasks as required. Regular water quality monitoring for *E. coli* would be performed to verify that the concentration of faecal bacterial contamination is acceptably low and that the management approach is working.

18.2.6 Public health status

Public health verification through surveillance would take place and would be designed to test for significant associations between consuming water and morbidity. In the study system, there was assumed to be no evidence of associations between water consumption and morbidity although it was assumed that monitoring found *E. coli* occasionally downstream of the open storage to suggest an average concentration above 1 per 100 ml. Since this leads to the exceedance of the water quality objectives, it was felt appropriate that an intervention should be undertaken. This was because it was reasoned that the concentration of *E. coli* demonstrated the presence of faecal material at a concentration that could foreseeably lead to a campylobacteriosis community infection rate greater than the public health target. It was assumed that this would occur should the population of animals or birds excreting the faeces into the storage become heavily infected with a human infectious *Campylobacter* strain. Options for intervention were put forward, such as disinfection downstream of the open storage or covering and vermin-proofing of the storage.

18.3 RECREATIONAL WATER

In contrast to the other two case studies, the recreational water group was asked to focus on the avoidance of acute gastrointestinal infection rather than a specific pathogen. The health target was set as 1 case of AGI per 80 exposures to a recreational water, along with no detectable outbreaks attributable to the recreational water during a summer bathing period. These levels were chosen to relate to present regulatory discussions and the draft guidelines relating to recreational water.

18.3.1 Trial study setting

The group decided that the best means of illustrating the framework was to apply it in a trial study format using real data. Beach A is a 4 km-long embayment with several compliance locations, set in a northern European location. One of the compliance locations passes the Guide standard and the others pass the Imperative criteria specified in the EC Bathing Water Directive. The bay takes most of the surface drainage from a community with a winter population of about 80,000 and a peak summer population of approximately 110,000. The local wastewater treatment works comprises an activated sludge plant with final settlement and ultra-violet (UV) disinfection of the treated wastewater producing an effluent of excellent microbiological quality. This effluent is discharged through a short outfall within the inter-tidal zone and represents nearly half of the freshwater input to the Bay.

In addition to the effluent treatment investments, considerable attention has been devoted to limiting the discharge of partially treated storm waters and untreated (but dilute) effluent from the combined sewerage system. The storage capacity has been designed to such a level that intermittent discharges following rainfall events have been virtually eliminated. Despite these measures, the beaches in the receiving waters do not all reliably achieve the Guide standards of the Bathing Water Directive and it has been proposed that stream inputs draining from agricultural catchments containing livestock and hinterland communities may be responsible.

18.3.2 Assess environmental exposure

In this example, data considered to be representative of the whole of Beach A are shown in Table 18.1.

Table 18.1. Microbiological data representative of Beach A

Environmental exposure data	
No. of samples	20
Geometric mean – faecal streptococci (/100ml)	12
Geometric mean – total coliform (/100ml)	215
Geometric mean – faecal coliform (/100ml)	71
Log_{10} standard deviation (faecal streptococci)	0.624
Log_{10} standard deviation (total coliform)	0.429
Log_{10} standard deviation (faecal coliform)	0.599
Bather number per year (i.e. no. of exposures)*	100,000

* Estimated based on the assumption that the average beach visitor population is 25,000 per fortnight over a 16-week summer period (i.e. 200,000 visitors per annum) and that 50% will bathe in the sea and swim once during their holiday.

Additional points, in relation to environmental exposure, are as follows:

- The effluent receives secondary (biological, activated sludge) treatment with tertiary UV disinfection. The geometric mean (GM) faecal coliform organism concentration in the discharged effluent after UV treatment is generally <50/100ml.
- The discharge point is in the inter-tidal zone via a short outfall.
- There is excellent storm-flow management of the combined sewerage system. High flow events are retained in the system through enhanced volume and storm retention. The spill frequency for this system is <1 in 5 years.
- There are significant inputs from diffuse sources causing episodic bacterial inputs to the bathing water which have been quantified. The potential exists for the application of a 'diffuse sources' prediction model to identify hot spots as part of a critical control point analysis.
- Streams draining an adjacent urban area through culverts may have cross connections causing minor but persistent microbiological loadings.
- There is an adjacent harbour with recreational craft that may produce intermittent bacterial discharges through inappropriate toilet discharges.

- The site has an average gull population and is not a major sea bird roosting area.

18.3.3 Assessment of risk

The assessment of risk is based upon the assumption that the visitor population is 25,000 per two-week period over a 16-week holiday period, with 50% of the visitors swimming in the recreational water. Previous epidemiological studies have investigated the risk of gastrointestinal infection from sea bathing and have demonstrated a dose–response relationship between faecal streptococci levels measured at chest depth and gastrointestinal illness (see Chapters 2 and 7). Using a disease burden approach (as outlined in Chapter 2) a risk of 43 cases of gastrointestinal infection/1000 population is derived. Over the summer season this equates to 4300 cases in 16 weeks or 269 cases of AGI attributable to sea bathing per week.

18.3.4 Acceptable risk and health targets

The health targets were based upon observing no outbreaks of illness attributable to a recreational water during a bathing season and one case (or less) of AGI per 80 exposures.

The assessment of risk, outlined in the previous section, found that the current level of illness was 4300 cases of illness from 100,000 bathing events. This equates to 1 case of illness in 23 exposures – clearly somewhat worse than the acceptable level.

In terms of outbreak detection, it is assumed that the local surveillance system will pick up a 15-fold increase above the background rate of gastrointestinal illness. The background rate of AGI is taken to be one case/person/year, which equates to 0.038 cases per two week period (or 0.019 cases/week) and hence a background rate of 480 cases per week in the visitor population (0.038 × 25,000 in a two-week period). The background rate in the local population is assumed to be the same and therefore adds an additional 1520 cases of illness/week (0.019 × 80,000), resulting in a total background rate of 2000 cases/week. The cases of illness attributable to sea bathing would, therefore, not be detected as an outbreak.

18.3.5 Risk management

An early stage in the risk management process is the setting of water quality objectives that are designed to allow the health target to be achieved. Following on from that, the harmonised framework requires that verifiable measures, interventions and key risk points (critical control points in HACCP terminology) should be defined. The approach taken by the recreational water group is outlined in the following sub-sections.

18.3.5.1 Water quality objectives

Based upon the desired health target, the water quality objective was set such that the faecal streptococci 95 percentile level should not exceed 50/100 ml during samples taken during the bathing season (for more details see Chapter 2).

18.3.5.2 Audit measures

- Microbiological concentrations in the bathing water and resultant compliance assessment.
- Final effluent quality monitoring for microbiological parameters and/or real time measurements of physico-chemical parameters in the effluent stream to facilitate instantaneous prediction of effluent microbiological quality.
- Combined sewer overflow (CSO) and storm spill volume monitoring and recording in real time.
- Diffuse source catchment modelling to predict the time and concentrations of diffuse source inputs.

18.3.5.3 Intervention measures

- Control of beach usage (time and/or space). Advisory notices could be posted to limit use to a specific area or restrict use for a specified time period.
- Adjustment of sewage treatment regime. The potential exists for plant optimisation and/or flow volume adjustment using the in-built storage to minimise faecal indicator loadings. It should be noted here that the science base describing the influence of management interventions within the activated sludge process on faecal indicator and enterovirus concentrations in the final effluent is very weak. It is stronger for interventions within the UV or microfiltration disinfection systems.
- Stream input quality. This may be adjusted by remedial solutions such as reed beds for small streams. Larger inputs would require some form of catchment management to control diffuse sources. Effort can best be targeted through the identification of pollution 'hot spots' informed by diffuse sources modelling.

- Compliance modelling. This could be used to predict the timing of elevated bacteriological concentrations for appropriate 'real time' intervention. This can take the form of simple statistical models that use antecedent conditions described by commonly available variables (such as sunlight, stream-flow, tidal state, wind speed and direction etc.) to provide a prediction of bacterial concentration at the compliance point.
- Removal of cross-connections between the sewage and storm-water systems. Such removal is as essential as the remediation of the catchment diffuse sources from agriculture. Almost universally, all surveys of inappropriate connections (i.e. of foul drains to surface-water drains and streams that are often culverted in urban areas) identify previously unknown problems.

18.3.5.4 Verification information

The need for the following verification information was identified. Points followed by an asterisk indicate that such data is currently already acquired in a number of countries.

- Compliance data.* Microbiological data acquired under the monitoring requirements of the Bathing Water Directive or other national/regional legislation or regulations.
- Spill volume data.* Acquired from telemetric monitors in the sewerage infrastructure. It is worth noting that some coastal sewerage systems are subject to marine water ingress causing siltation during high tides. This makes flow and level monitoring data difficult to interpret. In such circumstances, modelled flow data may be a more appropriate measure of CSO discharge, although such CSO modelling does require good spatial resolution and precision in the available rainfall data to drive the model.
- Effluent quality data.* Acquired though routine plant monitoring, this may not always include the microbiological parameters which should be placed on the suite of routine determinands.
- Stream water quality data. This is rarely available and, where data have been acquired, sampling is often biased towards low-flow conditions. The reason for this is the logistics of sampling within the working day and the requirement to get samples to a laboratory for analysis within the working week. However, samples collected under low flow conditions are almost worthless in characterising the impact of streams and catchment diffuse sources on bathing

waters because most of the bacterial delivery from streams and rivers occurs during high flow events.

- Beach usage rate data. Again, good quality data are rarely available for this parameter. Surveys offer one empirical means of data acquisition but broad estimates of usage from commonly acquired tourist data such as bed-night occupancy may be the best data available.

18.3.6 Public health status

Although there will be no detectable outbreaks of illness relating to the use of Beach A during a bathing season, the level of risk is currently greater than the acceptable level. The following table (Table 18.2) outlines a number of possible interventions along with the estimated health gain from each proposed measure. Such estimates could be used in a cost-benefit analysis, which may then lead to reconsideration of the level of acceptable risk.

Table 18.2. Interventions and health gain estimates

Intervention	Estimated health gain
Control of beach usage to prevent access to polluted water after episodic inputs from diffuse sources.	4000+ cases of AGI (assuming perfect prediction and control).
Adjustment of present sewage treatment regime.	Very little, as effluent quality is already very good.
Improvement of stream input quality to 'no effect' level.	Given the low effluent bacterial loadings – 4000+ cases of AGI. (A 'high flow' bacterial budget calculation is needed to underpin this calculation.)
Compliance model to predict the timing and/or spatial extent of peak bacterial indicator concentrations to facilitate appropriate advisory notices and/or beach zoning. (The utility of this approach should be judged on the basis of the model explained variance).	Given the low effluent bacterial loadings – 4000+ cases of AGI.
Remediation of all cross-connections in the hinterland catchments and adjacent urban areas.	Probably a small loading, maybe >300 cases of AGI.

18.4 WASTEWATER REUSE

The wastewater reuse group was asked to apply the framework to vegetable irrigation with wastewater. The reference pathogen was hepatitis A virus (for which there is no direct analytical method in environmental samples). The tolerable burden of disease was to be equivalent to 1 case of AGI per 10 people per year. Infection with hepatitis A was considered to be equivalent to 200 cases of AGI.

18.4.1 Trial study setting

The scenario chosen by the group related to furrow or flood irrigation of a lettuce crop with untreated wastewater.

18.4.2 Assessment of environmental exposure and risk

In order to determine environmental exposure a number of assumptions were made in relation to the concentration of hepatitis A virus in faeces and wastewater and also the residual level of wastewater on the lettuce crop. The approach taken was based upon inputs from epidemiological studies and risk assessment models. For the purposes of the exercise the assumptions shown in Table 18.3 were made (based loosely on the literature).

Table 18.3. Assumptions and data inputs

Data required	Assumptions
Concentration of hepatitis A in faeces	10^4/g of faeces
N_{50} (median infectious dose)	0.5 g of faeces
Wastewater production	150 litres/person/day 5.5×10^4 litres/person/year
Faeces production	250 g/person/day 9.1×10^4 g/person/year
Prevalence of hepatitis A shedding	2% of the population – i.e. 0.02
Duration of shedding	7 days/year – 0.0192
Residual water on the lettuce crop	0.11 ml/g of lettuce
Lettuce consumption	100 g/person/day

These assumptions lead to an estimate of the daily hepatitis A intake from lettuce consumption. The virus production can be estimated by multiplying faeces production by prevalence in the population, duration of shedding and by the concentration of hepatitis A in the faeces.

$$9.1 \times 10^4 \times 0.02 \times 0.0192 \times 10^4 = 3.5 \times 10^5 \text{ hepatitis A virus/person/year}$$

The concentration of hepatitis A virus in wastewater is, therefore, calculated from the amount of virus produced per person and amount of wastewater production.

$$\frac{3.5 \times 10^5}{5.5 \times 10^4} = 6.4 \text{ hepatitis A virus/litre (i.e. } 6.4 \times 10^{-3} \text{ /ml)}$$

The actual daily intake of hepatitis A virus by lettuce consumption is a function of the concentration of the virus in the wastewater per ml, the volume of wastewater in the lettuce consumed and the per capita lettuce consumption (assuming no removal of pathogens through washing of the lettuce prior to consumption).

$$6.4 \times 10^{-3} \times 0.11 \times 100 = 7 \times 10^{-2} \text{ hepatitis A virus/person/day}$$

18.4.3 Acceptable risk and health targets

The acceptable risk was defined to the group as being 1 case of AGI per 10 people per year, with the AGI equivalent for hepatitis A being 200. This, therefore, equates to 0.005 cases of AGI per 10 people/year or 5×10^{-4} cases/person/year.

Since exposure is based on intake, it is also necessary to convert the acceptable level of illness to an intake. The acceptable level of hepatitis A virus intake is a function of the acceptable risk, the N_{50} value and the concentration of hepatitis A virus in faeces. If N_{50} is expressed as concentration of hepatitis A virus (0.5×10^4), this can be related to acceptable daily intake (ADI) as follows:

$$ADI - -N_{50} - \ln\left[1 - \frac{5 - 10^{-4}}{365}\right]$$

where 5×10^{-4} is the acceptable annual risk and 365 is the number of days per year. Acceptable daily intake thus equals 6.9×10^{-3} hepatitis A virus/person/ day. It can be seen from this calculation that the assessment of exposure is an order of magnitude greater than the level of acceptable risk and therefore requires a risk management strategy that would yield at least a 10-fold reduction in hepatitis A virus intake.

18.4.4 Risk management

It has been found for this hypothetical, yet realistic, example that the actual risk of contracting hepatitis A infection is somewhat greater than that deemed acceptable. In human wastewater reuse there are four measures that may be implemented individually or combined to reduce the risk of transmitting excreta-related infections:

(1) Treatment of waste.
(2) Choosing suitable methods of waste application.
(3) Restricting certain crops.
(4) Improved personal and domestic hygiene.

It has been assumed in this example that there is no legal restriction on crops to be grown or, if such restrictions do exist, either farmers do not respect them and/or they are not being enforced. Hence, as is often the case, particularly in developing countries, peri-urban farmers have chosen to grow vegetables. In an urban setting, these crops are likely to yield the highest cash income and contribute greatly to food security, for both the farmer and his family as well as for the urban populace.

The thrust for risk management, therefore, rests on treating the wastewater and adopting irrigation methods that reduce the risk of contaminating the crop. Improved hygiene practices would primarily help the farmer and his family.

18.4.4.1 Drip irrigation

Drip irrigation is likely to lead to a 100-fold (2-log cycle) reduction in the pathogen load contaminating irrigated vegetables when compared with spray or flood irrigation. Hence, if technically and financially feasible for the farmer, this measure alone would lower the risk to the consumer to below the acceptable level.

18.4.4.2 Wastewater treatment

There exist several treatment options to achieve a reduction in exposure to hepatitis A virus. In reality, the choice of a particular option will depend upon socio-economic, financial, technical and institutional criteria. Partial treatment in a waste stabilisation pond scheme (consisting of a facultative pond or an aerobic pond followed by a facultative pond) as well as conventional secondary treatment are both likely to comfortably satisfy the stipulated 10-fold reduction in hepatitis A virus levels. An upflow anaerobic sludge blanket clarifier, a new treatment option, currently popular in Latin American countries, may also satisfy the required reduction, although it still requires research on specific pathogens removal. Irrespective of the treatment option chosen, the expected/required performance can only be achieved if the systems are adequately designed, and properly operated and maintained.

18.4.5 Public health status

The final stage in the first iteration of the framework is an examination of 'public health status' as a verification that the measures put in place are adequate and appropriate. In the case study scenario it was found that with the introduction of partial waste treatment and drip irrigation, levels of hepatitis A fell to within the acceptable level within the urban community. There was a suggestion, however, that levels of hepatitis A infection within the farming community remained high and additional measures may be required to target the health of this group.

18.5 DISCUSSION

The trial examples, drawn from each guideline area using realistic hypothetical scenarios demonstrated that the proposed harmonised framework is a valuable tool. Data needs and availability vary between the three guideline areas and this was clear from the types of data adopted and the specific approaches taken by the individual groups. However, in each case the framework was sufficiently inclusive to allow the use of the best data available and also acted to guide the groups through the process in a logical fashion. The need for the framework to be seen as a series of iterations, rather than simply a one-off exercise, was demonstrated by each example. Given the short period of time available for these group discussions, elaboration on the 'public health status' aspect was limited and none of the groups was able to consider their scenario in terms of public health more generally. In terms of hepatitis A infection, for example, it may have been constructive for the wastewater use group to 'examine' the likelihood of hepatitis A infection from consumption of contaminated shellfish or even recreational water use.

Index

(Note that tables are indicated by **bold** page numbers and illustrations by *italics*).

absolute risk 238
acceptable risk *see* risk, acceptable
acquired immunity, determinant of disease 236–7
Africa, sanitation (in 1990 and 2000) **99**
animal faeces (warm-blooded), microbiological water quality indicators 294, **295**
ascariasis
  Israel 141–2
  Mexico 150–5, 201
  quality audit (QA) **203**
  wastewater reuse, quality audit (QA) **203**
attributable risk 238
Australia
  Melbourne case studies 274–6
    drinking water quality 156
  Sydney Water case studies **267–8**, 272–4
    hazards and ranking scheme **273**

Bacillus spores 298
bacterial probability density function, recreational water quality 35–6
bacterial regrowth, endemic waterborne disease 72
bacteriophages 295–7
  defined 291
  limitations as indicators 298–9
  major groups of indicator coliphages **296**
  phages in water environments 296
Bacteroides fragilis
  bacteriophages 291
  detection 302
bargaining
  bureaucratic model 219
  principal agent model 218
  Sobel–Takahashi multi-stage model 218
bathing/beaches (sea/freshwater)
  bather shedding 280
  primary classification of beaches 281–2, **282**
  quality, monitoring and assessing, management strategies 281–2, **282**
  risk, economic approach 211–14
  see also recreational water quality
beneficiaries, identifying for cost recovery 343–4
benefits, economic evaluation of water and sanitation interventions 340–4, **342**

bifidobacteria 294
  defined 291
BMJ guidelines, economic evaluation **334–5**
Brazil, incidence of endemic GI disease **78**

California standards, coliforms 23–4
Campylobacter infection
  disease outbreak causes, UK 263
  hypothetical study, drinking water, guideline development in practice 394–401
  Netherlands, DALYs **55**
  Sweden 122–4
Canada
  drinking water case studies, epidemiology/risk assessment 155–6
  French–Canadian population, seroprevalence **68**
cancer, risk, acceptable 56
case-control studies, epidemiology 144
cause-effect 248–9
Chadwick, Edwin, on public health issues (historical) 228–9
chemical pollutants, TDI 19
chemical risk, vs microbiological risk 258–9
chemical risk paradigm *see* risk assessment
China, diarrhoeal disease study 79–80
cholera epidemic, Peru, economic evaluation 343
chromogenic substances 299
clostridia
  Clostridium perfringens, defined 290
  indicator development 294–5
  sulphite-reducing clostridia 281, 290
cohort studies 143–4
coliforms, faecal (FC) 289–93
  bacteriophages (phages) 291
  Bacteroides fragilis bacteriophages 291
  bifidobacteria 291
  California standards 23–4
  Campylobacter infection **55**, 122–4, 263
  case studies, recreational water 146–9
  coliphages 291, 296
  defined **290**
  defined substrate methods 293
  Escherichia coli (E. coli) 54, 281, 290
  Guidelines for Water Reuse, USEPA/USAID 370–3
  history 289
  identification schemes 289–93
  Klebsiella 290, 292
  membrane filtration method 292
  most probable number method 291–2
  survival in different media **93**
  TC quality standard 23
  thermotolerant coliforms 290
  wastewater, guidelines 28–31
coliphages 291, 296
  defined 291
  F-RNA 295, 297
  indicator **296**
communication *see* risk communication
community-managed drinking water supplies
  application of framework 386–8
  management by water source committee 387
community-managed waste management, application of framework 388–9
contingent valuation 349–50
coprostanol 297
cost-effectiveness, and willingness-to-pay **338–9**
Critical Control Points (CPs) 269
  corrective actions 385
  risk management 382–5
crops, pathogen and indicator survival **93**
cross-sectional studies 143
Cryptosporidium case studies in risk assessment 55, 166–73, **263**
  aetiology, management deficiencies **262**
  caveats 172
  identification 300
  input exposure variables 167–70
    mean oocyst levels estimated by different methods **170**
    oocyst levels, reservoir samples **168–9**
  Milwaukee episode (1993) 67, 72
  Monte Carlo simulation 166, 171–2
  opportunity cost analysis 213
  results and point estimates 171, **189**
    computed point estimates for daily risk **171**
  summary of trials, daily infection risk **172**
cultural theory
  and acceptable risk 215–17
  fright factors 215–16
  media interest 216–17

data collection, BMJ **334**
decision-making, environmental health 249–52
defined substrate methods 293
developing countries
  DALYs (1990) **50**
  diarrhoeal disease 62–3, 78–80
  drinking water treatment systems **362**
  economic evaluation
    interventions in water and sanitation 331–57

water and sanitation improvements **98**
global burden of disease (1990) study **50**
guideline implementation 360–3, **361**
problems with setting standards 363–5, **364–5**
incidence of endemic GI disease **76–7**, 74–80
tourism 379
wastewater treatment systems **362**
diarrhoeal disease
studies **62–3**, 79–80
travellers' diarrhoea **78**
see also coliforms; infections; streptococci; *specific organisms*
disability
classes and indicator diseases **47**
measuring 46–7
see also DALYs
Disability Adjusted Life Years (DALYs) 52–6
causes of DALYs 96–88
definitions 45
level of acceptable risk 56
lost years of life (YLL) 45
years lived with disability (YLD) 45–6
developed and developing regions (1990) **50**
GI disease 52–4
chain model 52–4
global burden of disease (GBD) study **50**, 43–59
guidelines, use in derivation 52–6
infection with thermophilic *Campylobacter*, Netherlands **55**
integrating health effects of exposure
one agent 54–5
several agents 55
disease burden approach
risk, acceptable 35–8, 56, 211
see also global disease burden; infectious disease
disease process, conceptual model **176**, 181
DNA sensing 303
DNA/RNA probe-based rRNA targets 303
dose–response analysis 163–6, 187–8
best-fit dose–response parameters **165**
chemical risk paradigm 163–5
classical risk assessment framework 257
exponential and beta-Poisson functions **164**
population 5
drinking water quality
case studies
epidemiology, risk assessment 155–6
rotavirus disease **180**
Cryptosporidium oocysts, reservoir raw water samples **168**
hypothetical study of *Campylobacter* sp. 394–401
acceptable risk and health targets 396–7
assessment of environmental exposure 394–6
public health status 400–1
risk management 397–400
trial study setting 394
inequality of risk **221**
ingestion, lognormal distribution model 167
and microbiological risk **259**, **263**
pollution by storm run-off **260**
reverse-osmosis filters, in intervention studies 144–5, 155
under-reporting of infectious disease **131**
USA (1995-1996) **127**
see also quality audit (QA)
drinking water quality guidelines (GDWQ) 18–22
European guidelines 18
faecal indicator organisms 20–1
indicators **4**
and international/national guidelines 132–3
operational/national guidelines 21–2
pathogens, reviewed in GDWQ **20**
drinking water treatment systems
application of framework 386–8
developed vs developing countries **362**

ecological/correlational studies 142–2
economic evaluation and priority setting, interventions in water and sanitation 331–57
application of guidelines 340–55
benefit inclusion 340–4
benefits to society of interventions **342**
identifying beneficiaries for cost recovery 343–4
benefit valuation 346–50
contingent valuation 349–50
household production function 348
market price of goods and activities 346–8
methods of valuation **347**
revealed preferences 348–9
BMJ guidelines **334–5**
comparisons with health interventions 332
cost effectiveness and willingness to pay for services **338–9**

economic evaluation and priority setting, interventions in water and sanitation
cost effectiveness and willingness to pay for services (*cont'd*)
cost inclusion 344–5
categorisation of health interventions **345**
discounting future costs and benefits 350–2
present value of future incomes and discount rates **351**
disease costs to society 82–3
effectiveness 352–4
framework 333–6, 340–54
identifying beneficiaries for cost recovery 343–4
risk of bathing vs costs of treatment 211–14
uncertainty 352–4
willingness-to-pay 336–7, **338–9**
ELISA 300
endemic level of disease, defined 234
enterococci 290
defined 290
environmental exposure to disease
assessment 10, 188
chemical risk paradigm 162–3
drinking water, avoidance of acute GI infection 394–401
harmonised risk assessment 10
measured 395–6
predictive 395
risk 396
environmental health, decision-making 249–52
epidemic, defined 234
epidemiology and risk assessment 135–60
analytical studies
case-control studies 144
cohort studies 143–4
cross-sectional studies 143
ecological or correlational studies 142–2
relationship between exposure and disease 139
case studies 146–56
drinking water 155–6
non-exposure-related risk factors for gastroenteritis **148**
recreational water 146–9
wastewater reuse
exposure and degree of storage **152**
Mexico 150–5, 201
descriptive and analytical 241–2
elements of study 136–41
epidemiological risk, types 238–9
evaluation of chance and bias 139–41
formulation of question or hypotheses 137
measurements of exposure and disease status 138–9
policy-making 242
practitioners' skills 238–42
selection of exposure indicators 138
selection of study populations 137
setting or evaluating microbiological guidelines 145–6
summary/discussion 157–8
surveillance, public health 240–1
types
analytical 142–4
descriptive 141–2
experimental or intervention studies 144–5
Escherichia coli (E. coli) 54, 281, 290
confirmation 291–2
defined 290
O157:H7 300
toxins 54
EU, disease outbreaks, (1986-1996), under-reporting **131**
EU guidelines
drinking water quality 18
recreational water quality, microbiological guideline design 33
excreta disposal/excreta management *see* sanitation
excreta-related infections *see* faecal indicator microorganisms; infections
experimental studies, epidemiology, risk assessment 144–5
extrapolation uncertainty 218

faecal indicator organisms 376–8
animal faeces (warm-blooded) 294, **295**
current applicability 303–4
drinking water quality guidelines (GDWQ) 20–1
faecal–oral pathogens, transmission routes **95**
key 290
transmission routes, pathogen and indicator survival in different media **93**
see also coliforms, faecal (FC); streptococci, faecal (FS)
faecal sludge treatment, technical options 106
faecal sterol biomarkers 297
faeces of warm-blooded animals, microbiological water quality indicators 294, **295**
FISH (fluorescence in situ hybridisation) 302

framework for guideline development
community-managed drinking water supplies 386–8
community-managed waste management 388–9
economic evaluation, interventions in water and sanitation 333–6, 340–54
elements and implementation, harmonised assessment of infection risk 9–16
hypothetical studies, Stockholm 393–410
drinking water, study of *Campylobacter* sp. 394–401
recreational water, study of avoidance of acute GI infection 401–6
wastewater reuse, study of hepatitis A infection 406–10
quality audit (QA) **191**, 191–9
risk assessment 256–8
`framing effect', acceptable risk 215
France, waterborne disease study 69, 72–3
freshwater, pathogen and indicator survival **93**
fright factors, cultural theory 215–16

gastrointestinal disease *see* infections
gene sequence based methods 301–2
giardiasis
disease outbreak causes, UK 263
identification 300
risk, US goal 11, 209
transmission 173
global burden of disease (GBD) study 43–59
GBD estimate, applications 51–6
international guidelines 57–8
major outcomes of study 47–51
causes of DALYs by developed and developing regions (1990) **50**
causes of death/GBD estimates (1990 and 1998) **48–9**
disease and injury attributable to selected risk factors **50**
use of DALYs in guideline derivation 52–6
measuring population health 44–7
measuring disability 46–7
years of life lost 45
years lived with a disability 45–6
problems of assessing disease burden in relation to water quality 56–7
groundwater pollution, risks from sanitation 105
guideline development framework
hypothetical studies 393–410
discussion 410
drinking water, study of *Campylobacter* sp. 394–401
recreational water, study of avoidance of acute GI infection 401–6
wastewater reuse, study of hepatitis A infection 406–10
guideline implementation 359–74
case study 370–3
background 370
methodology 371
results 372
risk of infection and disease from various pathogens **372**
cost implications 369
current position 17–41
and international/national guidelines 38
developing countries
compliance with standards **361**
vs developed 360–3
problems with setting standards 363–5, **364–5**
selecting water and wastewater treatment systems **362**
stepwise implementation of standards 366–9
and international guidelines/national regulations 373
principle of equity 369
standards, improvement of water quality **367**
wastewater, faecal coliforms (FC) 28–31
see also drinking water; recreational water; wastewater and excreta use
Guidelines for Water Reuse, USEPA/USAID 370–3
Guillain–Barré syndrome 54, 55

harmonised assessment of risk 1–16
harmonised framework for guideline development *see* framework for guideline development
hazard analysis
and acceptable risk, microbiological quality 378–9
risk management plan 381–5
Hazard Analysis and Critical Control Points (HACCP) 3, 162, 187, 265–70
principles **265**
scoring risks, Sydney Water, Australia **267–8**, 272–4
worksheet **267**, **275**

health education
  and behavioral modification 246
  example of risk communication 319
health impact assessments (HIAs) 250–1
health targets, benefits **13**
hedonic pricing 348
helminths
  standards 28, 32
  survival in different media **93**
hepatitis A
  case study in wastewater irrigation 372
  wastewater reuse, guideline development framework, hypothetical study 406–10
hepatitis E 378
hypothetical studies *see* risk, acceptable

immunity, acquired 236–7
immunomagnetic separation, and other rapid culture based methods 301
index organism, defined 288
indicator *see* faecal indicator organisms
infections (general/gastrointestinal) 61–89
  aetiology 62–3
  chain model **53**
  control by sanitation 91–115
  costs to society 82–3
    see also economic evaluation
  endemic waterborne disease in industrialized countries 68–74, 81–2
    health significance of bacterial regrowth 72
    intervention studies 70–2
  environmental health decision-making 249–52
  exposure indicators, epidemiology and risk assessment 138
  faecal sludge, *see also* wastewater and excreta in agriculture and aquaculture
  faecal streptococci and enterococci 293–4
  harmonised assessment of risk 1–16
    acceptable risk 10–13, 207–26
    expanded version **9**
    framework elements and implementation 9–16
    further development 16
    future guidelines 6–9
    indicators and good practice requirements by guideline area **4**
    necessity 4–6
    public health status 15–16
    risk management 13–14
    World Health Organization guidelines on water quality 2
  incidence
    developing countries 74–80
    endemic disease 74–9
    industrialized countries 63–74, 81–2
  indicator diseases **47**
  inequality of risk **221**
  and international guidelines/national regulations 112
  and international/national guidelines 83, 158
  interventions 247–50
  outbreak, defined 234
  risk factors, non-exposure-related **148**
  routes of transmission 231–4
  surveillance and waterborne outbreaks 117–34
    Sweden 118–24
    USA 124–8
  transmission routes 92–6, 231–2
    faecal–oral pathogen transmission routes **95**
  under-reporting 130–2
  water-related diseases **232**
information
  trusted sources **324**
  see also risk communication
Inoviridae 296 **298**
International Life Sciences Institute (ILSI) 177–8
intervention measures
  epidemiology, risk assessment 144–5
  public health 245–50
  risk management 404–5
Israel, *Ascaris* infection 141–2

Klebsiella infection 290, 292
  detection 300

latrines *see* sanitation, technical options
Legionella infection, detection 300, 302
Leviviridae 296 **298**
Lubbock, health effect study 68

mathematical modelling (quantitative risk assessment) 242–3
media interest
  cultural theory 216–17
  triggers **321**
membrane filtration method, coliforms (FC) 292
Mexico, wastewater reuse case studies, epidemiology 150–5, 201
microbiological methods 299–303
  fast detections using chromogenic substances 299

microbiological methods (*cont'd*)
  future developments 302–3
  gene sequence based methods, FISH and PCR 301–2
  immunomagnetic separation (IMS)/culture and other rapid methods 301
  monoclonal and polyclonal antibodies 300
  most probable number (MPN) method 291–2
microbiological risk, drinking water 259–64
  vs chemical risk 258–9
  multiple barriers 261
  origins 259–60
  outbreak aetiology 262–4
    cryptosporidiosis outbreaks, management deficiencies **262**
    scenarios **263**
  pollution by storm run-off **260**
  quantitative assessment (QMRA) 162
  sources **259**
microbiological water quality indicators 287–314
  current applicability of faecal indicators 303–4
  emerging microbiological methods 299–303
  indicator development 289–97
    bacteriophages 295–7
    definitions, indicator and index microorganisms **288**
    faecal sterol biomarkers 297
    sulphite-reducing clostridia and other anaerobes 281, 190, 294–5
  and international/national guidelines 304
  pathogen models and index microorganisms 297–9
  see also coliforms; streptococci
microbiological water quality regulation 378–89
  defining hazards and acceptable levels of risk 378–9
  drinking water supply 386–8
  and international guidelines/national regulation 377–8, 389
  objectives 376–8
  risk management 380–5
    critical control points 269, 382–5
    verification and auditing 385
  wastes management 388–9
Microviridae 296 **298** , 292
monitoring systems, matching critical control points 384–5
monoclonal and polyclonal antibodies 300
Monte Carlo simulation, risk assessment 166, 171–2
most probable number (MPN) method 291–2
multiple- tube fermentation (most probable number method) 291–2
Myoviridae 296 **298**
nematodes
  cohort studies 143–4
  standards 28
  survival in different media **93**
  see also ascariasis
Netherlands, *Campylobacter* infection, DALYs **55**
null hypothesis 137

opportunity cost, defined 213
outbreak
  defined 234
  types 234–5

pathogens
  inactivation 108–9
  pathogen–host properties **25**
  survival in different media **93**
  survival on soils and crops, warm climate **26**
PCR (polymerase chain reaction) 301–2
Peru, cholera epidemic, economic evaluation 343
phages *see* bacteriophages; coliphages
Philippines, diarrhoeal disease study 80
Plesiomonas shigelloides 127–8
Podoviridae 296**298**
policy-making, epidemiology, risk assessment 242
political resolution of risk issues 217–22
  bargaining, models 218–19
  pre-defined probability approach 208–10
  public acceptance 214–17
  satisficing 220
  stakeholder inequality 220–2
    inequality of health risks **221**
political will, interventions in public health 246
polyclonal antibodies 300
population dose–response analysis 5
population health measurement 44–7
  see also DALYs; risk, acceptable
poverty, as determinant of disease 236
preventive medical care, public health 245
process adequacy (validation), critical control point, risk management 383–4
process indicator, defined 288
protozoan oocysts
  drinking water ingestion case studies 166–73
  see also Cryptosporidium

public health 227–54
acceptance of risk 214–17
see also risk, acceptable
biological and physical sciences 243–4
defined 228
demography 244
and international/national guidelines 252–3
interventions 245–50
classification 245–6
control of environment 246
cultivating political will 246
health education and behavioral modification 246
preventive medical care 245
waterborne disease, cause-effect 248–9
metaphor for surveillance 229–30
nature and determinants of disease 230–7
acquired immunity 236–7
determinants of ill health 235–7
endemic/epidemic disease and outbreaks 234–5
environmental exposure 235–6
poverty 236
pre-existing health 236
routes of transmission 231–4
classification of water-related diseases **232**
practitioners' skills and tools 237–44
epidemiology, descriptive and analytical 238–42
mathematical modelling (quantitative risk assessment) 242–3
policy making 242
social and behavioural sciences 244
surveillance 240–1
risk acceptance, political resolution 15–16, 214–17
setting standards 223–4, 250–2
public information, trusted sources **324**

Quality Adjusted Life Years (QALYs), costs 213
quality audit (QA) 185–206
applications 201–3
case study 201–3
stages, water supply pathway **202**
wastewater reuse, and ascariasis **203**
and international/national guidelines 204–5
outline QA, studies on drinking water consumption **200**
proposed framework **191**, 191–9
method 195
observation 194–5
outline **193–4**
output 196–7, 199–201
peer review 197
validity 197–8
science in risk estimates 189–90
uncertainty 187–9
Cryptosporidium in tap water **189**
quality indicators *see* microbiological water quality indicators
quantitative microbiological risk assessment (QMRA) 162
quantitative risk assessment (QRA), public health 242–3

recreational water quality
bacterial probability density function 35–6
case studies 146–9
coliforms, faecal (FC) 146–9
epidemiology, risk assessment 146–9
hypothetical study of avoidance of acute GI infection
acceptable risk and health targets 403
assessment of environment exposure 402
microbiological data **402**
public health status **406**
risk management 403–5
trial study setting 401
indicators **4**
marine 280–1
microbiological guideline design 33–8
acceptable risk 35–8
combining epidemiological and environmental data 35–7
current position 32–8
dose–response curve, faecal streptococci and gastroenteritis **35**
epidemiology 33–4
estimated disease burden **37**
probability density function of faecal streptococci **36**
water quality data 34–5
monitoring and assessment
classes of health risk **277**
human faecal contamination 278–80
management strategies 276–84
microbiological quality 33, 280–1
new approach 276–8
principal sources of human faecal contamination 278–80
riverine discharges 279–80
sewage discharges 278–9, **279**

monitoring and assessment (*cont'd*)
under-reporting of disease, outbreaks (1986–1996) **131**
and wastewater contamination 276–8
see also bathing/beaches
relative risk 238
reverse-osmosis filters, in intervention studies 144–5, 155
ribosomal RNA (16S rRNA), FISH 302
risk
absolute, attributable, relative 238
defined 256
pathways, omission, examples 198
risk, acceptable 10–13, 207–26
vs accepted 210–11
bargaining, Sobel–Takahashi multi-stage model 218
of cancer
`essentially zero' level as gold standard 208
vs infection 56
and cultural theory 215–17
currently tolerated approach 210–11
definitions 56, 208–9
disease burden approach 35–8, 211
economic approach 211–14
`framing effect' 215
and hazards, microbiological quality 378–9
hypothetical studies
drinking water, study of *Campylobacter* sp. 394–401
recreational water, study of avoidance of acute GI infection 401–6
wastewater reuse, study of hepatitis A infection 406–10
and international/national guidelines 225
political resolution of issues 217–22
pre-defined probability approach 208–10
public acceptance 214–17
setting standards 223–4, 250–2
uncertainty **186**, 187–9, 217
risk assessment 161–83
background 161–2
chemical risk paradigm 162–6, 247
best-fit dose–response parameters **165**
dose–response analysis 163–5
exponential and beta-Poisson dose–response functions **164**
exposure assessment 162–3
hazard assessment 162
risk characterisation 165–6
risk management 166
classical framework 256–8
dose–response assessment 257
elements and implementation 9–16
hazard identification 256
risk characterisation 257–8
Cryptosporidium case studies 55, 166–73
dynamic epidemiologically-based model 173–4
environmental exposure assessment 10
expanded framework **9**
Health & Safety Executive (HSE), UK 208–9
health targets, benefits **13**
and international/national guidelines 181
rotavirus disease process case study 174–80
summary/discussion 180–1
see also quality audit; epidemiology
risk characterisation
chemical risk paradigm 165–6, 247
management strategies 257–8
Monte Carlo approach 166, 171–2
rotavirus disease process case study 179–80
risk communication 317–30
communication techniques 325–27
empathy 326
silence 327
uncertainty 326–7
educuation example 319
evaluation 327–8
functions 317–18
long term trust 324–5
media interest
cultural theory 216–17
triggers **321**
risk management cycle **317**
risk perception **323**
situation management 319–24
and WHO guidelines 328–9
risk management 319–24, 255–86, 397–400
analytical verifications 399–400
anticipating concerns 323
approaches 380–1
assessment, recreational water quality 276–84
audience focused communication 321
audit measures 404
Australia 272–6
Melbourne 274–6
Sydney **267–8**, 272–4
basic control 397
chemical vs microbiological risk 258–9

risk management (*cont'd*)
  chemical risk paradigm 166, 247
  classical risk assessment framework
    dose–response assessment 257
    exposure assessment 257
    hazard identification 256
    risk characterisation 257–8
  critical control point 382–5
    corrective actions 385
    identification 382–3
    monitoring 384–5
    process adequacy (validation) 383–4
    verification and auditing 385
  Cryptosporidium case study 166–73
  current condition 398
  cycle, risk communication **317**
  definitions 256–8
  harmonised assessment 13–14
  and international/national guidelines 284
  intervention measures 404–5
  key risk points and audit procedures 398–9
  long term trust 324–5
  managing people and processes 271–2
  media triggers **321**
  microbiological quality regulation 380–5
  microbiological risk 259–64
  negative feedback and outrage 322–3
    risk perception **323**
  objectives 404
  origins 258–64
  public information sources 324
    trusted sources of impartial advice in UK **324**
  recreational water quality assessment
    classes of health risk **277**
    microbiological quality 280–1
    primary classification of beaches 281–2, **282**
    principal sources of human faecal contamination 278–80
  systems approach 265–71
    critical control points 268–9
    critical limits 269–70
    flow chart and flow chart verification 266
    hazard analysis **265**, 267–8
    monitoring and corrective actions 270–1
    record keeping, validation and verification 271
    team skills and resources 266
    water description and use 266–7
  theory/reality 264–5
  verification information 405
  water quality objectives 397–8
  see also Hazard Analysis and Critical Control Points; risk communication
riverine discharges, and recreational water quality 279–80
rotavirus disease process case study 174–80, 372
  average daily prevalence, children exposed to drinking water contamination **180**
  conceptual model **176**
  implementation 177–80
    International Life Sciences Institute (ILSI) 177–8
    problem formulation and analysis 178
    risk characterisation 179–80
    schematic application of ILSI framework **178**

Salmonella infection
  case study in wastewater irrigation 372
  detection 300
  disease outbreak causes, UK 263
saltwater, pathogen and indicator survival **93**
sanitation 96–111
  containment 109–11
    selected excreta management and treatment options **110–11**
  diarrhoeal disease, water and sanitation improvements **98**
  global coverage (in 1990 and 2000) and Africa **99**
  health and poor sanitation 96–8
  pathogen inactivation 108–9
    organism survival in faecal sludge **109**
  sanitation coverage 98–100
  scenario, low-income neighbourhood 99–100
  sewerage system, costs 107
  technical options 102–8
    containment efficiency **109**
    conventional waterborne, disadvantages 107
    faecal sludge treatment 106
    groundwater pollution risks from on-site sanitation 105
    off site (sewered) sanitation 106–7
    on-site installations 102–4
    septic tanks **104**
    VIP latrine and double vault no-mix latrine **103**
    wastewater treatment 107–8
satisficing, political resolution of risk issues 220
selection bias 137
seroprevalence, selected enteric pathogens, French–Canadian population **68**

sewage discharges, recreational water quality 278–9, **279**
sewage treatment *see* wastewater treatment
shellfish, faecal–oral pathogen transmission routes **95**
small round structure virus (SRSV) 128
Snow, John, cholera in London 228
Sobel–Takahashi multi-stage model, bargaining 218
soils, pathogen and indicator survival **93**
soils and crops, pathogens **26**
stakeholder inequality, political resolution of risk issues 220–2
sterol biomarkers 297
storm run-off, microbiological risk **260**
streptococci, faecal (FS) 290
 absolute numbers 37
 defined 290
 and enterococci 281–2
 gastroenteritis **35–6**
 thermotolerant/faecal streptococci ratio 294
study design, BMJ **334**
sulphite-reducing clostridia, indicator development 281, 290, 294–5
surveillance systems
 epidemiological, public health 240–1
 waterborne infections (general) 117–34
Sweden 118–24
 clinical and laboratory surveillance 121–4
  diagnosis **122**
  initially reported and actual numbers **123**
  risk factors identified **124**
 recognition of outbreaks 120–1
 water sampling 121
 waterborne disease 1980-99 118–20, **119**
Siphoviridae 296, **298** , 292

tap water *see* drinking water
tolerable daily intake (TDI), chemical pollutants 19
tourism
 developing countries 379
 travellers' diarrhoea **78**
transmission routes
 GI infections 231–4
  giardiasis 173
travellers' diarrhoea, developing countries **78**
turbidity
 bacterial, endemic waterborne disease, industrialized countries 72
 Milwaukee, USA 72

UK
 GI disease, studies 64
 Health & Safety Executive (HSE), risk assessment 208–9
uncertainty **186**, 187–9
 extrapolation uncertainty 218
 risk communication 326–7
 types 217
urine as crop fertilizer 102
USA
 drinking water quality, (1995-1996) **127**
 FoodNet site 65–6
 recreational water quality, microbiological guideline design 33
USA disease outbreaks
 Cleveland study of GI disease 63–4
 Cryptosporidium
  Milwaukee (1993) 67, 72
  New York 166–73
 drinking water (1995-1996) 132–3
 management 128–30
 Minnesota study 127
 Philadelphia study 73
 Tecumseh study of GI disease 63–4
 waterborne outbreaks (WBDOs) 124–8, **127**
USEPA (Environmental Protection Agency)
 standards 199, 208, 210
 and World Bank
  case study in wastewater irrigation 370–3
  recommendations 372
utility, defined 213

validation, critical control point, risk management 383–4
Vibrio cholerae, case study in wastewater irrigation 372
viruses
 enteric 67–8
 human enteric 297–9
 seroprevalence, French–Canadian population **68**
 small round structure virus (SRSV) 128

wastewater and excreta in agriculture and aquaculture 22–32
 as benefit 24–5

wastewater and excreta in agriculture and aquaculture (*cont'd*)
case studies
epidemiology 149–55
Mexico 149–55, 201
community management, application of framework 388–9
derivation of WHO (1989) guidelines 27–32
controversy on wastewater reuse 32
incorporation into standards 31–2
main features 30–1
model of reducing health risks **30**
recommended microbiological quality **28**
effect of exposure and degree of storage **152**
indicators **4**
irrigation, World Bank and WHO, case study 370–3
pathogen–host properties **25**
and recreational water quality 276–8
reuse case studies, epidemiology, risk assessment 149–55
WHO, reuse guidelines, history 23–7
see also quality audit (QA); wastewater treatment
wastewater reuse, hypothetical study of hepatitis A infection
acceptable risk and health targets 408
assessment of environmental exposure and risk 407–8
assumptions and data inputs **407–8**
drip irrigation 409
public health status 409–10
risk management 408–9
trial study setting 407
wastewater treatment 107–8, 409, 278–9, **279**
systems, developed vs developing countries **362**
waste stabilization ponds (WSP) 24, 27
organism survival periods **109**
recommended microbiological quality guidelines 24, 27
see also wastewater and excreta in agriculture and aquaculture
Water Decade 5
water quality indicators *see* microbiological water quality indicators
water source committee, community-managed drinking water supplies 387
waterborne outbreaks (WBDOs) *see* USA
WHO
carcinogens, acceptable risk defined 208
drinking water quality guidelines (GDWQ) 18–22
estimates of world disease (1996) 62
health, defined 230
Health Guidelines for the Use of Wastewater 370–3
Stockholm, hypothetical studies, harmonised framework for guideline development 393–410
wastewater reuse guidelines 23–7
water quality, harmonised assessment **2**
and World Bank, case study in wastewater irrigation 370–3
willingness-to-pay, studies 336–7, **338–9**
World Bank, and WHO, case study in wastewater irrigation 370–3

years of life lost (YLL), definitions of DALYs 45
years lived with disability (YLD) 45–6